Strategies for the Synthesis of Heterocycles and Their Applications

Edited by Premlata Kumari and Amit B. Patel

Published in London, United Kingdom

Strategies for the Synthesis of Heterocycles and Their Applications
http://dx.doi.org/10.5772/intechopen.104135
Edited by Premlata Kumari and Amit B. Patel

Contributors
Aravazhi Amalan Thiruvalluvar, Abha Kathuria, Anwar Jahan, Poonam Dwivedi, Amit B. Patel, Premlata Kumari, Zebabanu Khalifa, Rachana Upadhyay, Yousef Najajreh, Maha Awwad Khoury, Rohit Singh, Avneet Kaur, Kishor Danao, Vijayshri Rokde, Deweshri Nandurkar, Ruchi Shivhare, Ujwala Mahajan, Warjeet S. Laitonjam, Nimalini Moirangthem, Lakshmi P. P. Kotra, Melissa M. Lewis-Bakker, Erika Lozano, Monica Dinodia, Vasudevan Dhayalan, Chitrarasu Manikandan, Deepika Sharma, Rambabu Dandela, Ramarajan Rajalakshmi, Subramaniyan Ramkumar

First published in London, United Kingdom, 2023 by IntechOpen
IntechOpen is the global imprint of INTECHOPEN LIMITED, registered in England and Wales, registration number: 11086078, 5 Princes Gate Court, London, SW7 2QJ, United Kingdom

British Library Cataloguing-in-Publication Data
A catalogue record for this book is available from the British Library

Additional hard and PDF copies can be obtained from orders@intechopen.com

Strategies for the Synthesis of Heterocycles and Their Applications
Edited by Premlata Kumari and Amit B. Patel
p. cm.
Print ISBN 978-1-83768-241-6
Online ISBN 978-1-83768-242-3
eBook (PDF) ISBN 978-1-83768-243-0

For EU product safety concerns:
IN TECH d.o.o., Prolaz Marije Krucifikse Kozulić 3, 51000 Rijeka, Croatia, info@intechopen.com or visit our website at intechopen.com.

Meet the editors

Dr. Premlata Kumari is an associate professor in the Department of Chemistry, Sardar Vallabhbhai National Institute of Technology (SVNIT), Surat, India. She has more than 16 years of teaching and research experience. She received her PhD from the University of Allahabad, India, in 2007. In 2016–2017, she was the head of the Applied Chemistry Department, SVNIT, Surat. She has published more than fifty-five peer-reviewed papers and fourteen book chapters. She has supervised twenty-one postgraduate students and five PhD students. At present, she is supervising six Ph.D. students and three master's students. Dr. Kumari's major study areas include the design and synthesis of bioactive drug candidates as well as their in vitro and in silico analyses. She is also interested in designing more potent drug candidates via the examination of the structure–activity relationship.

Dr. Amit B. Patel has 10 years of teaching and research experience. He has published twenty research papers in peer-reviewed journals and seven books/book chapters. He is a reviewer for many reputed journals. His research interest focuses on sustainable and environmentally friendly synthetic approaches in developing new transition metal-mediated syntheses of various novel organic and heterocyclic scaffolds.

Contents

Preface **IX**

Chapter 1 **1**
Substituted Azoles as Non-Nucleoside Reverse Transcriptase Inhibitors Activity
by Rohit Singh and Avneet Kaur

Chapter 2 **15**
Pyrazole Scaffold: Strategies toward the Synthesis and Their Applications
by Deweshri Nandurkar, Kishor Danao, Vijayshri Rokde, Ruchi Shivhare and Ujwala Mahajan

Chapter 3 **39**
Synthesis and Biological Applications of Thiazolidinone
by Ramarajan Rajalakshmi and Subramaniyan Ramkumar

Chapter 4 **63**
Synthetic Approaches for Pharmacologically Active Decorated Six-Membered Diazines
by Yousef Najajreh and Maha Awwad Khoury

Chapter 5 **103**
Selected Imidazole Derivatives: Synthesis and X-Ray Crystal Structure – A Review
by Aravazhi Amalan Thiruvalluvar

Chapter 6 **125**
Synthesis of Five-Membered Heterocycles Using Water as a Solvent
by Abha Kathuria, Anwar Jahan and Poonam Dwivedi

Chapter 7 **139**
Current Development in the Synthesis of Benzimidazole-Quinoline Hybrid Analogues and Their Biological Applications
by Zebabanu Khalifa, Rachana Upadhyay, Premlata Kumari and Amit B. Patel

Chapter 8 159
Construction of Biologically Active Five- and Six-Membered Fused Ring Pyrimidine Derivatives from 1,3-Diarylthiobarbituric Acids (DTBA)
by Warjeet S. Laitonjam and Nimalini Moirangthem

Chapter 9 177
Synthetic Strategies and Biological Activities of 1,5-Disubstituted Pyrazoles and 2,5-Disubstituted Thiazoles
by Erika Lozano, Melissa M. Lewis-Bakker and Lakshmi P. Kotra

Chapter 10 197
Greener Approach towards the Synthesis of Nitrogen Based Heterocycles
by Monica Dinodia

Chapter 11 211
Recent Methods for Synthesis of Coumarin Derivatives and Their New Applications
by Deepika Sharma, Vasudevan Dhayalan, Chitrarasu Manikandan and Rambabu Dandela

Preface

Heterocycles have diverse applications in various fields, such as pharmacy, medicine, dyestuffs, and polymers. As such, there is a need to continually upgrade the information related to heterocyclic compounds, which play a significant role in biochemical reactions in living cells. There is a whole range of heterocycles/fused heterocycles with one/two heteroatoms with a variety of applications. The momentum gained for the synthesis of new heterocycles testifies to their significance. The challenges in synthesizing new heterocycles and heterocycle properties that expand their use have stimulated research in the field. This book discusses various chemical strategies for the synthesis of the spectrum of heterocycles and their relevant applications in pharmaceuticals, agriculture, and material science. The green synthetic approach is an eco-friendly method to minimize/eliminate waste, decrease the consumption of energy, and promote the usage of green solvents. The book examines these green synthetic approaches for synthesizing the spectra of heterocyclic scaffolds. It includes eleven chapters.

Chapter 1, "Substituted Azoles as Non-Nucleoside Reverse Transcriptase Inhibitors Activity", discusses five-member azoles as anti-HIV medicines and, in particular, as non-nucleoside reverse transcriptase inhibitors (NNRTIs). Chapter 2, "Pyrazole Scaffold: Strategies toward the Synthesis and Their Applications", addresses the broad-spectrum bioactivity of pyrazoles and the green synthetic methodology used for synthesizing them. Chapter 3, "Synthesis and Biological Applications of Thiazolidinone," discusses the synthesis of thiazolidinone derivatives and their biological effectiveness, including their anti-diabetic, antioxidant, and antibacterial activities. Chapter 4, "Synthetic Approaches for Pharmacologically Active Decorated Six Membered Diazines", elaborates on the various synthetic techniques used to synthesize pharmacologically active decorated diazines, paying particular attention to non-fused substituted pyrimidines that have clinical applications. According to reports, diazines have numerous beneficial effects on human health, including anti-metabolite, anticancer, antibacterial, anti-allergic, tyrosine kinase, antimicrobial, anti-inflammatory, analgesic, antihypertensive, antileishmanial, antituberculostatic, anticonvulsant, diuretic potassium-sparing, and anti-aggressive activities. Chapter 5, "Selected Imidazole Derivatives: Synthesis and X-Ray Crystal Structure – A Review", describes the synthesis and X-ray crystal structure analysis of selected significant imidazole derivatives, including DMDPIMH: 4,5-Dimethyl-1,2-diphenyl-1H-imidazole monohydrate; DMPPTI: 4,5-Dimethyl-2-phenyl-1-(p-tolyl)-1H-imidazole; FPDMMPI: 2-(4-Fluorophenyl)-4,5-dimethyl-1-(4- methylphenyl)-1H-imidazole; DMPDMPIHH: 1-(3,5-Dimethylphenyl)-4,5-dimethyl-2-phenyl-1H-imidazole hemihydrate; DMPFPDMI: 1-(3,5-Dimethylphenyl)-2-(4-fluorophenyl)-4,5-dimethyl-1H-imidazole; FPMOPDMI: 2-(4-Fluorophenyl)-1-(4-methoxyphenyl)-4,5-dimethyl-1H-imidazole; DMOPFPDMI: 1-(3,5-Dimethoxyphenyl)-2-(4-fluorophenyl)-4,5-dimethyl-1H-imidazo; and FPTPI: 2-(4-Fluorophenyl)-1,4,5-triphenyl-1H-imidazole.

Chapter 6, "Synthesis of Five-Membered Heterocycles Using Water as a Solvent", discusses three typical five-membered heterocyclic compounds (furan, thiophene, and pyrrole) that can be synthesized using water as a solvent, as an aqueous medium has attracted tremendous interest from chemists for various organic synthesis. Chapter 7, "Current Development in the Synthesis of Benzimidazole-Quinoline Hybrid Analogues and their Biological Applications", presents synthetic methodologies for quinolone benzimidazole hybrid scaffolds and their potential pharmacological activity. Quinoline and benzimidazole are crucial scaffolding for drug discovery leads and have significance in medicinal chemistry. Chapter 8, "Construction of Biologically Active Five- and Six-Membered Fused Ring Pyrimidine Derivatives from 1,3-Diarylthiobarbituric Acids (DTBA)", highlights the synthesis and biological activity of certain fused pyrimidines, such as pyrazolo[3,4-d]pyrimidine-6-thiones; 5,7-diaryl-4-oxo-isoxazolo[5,4-d]pyrimidine-6-thiones; 5-oxo-pyrimido[4,5-d] pyrimidine-7-thiones; 2-thioxo-pyrano[2,3-d]pyrimidine-4-ones; pyrido[2,3-d] pyrimidines; quinazoline-4-oxo-2-thiones.Pyrimidines and fused pyrimidines have attracted a lot of attention because of the diverse range of biological functions they exhibit. Moreover, the pyrimidine ring with the thiourea linkage (-NH-C(S)-NH) is significant for pharmacological purposes because heterocyclic molecules with nitrogen and sulphur atoms were a major contributor to the development of medicine. Pyrimidine derivatives hold a special position in biological and medicinal chemistry because of the diverse spectrum of biological activities they display.

Chapter 9, "Synthetic Strategies and Biological Activities of 1,5-Disubstituted Pyrazoles and 2,5-Disubstituted Thiazoles", focuses on the strategies and various viewpoints on the synthesis of thiazoles and pyrazoles, including derivatives at the 2,4,5 positions and 1,5 positions, respectively. The chapter also explores their biological and therapeutic evaluations. Chapter 10, "Greener Approach towards the Synthesis of Nitrogen Based Heterocycles", covers the synthesis of N-based heterocycles using environmentally friendly and greener approaches. Chapter 11, "Recent Methods for Synthesis of Coumarin Derivatives and Their New Applications," discusses the synthesis of coumarins during which equivalent coumarin analogues are produced from substituted alkyne substrates.

Overall, this book offers an overview of synthetic methods for the synthesis of a variety of heterocycles, including green synthetic methods and heterocycle applications in a variety of fields.

Premlata Kumari
Associate Professor,
Department of Chemistry,
Sardar Vallabhbhai National Institute of Technology,
Surat, India

Amit B. Patel
Department of Chemistry,
Government College Daman,
Daman, India

Chapter 1

Substituted Azoles as Non-Nucleoside Reverse Transcriptase Inhibitors Activity

Rohit Singh and Avneet Kaur

Abstract

This Review article gives an overview of substituted azole based synthetic medicine and their significance application in treating various ailments. The chemical reactions of azoles with other heterocyclic compounds/chemical reagents produce a lot of new substituted molecules, which have plentiful biological actions and potential pharmaceutical applications. Azoles is considered a major class of heterocyclics and worldwide researchers have put great efforts in studying this nucleus in order to design and synthesize various new derived of imidazole and hybrid molecules with the identification of their significant role in many classes of drugs such as antiviral including anti-HIV. Azoles has inspiring molecular geometric structure that offers a suitable skeleton to build newer chemical entities which has encouraged medicinal chemists to design and synthesis of novel and effective molecule as potential therapeutic agents. However, there is an urgent need to study the mechanism of action at molecular level of several pharmacological actions resulting from azoles scaffold through modern methods to furnish safer and effective new molecules for the treatment of various diseases.

Keywords: azoles, heterocyclic compounds, hybrid compounds, synthetic reactions and biological application

1. Introduction

AIDS (Acquired immune deficiency syndrome) is the major cause of death which leads to pandemic worldwide. In today's scenario combination of Nucleoside reverse transcriptase inhibitors (NRTI), Non nucleoside reverse transcriptase inhibitors (NNRTI) and Protease inhibitors (PI). Highly active antiretroviral therapy (HAART) also called as anti retroviral therapy which is a combination of two or more combination of antiviral drugs become unsuccessful because of mutational change and virus getting resistant against anti viral drugs. Non nucleoside reverse transcriptase Inhibitors (NNRTI) binds to specific active site of reverse transcriptase of HIV are the important cascade of anti HIV-1 drugs mechanism. Molecular modeling studies of HIV enzyme complexes and chemical generation of second and third generation of

NNRTI's led to higher specificity of NNRTI's against HIV-1 and leads to major discovery in the generation of anti-viral drugs [1].

Non nucleoside reverse transcriptase inhibitor binding pocket were found to play an important role in binding site with drugs and found less prone against mutation. Here we also describe the chronological history of non nucleoside reverse transcriptase inhibitors development and basically focused on small molecules belonging to NNRTI class against AIDS management therapy [2].

In this review the chronological advancement of non nucleoside reverse transcriptase inhibitors and their belongings to NNRTI class in the management of drugs used in treatment of AIDS have been focused. The generation of resistance against NNRTI drugs and pharmacokinetic problems are the serious concern in management therapy.

2. Reverse transcriptase

RT is the replicative enzyme of HIV and other retroviruses. RT copies the single-stranded viral genomic RNA into double-stranded DNA (**Figure 1**), which is subsequently integrated into host cell DNA.

Reverse transcriptase (RT) is a key enzyme which plays an essential and multifunctional role in the replication of the human immunodeficiency virus type- 1 (HIV-1) and thus constitutes an attractive target for the development of new drugs useful in AIDS therapy [3].

Reverse Transcriptase is a major target for anti HIV drug development along with two classes of inhibitors, the nucleoside or nucleotide reverse transcriptase inhibitors and non nucleoside reverse transcriptase inhibitors (NNRTIs) have been approved by US Food and Drug Administration (FDA) for the treatment of HIV-1 infection. Reverse Transcription (RT) is formation of double stranded DNA from single stranded RNA genome which is an essential step in life cycle of HIV-1 replication. This is one is very important and complex step required for both DNA polymerase and ribonuclease active sites of HIV-1 reverse transcriptase [4, 5].

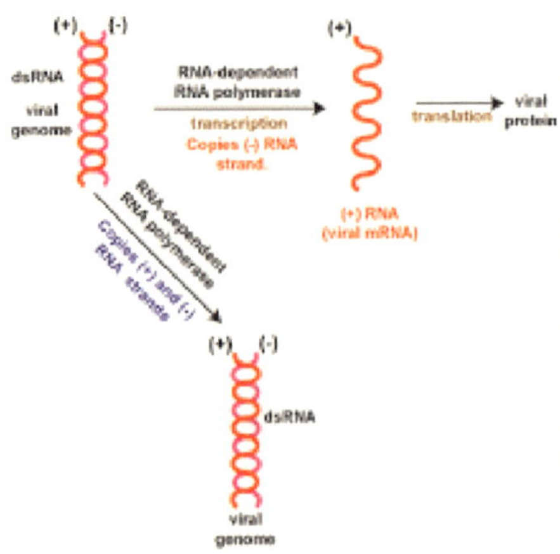

Figure 1.
Reverse transcriptase copies the single-stranded viral genomic RNA into double-stranded DNA (reverse transcription).

3. Non nucleoside reverse transcriptase inhibitors (NNRTIs)

The era of NNRTIs began before two decades, with the discovery of TIBO [4, 6] and HEPT [7] as specific inhibitors of reverse transcriptase. NNRTIs are highly specific and potent inhibitors of HIV-1 RT, and do not interfere with cellular or mitochondrial DNA synthesis [8]. However, the rapid emergence of resistant virus variants and the problem of resistance have limited their clinical use [9].

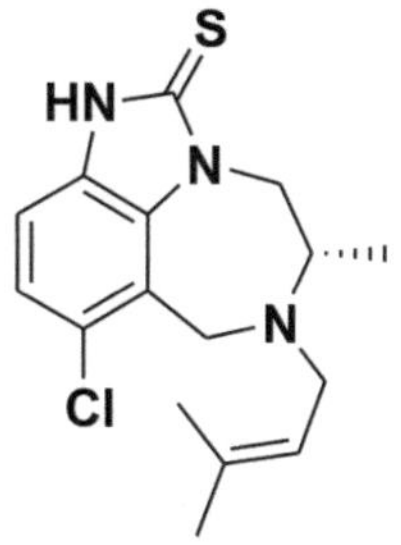

(1)

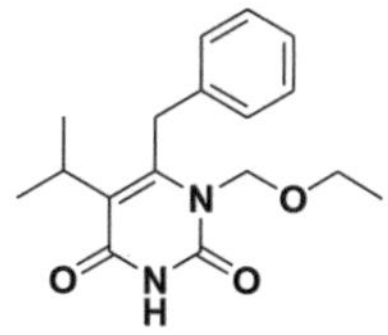

(2)

Nevirapine, Delavirdine and Efavirenz are the three Non nucleoside reverse transcriptase inhibitors which were approved as anti-HIV drugs in the year 1996, 1997 and 1998. Out of which Nevirapine is one the agent used to inhibit mother to child transmission of HIV [10–12].

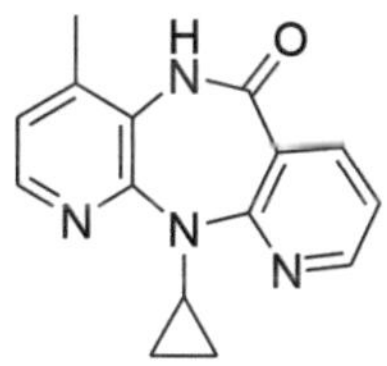

NEVIRAPINE

(3)

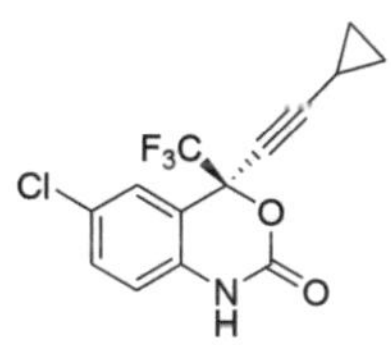

EFAVIRENZ

(4)

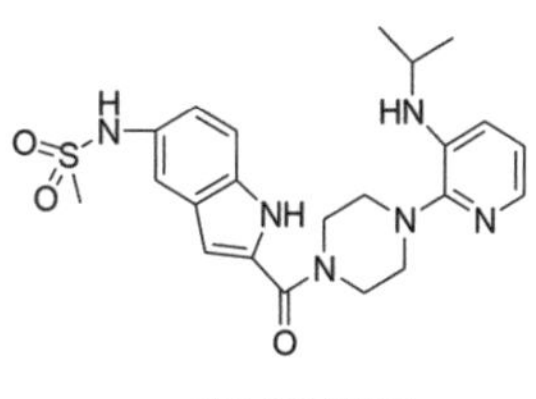

DELAVIRDINE

(5)

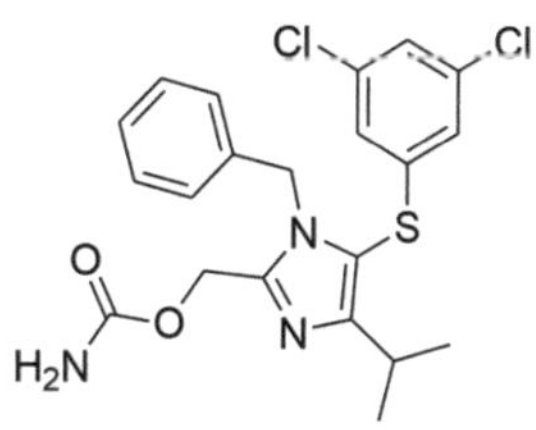

CAPRAVIRINE

(6)

Non Nucleoside Reverse Transcriptase Inhibitors (e.g. Nevirapine, Delavirdine, Efavirenz and Etravirine) are used as antiviral compounds which prevent the possibility transmission from mother to child. The interaction between NNRTI and NNRTI binding pocket mechanism from DNA polymerase active site has shown in **Figure 2.**

Basically in cytoplasm with the process of reverse transcription begins with entry of viral particle to the target cell of viral genome as a part of nucleoprotein conjugation whose structure is not well defined. This DNA is formed on the template of RNA but its terminal contains duplications which is known as long terminal repeats (LTRs)

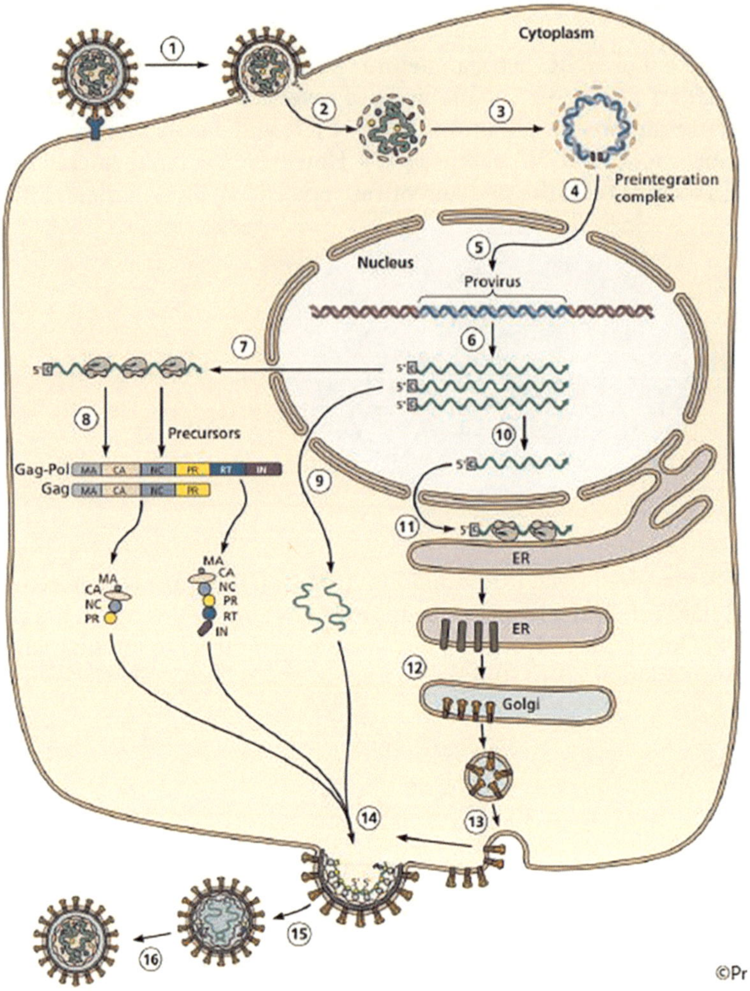

Figure 2.
Mechanism of NNRTI inhibition.

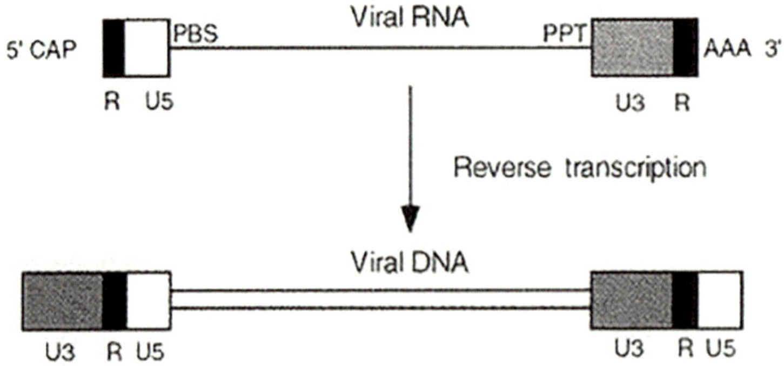

Figure 3.
Strand transfer reaction or jumps required to generate long terminal repeats (LTR's).

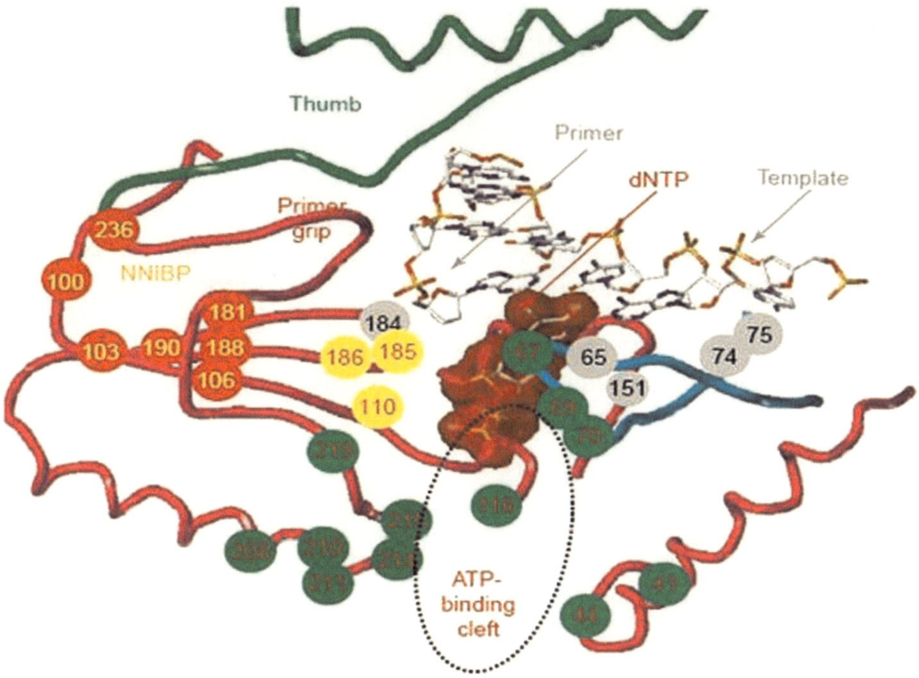

Figure 4.
The structure of HIV-1 RT in the region near the polymerase active site.

which is not present in viral RNA. (**Figure 3**) Strand transfer reactions or jumps are required to generate long terminal repeats (LTRs) which is perfect model for reverse transcription propose two specialized template switches [13, 14].

Reverse transcriptase enzyme is a type of replicative enzyme of HIV-1 and other type of retroviruses where retro viruses generates double stranded DNA by copying from single stranded genomic viral RNA and subsequently integrated with host cell DNA. Reverse transcriptase (RT) has two enzymatic activities: First one is polymerase which copy either DNA or RNA and second one is RNAase that degrade RNA strand of RNA-DNA intermediates formed during viral DNA synthesis. Human immunodeficiency syndrome virus (HIV-1) has two subunits p66 and p51. Polymerase chain of domain site p66 and p51 contain four subdomain namely fingers, palm, thumb and connection. Folding and connections of both subdomain p66 and p51 are similar in nature but their spatial arrangements are different. Domain p66 contains active site for both polymerase as well as RNAse while domain p51 plays primarily structural role for DNA and RNA.

Fingers and palm size of domain p66 are highly conserved region and together with two helices of thumb subdomain acts as a clamp that holds the position template primer (**Figure 4**). One of these region which is part of palm subdomain act as a primer group of DNA. The primer group is the candidate solely responsible for the exact placement of primer terminus at the active site of polymerase enzyme and involved in translocation of DNA template primer followed by 4–7 nucleotide incorporation. Exact positioning or binding of template primer is also very much important for cleavage of DNA/RNA template by the RNase activity of reverse transcription enzyme. Now a days HIV-1 reverse transcriptase inhibitors currently available as anti HIV drugs in the management of AIDS by targeting enzymatic activity of DNA polymerase activity [15–17].

3.1 Capravirine

Capravirine (formerly known as S-1153 or AG1549) is a 1,2,4,5-substituted imidazole derivative (**Figure 2**) that inhibits HIV-1 replication in a standard $CD4^+$.

Capravirine exhibited significant expansion as antiviral spectrum as anti HIV agent and covered many typical resistant mutant strain of NNRTI including single K103N mutation and double K103N-L100I and K103N-P225H mutations [18–20]. Y181C was also other common mutation which arises under nevirapine pressure and also inhibited by capravirine [21].

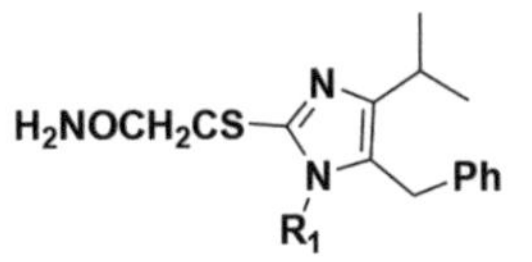

(7)

Barreca et al. were carried out their research in the fields of Non nucleoside reverse transcriptase inhibitors (NNRTIs).

4. Nucleoside reverse transcriptase inhibitors (NRTI's)

Out of the 13 compounds that have been formally approved for the treatment of human immune deficiency virus (HIV) infections, nine are targeted at the viral reverse transcriptase (RT), the other four (saquinavir, ritonavir, indinavir and nelfinavir) being targeted as the viral protease. Of the nine RT inhibitors that have been licensed, six can be considered as nucleoside reverse transcriptase inhibitors (NRTI's) namely zidovudine (AZT) **(1)**; Didanosine (ddI) **(2)**; Zalcitabine (ddC) **(3)**; Stavudine (d4T) [4]; Lamivudine (3TC) **(5)** and Abacavir (ABC) **(6)**.

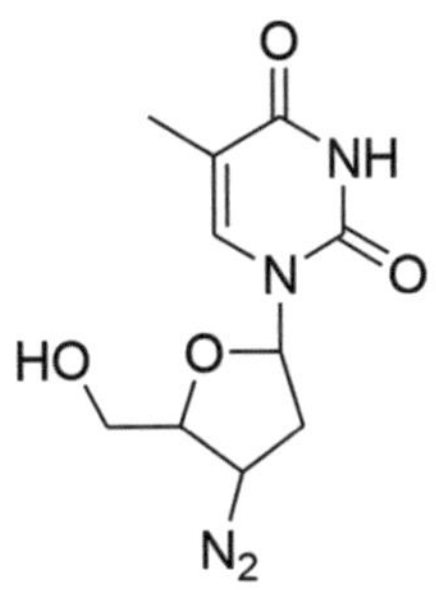

8. Zidovudine

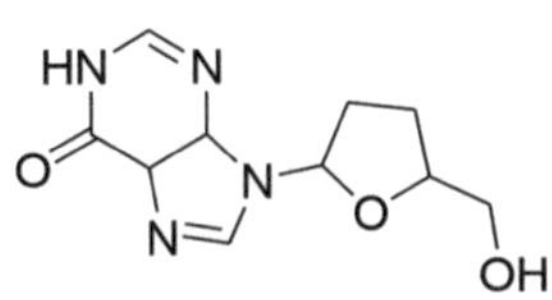

9. Didanosine

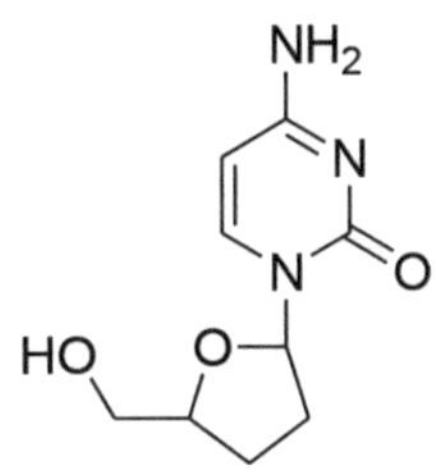

10. Zalcitabine

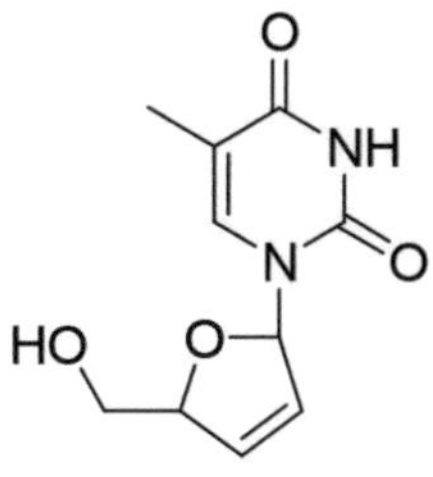

11. Stavudine

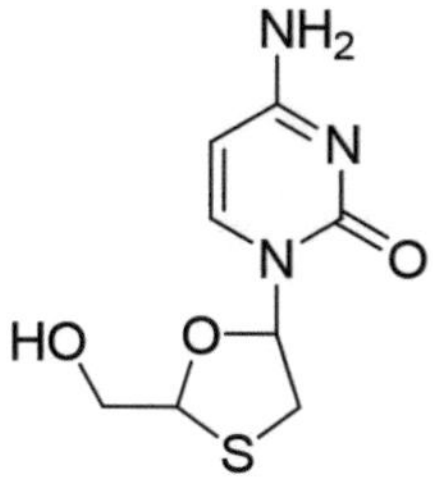

12. Lamivudine

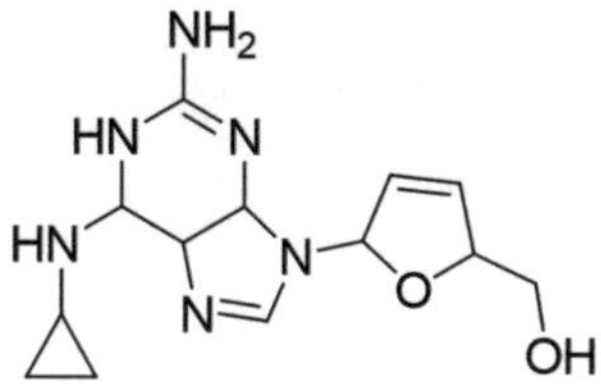

13. Abacavir

Some of these compounds are still in pre clinical stage of development but others have already proceeded to phase I, II or III clinical trial. Most advanced among new RT inhibitors are (−) FTC (emitricitabine), F-ddA (lodenosine), PMEA (adefovir) and its oral prodrug form bis (POM)-PMEA [adefovir dipivoxil], PMPA (tenofovir) and its oral prodrug form bis (POC)-PMPA (tenofovir disoproxyl fumarate), which can be considered as NRTI's.

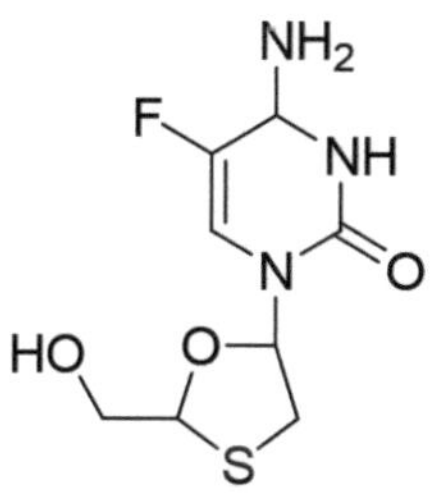

14 . (-) FTC Emitricitabine

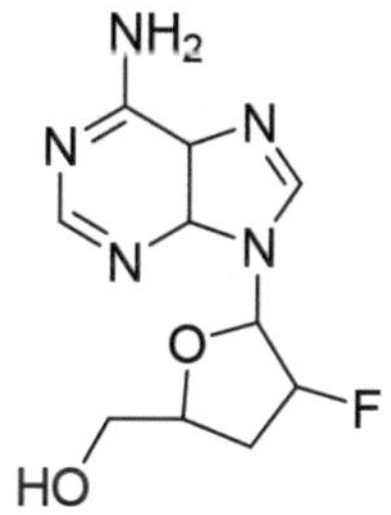

15. F-ddA (Lodenosine)

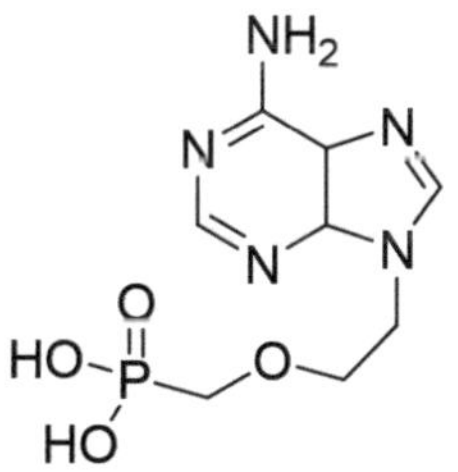

16. Adefovir

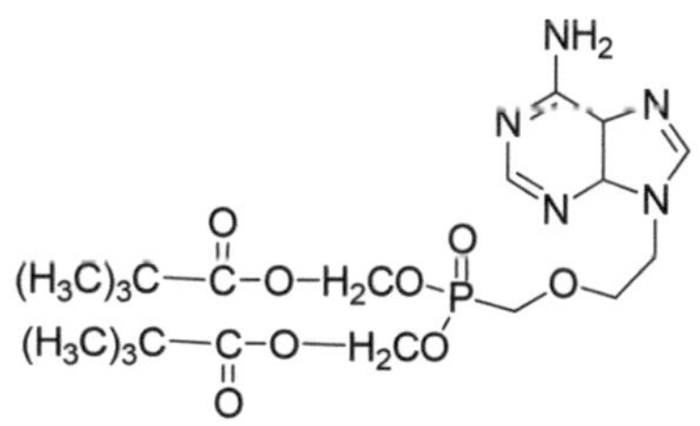

17. Adefovir dipivoxil

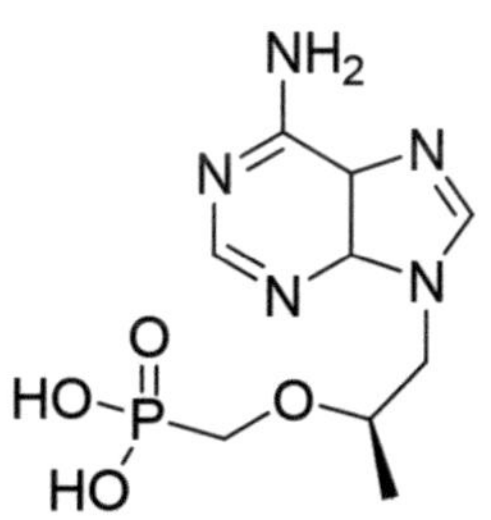

18. Tenofovir

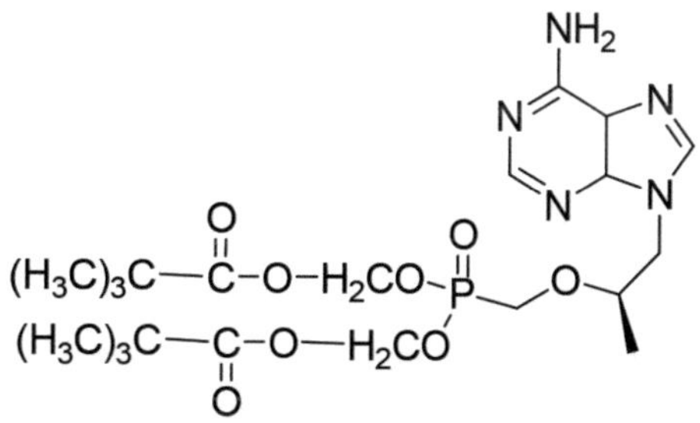

19. Tenofovir disoproxyl fumarate.

Agents		Major side effects
Nucleoside reverse transcriptase inhibitors (NRTI'S)		
1.	Zidovudine (AZT, Retrovir)	Osteoporosis, bone marrow anemia, GI disturbances, insomnia, neutropenia
2.	Didanosine (ddI, Videx)	Pancreatitis, Peripheral neuropathy, GI disturbances.
3.	Zalcitabine (ddC, Hivid)	Peripheral Neuropathy, stomatitis.
4.	Stavudine (d4T, Zerit)	Peripheral Neuropathy, pancreatitis.
5.	Lamivudine (3TC, Epivir)	Nasal symptoms, lactic acidosis, liver problems etc.
6.	Abacavir (ABC, Ziagen)	Hypersensitivity, Skin Rash (Stevens-johnson syndrome), abdominal and stomach pain.
Non nucleoside reverse transcriptase inhibitors (NNRTI'S)		
7.	Nevirapine (NVP, Viramune)	Rash (Stevens-johnson syndrome), tingling, burning & prickly sensations.
8.	Delavirdine (DLV, Rescriptor)	Rash (Stevens-johnson syndrome), excessive tiredness.
9.	Efavirenz (EFV, Sustiva)	Convulsions, false-positive test for urine, dizziness, drowsiness.
Protease inhibitors		
10	Saquinavir (SQV, Invirase)	GI symptoms, hepatitis, increased TGC, dry lips or skin.
11.	Ritonavir (RTV, Norvir)	Muscles or joint pains, itching, numbness of mouth area, headache, dizziness, pancreatitis, heart problem.
12.	Indinavir (IDV, Crixivan)	Itchy skin, blurred vision, headache, dizziness, GI symptoms rashes.
Integrase inhibitors		
13.	Raltegravir (RAL, Isentress)	Blue in urine, joint pain, stiffness, headache.
Fusion and entry inhibitors		
14.	Enfuvirtide (T-20, Fuzeon)	Nervousness, muscle pain, nausea, depression.
15.	Maraviroc (MVC, Selzentry)	Selzentry, diarrhea, constipation, dizziness.

Table 1.
Side effects of anti-HIV compounds [23, 24].

The description of the RT inhibitors as anti-HIV agents should be viewed in the broad scope of anti-HIV therapy (1) and therapeutic approaches for intervention with HIV infections (2), and strategies to overcome or prevent the problem of HIV resistance development to anti-HIV agents in general (3) and NNRTI's particular (4).

Nucleoside reverse transcriptase inhibitors (NRTIs) were the first class of drugs as antiretroviral drugs to gain regulatory approval in 1987. Zidovudine (AZT) was the first drug to be licensed for the treatment of HIV infection [22], a mere four years after the identification of HIV as the etiology agent for AIDS (**Table 1**) [23, 24].

5. Non nucleoside reverse transcriptase inhibitors (NNRTIs)

Non nucleoside reverse transcriptase inhibitors (NNRTI's) causing an allosteric inhibition of reverse transcriptase enzyme by binding directly to enzyme at different site of nucleoside binding component. Nevirapine, delavirdine and efavirenz are the three members of this group which are used clinically. These agents are given in combination with other antiviral compounds for the treatment of AIDS. All class of NNRTI's are rapidly absorbed from gastrointestinal tract and with highly bound plasma proteins. These NNRTI's are extensively metabolized by cytochrome P450 enzyme along with variable amount of administered drug which are being recovered from urine as a metabolite. Skin rashes is the most common adverse effect associated with NNRTI class of drugs [25].

6. Protease inhibitors

Protease inhibitors (PI's) interfere with replication of viruses, maturation and infection of new viral cells by inhibiting the cleavage of enzyme necessary for viral protein precursors. Like NNRTI's, protein inhibitors are also metabolized by cytochrome P450 enzymes which leads to drug-drug interaction. PI's can also cause gastrointestinal side (GI's) effects and elevation in liver transaminase enzyme. Some protease inhibitors like saquinavir have first pass metabolic effect which results poor bioavailabilty which may be improved by administrating with food.

Physiological factors like achlorhydria, malabsorption and poor hepatic dysfunction may impair the bioavailability of protease inhibitors in HIV infections which may influence their antiHIV activity *in vitro*. PI's like ritonavir and indinavir may displace other protease inhibitors like saquinavir, from plasma protein and inhibit their metabolism. In addition PI's can worsen diabetes, insulin resistance, hypercholesterolemia, hypertension and hyperlipidemia [26].

7. Integrase inhibitors

Integrase is the viral enzyme which catalyzes the integration of DNA derived from virus into the DNA of host cell present in the nucleus and form proactive virus that can be activated to produce viral proteins.

Integrase inhibitors are the type of antiretroviral therapy agents and they rely on the fact that HIV needs integrase enzyme to replicate and the integrase inhibitor drugs blocks the enzymatic action of integrase enzyme and without the help of this enzyme HIV cannot take over T cells to copy itself. The US food and drug administration (FDA) approved the use of integrase inhibitors in year 2007. Raltegravir, dolutegravir, elvitegravir, bictegravir etc. are some marketed drugs of integrase inhibitors are in use. Raltegravir is the first approved integrase inhibitors, is a valuable addition to the drug therapy against multidrug resistance. Dolutegravir and elvitegravir are also present in combination medication such as genvoya (elvitegravir + emtricitabine + tenofovir alafenamide fumarate + cobicistat), stribild (elvitegravir + emtricitabine + tenofovir disoproxil fumarate + cobicistat), triumeq (Dolutegravir + abacavir + lamivudine), juluca (Dolutegravir + rilpivirine) and biktarvy (biktegravir + emtricitabine + tenofovir alafenamide fumarate) [27, 28].

8. Fusion and entry inhibitors

Fusion inhibitors are also known as entry inhibitors is a class of antiviral drugs which prevent the entry of virus to host cell by blocking the receptors. They are used in combination therapy for the treatment of HIV infections by blocking the steps involved in HIV replication cycle. Once virus enters into the host cell the viral surface protein gp120 form complex with CD4 receptor with co-receptors like CCR5 and CXCR4 which is necessary step for viral entry into the host cell.

Enfuvirtide, the first approved drug of this class which interfere with viral attachment with host cell membrane by inhibiting the necessary conformational change in particular viral envelop protein (gp41) which trigger the formation of transmembrane pore through which virus would enter the host cell. Maraviroc is also a new chemokine co-receptor antagonist (CCR5 antagonist) drugs that blocks the binding of HIV to CCR5 co-receptor in HIV infection [29, 30].

9. Conclusion

Drugs like Nucleoside reverse transcriptase inhibitors, Non nucleoside reverse transcriptase inhibitors, Protease inhibitors, Integrase inhibitors and Fusion or Entry inhibitors are the potent candidates as anti HIV and efforts are still continuing to develop potent molecules which will be effective in the management of AIDS, and search is still going on as viruses exhibited rapid mutation which makes the drug ineffective against the treatment of AIDS and develop resistance against multiple drug therapy and drug become active only for short span of life.

As most of anti HIV agents have diverse chemical structures showed in this book chapter which exhibited potent action against many viruses and considered as best candidate as anti HIV.

The current Covid-19 problem is an excellent example of mutant variation of viruses which exhibited variant strain rapidly and for the reason of developing resistance against many anti-viral drugs. Combination of NRTI and NNRTI along with PI are used administered to combat with drug resistance and be a part of multi drug resistance therapy.

Author details

Rohit Singh[1]* and Avneet Kaur[2]

1 Department of Pharmaceutical Chemistry, Institute of Pharmaceutical Sciences, University of Lucknow New Campus, Lucknow, Uttar Pradesh, India

2 Department of Pharmaceutical Chemistry, SGT College of Pharmacy, Shree Guru Gobind Singh Tricentenary University, Gurugram, Haryana, India

*Address all correspondence to: rohitsingh20485bitmesra@gmail.com

References

[1] UNAIDS, Global AIDS report. 2022

[2] Sarafianos SG, Das K, Hughes SH, Arnold E. Taking aim at a moving target: Designing drugs to inhibit drug-resistant HIV-1 reverse transcriptases. Current Opinion in Structural Biology. 2004;**14**:716

[3] Jonckheere H, Anne J, Clercq E. De. The HIV-1 reverse transcription (RT) process as target for RT inhibitors. Medicinal Research Reviews. 2000;**20**: 129-154

[4] Kohlstaedt LA, Wang J, Friedman JM, Rice PA, Steitz TA. Crystal structure at 3.5 a resolution of HIV-1 reverse transcriptase complexed with an inhibitor. Science. 1992;**26**:1783-1790 [PubMed:1377403]

[5] Esnouf R, Ren J, Ross C, Jones Y, Stammers D, Stuart D. Mechanism of inhibition of HIV-1 reverse transcriptase by non-nucleoside inhibitors. Nature Structural Biology. 1995;**2**:303-308 [PubMed: 7540935]

[6] Telesnitsky A, Goff SP. In: Coffin JM, Hughes SH, Varmus HE, editors. Retroviruses. Cold Spring Harbor, NY: Cold Spring Harbor Laboratory Press; 1997. pp. 121-160

[7] Jacobo A, Ding J, Nanni RG, Clark AD, Lu X, Tantillo C, et al. Crystal structure of human immunodeficiency virus type 1 reverse transcriptase complexed with double-stranded DNA at 3.0 A resolution shows bent DNA. Proceedings of the National Academy of Sciences of the United States of America. 1993;**90**:6320

[8] Ding J, Das K, Hsiou Y, Sarafianos SG, Clark AD, Jacobo MA, et al. Structure and function of HIV-1 reverse transcriptase: Molecular mechanisms of polymerization and inhibition. Journal of Molecular Biology. 1998;**284**:1095

[9] Ghosh M, Jacques PS, Rodgers DR, Ottman M, Darlix JL, Le Grice SF. Analysis of human immunodeficiency virus type 1 reverse transcriptase subunit structure/function in the context of infectious virions and human target cells. Biochemistry. 1996;**35**:8553

[10] Sarafianos SG, Das K, Tantillo C, Clark AD, Ding J, Whitcomb JM, et al. Crystal structure of HIV-1 reverse transcriptase in complex with a polypurine tract RNA:DNA. The EMBO Journal. 2001;**20**:1449

[11] Julias JG, McWilliams MJ, Sarafianos SG, Arnold E, Hughes SH. Mutations in the RNase H domain of HIV-1 reverse transcriptase affect the initiation of DNA synthesis and the specificity of RNase H cleavage in vivo. Proceedings of the National Academy of Sciences of the United States of America. 2002;**99**:9515

[12] Julias JG, McWilliams MJ, Sarafianos SG, Alvord WG, Arnold E, Hughes SH. Mutations in the RNase H domain of HIV-1 reverse transcriptase affect the initiation of DNA synthesis and the specificity of RNase H cleavage in vivo. Journal of Virology. 2003;**77**:8548

[13] Pauwels R, Andries K, Desmyter J, Schols D, Kukla M, Breslin H, et al. In vitro activity of acyclic nucleoside phosphonate derivatives against feline immunodeficiency virus in crandell feline kidney cells and feline peripheral blood lymphocytes. Nature. 1990;**343**: 470

[14] Kukla M, Breslin H, Pauwels R, Fedde C, Miranda M, Scott M, et al. Synthesis and anti-HIV-1 activity of 4,5,6,7-tetrahydro-5-methylimidazo [4,5,1 jk][1,4]benzodiazepin-2(1H)-one (TIBO) derivatives. Journal of Medicinal Chemistry. 1991;**34**:746

[15] Miyasaka T, Tanaka H, Baba M, Hayakawa H, Walker RT, Balzarini J, et al. Palladium-catalyzed synthesis of [E]-6-(2-acylvinyl)uracils and [E]-6-(2-acylvinyl)-1-[(2-hydroxyethoxy) methyl]uracils—their antiviral and cytotoxic activities. Journal of Medicinal Chemistry. 1989;**32**:2507

[16] De Clercq E. The role of non-nucleoside reverse transcriptase inhibitors (NNRTIs) in the therapy of HIV-1 infection. Antiviral Research. 1998;**38**:153

[17] Richman DD, Havlir D, Corbeil J, Looney D, Ignacio C, Spector SA, et al. Nevirapine resistance mutations of human immunodeficiency virus type 1 selected during therapy. Journal of Virology. 1994;**68**:1660

[18] Sorbera LA, Castaner J, Bayes M. A comprehensive review of capravirine, covering chemistry, virology and all available clinical data. Capravirine. Drugs of the Future. 2003;**28**:1149-1158

[19] Blair WS, Sata A, Sugimoto H, Fujiwara T, Alexander TN, Bahn V, et al. Resistance and cross-resistance profiles of capravirine, a novel NNRTI of HIV-1. Antiviral Research. 2001;**50**:abstract 45

[20] Alexander TN, Leavitt MC, Rudy JJ, Isaacson JS, Hertogs K, Larder BA, et al. Antiviral activity of the HIV-1 nonnucleoside reverse transcriptase inhibitor (NNRTI) capravirine against HIV-1 variants from NNRTI-experienced patients. Antiviral Therapy. 2001;**6**(Suppl 1):abstract 10

[21] Alexander TN, Graham JP, Sato A, Sugimoto H, Fujiwara T, Banh VN, et al. Resistance and cross-resistance of capravirine, a novel non-nucleoside reverse transcriptase inhibitor of HIV-1. In: 14th International Conference on Antiviral Research. Seattle; 8-12 April 2001

[22] Ezzell C. AIDS drug gets green light. Nature. 1987;**329**:751

[23] Barre-Sinoussi F, Chermann J, Rey F, Nugeyre M, Chamaret S, Gruest J, et al. Isolation of a T-lymphotropic retrovirus from a patient at risk for acquired immune deficiency syndrome (AIDS). Science. 1983;**220**:868-871

[24] Gallo RC, Sarin PS, Gelmann EP, Robert-Guroff M, Richardson E, Kalyanaraman VS, et al. Isolation of human T-cell leukemia virus in acquired immune deficiency syndrome (AIDS). Science. 1983;**220**:865-867

[25] Balzarini J, Holy A, Jindrich J, Naesens L, Snoeck R, Schols D, et al. Differential antiherpesvirus and antiretrovirus effects of the (S) and (R) enantiomers of acyclic nucleoside phosphonates: Potent and selective in vitro and in vivo antiretrovirus activities of (R)-9-(2-phosphonomethoxypropyl)-2,6-diaminopurine. Antimicrobial Agents and Chemotherapy. 1993;**37**:332-338

[26] The DAD Study Group. Class of antiretroviral drugs and the risk of myocardial infarction. The New England Journal of Medicine. 2007;**356**: 1723-1735

[27] Hammer SM, Eron JJ, Reiss P, Schooley RT, Thompson MA, Walmsley S, et al. Antiretroviral treatment of adult HIV infection: 2008 recommendations of the international AIDS society–USA panel. Journal of the

American Medical Association. 2008; **300**(5):555-570

[28] Markowitz M, Nguyen BY, Gotuzzo E, Mendo F, Ratanasuwan W, Kovacs C, et al. Rapid and durable antiretroviral effect of the HIV-1 integrase inhibitor raltegravir as part of combination therapy in treatment-naive patients with HIV-1 infection: Results of a 48-week controlled study. JAIDS Journal of Acquired Immune Deficiency Syndromes. 2007;**46**(2):125-133

[29] The Medical Letter, Inc. Two new drugs for HIV infection. The Medical Letter on Drugs and Therapeutics. 2008; **50**(1277):2-4

[30] Saag MS. A multicenter, randomized, double blind, comparative trial of a novel CCR5 antagonist, maraviroc versus efavirenz, both in combination with combivir (zidovudine/lamivudine), for the treatment of antiretroviral-naive subjects infected with R5 HIV: Week 48 results of the MERIT study. In: IAS Conference on HIV Pathogenesis; Treatment and Prevention; July 2007. Sydney, Australia; 2007. pp. 22-25

Chapter 2

Pyrazole Scaffold: Strategies toward the Synthesis and Their Applications

Deweshri Nandurkar, Kishor Danao, Vijayshri Rokde, Ruchi Shivhare and Ujwala Mahajan

Abstract

Pyrazoles have a wide range of applications in medicinal chemistry, drug discovery, agrochemistry, coordination chemistry, and organometallic chemistry. Their popularity has skyrocketed since the early 1990s. Basically, Pyrazole ($C_3H_3N_2H$) is a simple doubly unsaturated five membered heterocyclic aromatic ring molecule comprising two nitrogen (N) atoms at positions 1- and 2- and three carbon (C) atoms. Pyrazole nucleus is synthesized with various strategies such as multicomponent approach, dipolar cycloadditions, cyclocondensation of hydrazine with carbonyl system, using heterocyclic system and multicomponent approach. A special emphasis is placed on a thorough examination of response processes. Furthermore, the reasons for the increasing popularity of pyrazoles in several fields of science are examined. Pyrazoles have recently been the focus of many techniques, mostly because of how frequently they are used as scaffolds in the synthesis of bioactive chemicals and reactions in various media. The goal of this chapter is to discuss the current developments in synthetic techniques and biological activity related to pyrazole derivatives. The many pharmacological functions of the pyrazole moiety and different synthesis techniques were discussed. This chapter has summarized novel strategies and wide applications of pyrazole scaffold.

Keywords: Pyrazole, scaffold, synthesis, application, green synthesis, microwave

1. Introduction

Ludwig Knorr coined the term “Pyrazole” in 1883. A 5-membered ring structure made up of three carbon atoms and two nitrogen atoms in close proximity defines the family of simple aromatic ring organic compounds known as Pyrazoles. These compounds belong to the heterocyclic series. Although being rarely in nature, they are categorized as alkaloids due to their structure and pharmacological effects on humans. Watermelon seeds yielded the first natural Pyrazole, 1-Pyrazolyl-alanine, in 1959 [1, 2]. Pyrazole refers to both an unsaturated parent chemical and a family of simple aromatic ring organic compounds of the heterocyclic diazole series, which are distinguished by a 5-member ring structure made up of two nitrogen atoms in the neighboring position and three carbon atoms in the central position (**Figure 1**). Pyrazoles have tautomerism because of the moving C-N double bond inside the heterocycle [3, 4]. Pyrazole's 1H-tautomer is known as 1H-pyrazole [5]. It is a base of a pyrazolium conjugate. It is an acid conjugate of pyrazol-1-ide. It is a tautomer of the 3H and 4H Pyrazoles (**Table 1**).

Two techniques have been used to synthesize substituted Pyrazoles (**Figure 2**):

a. Cyclocondensation of hydrazines with 1,3-dicarbonyl compounds or their synthetic 1,3-dielectrophilic equivalents and

b. Cycloaddition of 1,3-dipoles to dipolarophiles

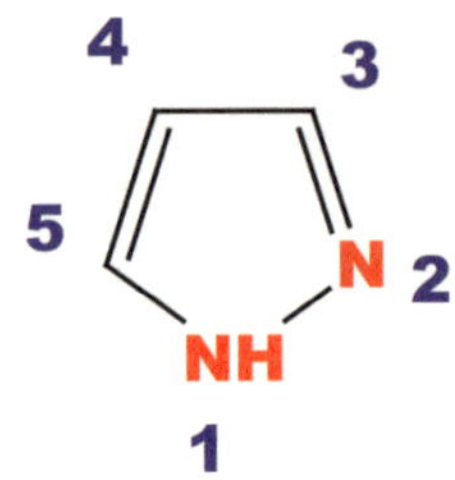

Figure 1.
Structure of Pyrazole.

Molecular Formula	$C_3H_4N_2$
IUPAC name	1*H*-pyrazole
Molecular Weight	68.08
Composition	C (52.93%) H (5.92%) N (41.15%)
Boiling Point	187°C
Melting Point	68°C
Hydrogen Bond Donor Count	1
Hydrogen Bond Acceptor Count	1
Rotatable Bond Count	0
Dissociation Constants (pKa)	2.48 (at 25°C)
Molar Refractivity	18.77 ± 0.3 cm^3
Molar Volume	60.9 ± 3.0 cm^3
Parachor	161.0 ± 4.0 cm^3
Index of Refraction	1.528 ± 0.02
Surface Tension	48.6 ± 3.0 dyne/cm
Dielectric Constant	Not available
Polarizability	7.44 ± 0.5 10-24 cm^3
Monoisotopic	68.037448 Da
Nominal Mass	68 Da

Table 1.
Chemistry and properties of Pyrazole [6].

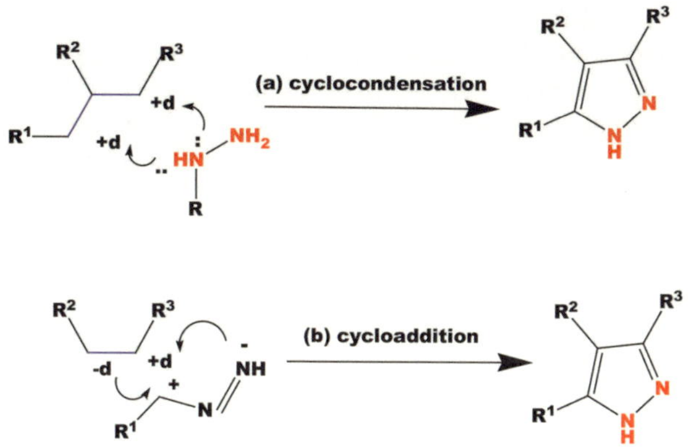

Figure 2.
Traditional methods for synthesizing Pyrazoles are shown in scheme (a) cyclocondensation; cycloaddition (b) [7].

Recently, new methods such multicomponent one-pot procedures, photoredox reactions, and transition-metal catalyzed reactions have been added to these two standard tactics to improve them. In this part, cyclocondensation will be covered first, then cycloaddition [7].

As [NN] synthons, substituted or unsubstituted hydrazines are easily accessible for the production of pyrazole derivatives. When hydrazines react with 1,3-dielectrophilic compounds like 1,3-dicarbonyl or with structures which are carbonyl, such as enones, ynones, and vinyl ketones possessing a leaving group, pyrazole molecules can be produced [8]. Hydrazines and comparable synthetic equivalents can efficiently condense to form substituted pyrazoles from 1,3-diketones, β-ketoesters, 2,4-diketoesters, and related compounds. The 1, 3-diketone and the appropriate hydrazine were cyclocondensed to create a series of powerful carbonic anhydrase, α-glycosidase, and cholinesterase enzyme inhibitors 1 [9]. Wang and co-workers reported a moderate and acid-free condensation of 1,3-diketones with substituted hydrazines to produce the 1,3,5-trisubstituted and completely substituted Pyrazoles [10]. Hydrazines and α-enones can be combined to produce pyrazolines, which can then be oxidized to produce the equivalent Pyrazoles. Iodine was used by Zhang *et al.* to mediate the creation of oxidative intramolecular C-N bonds, and the intermediate hydrazones were then cycled to produce Pyrazoles [11]. Ding *et al.* also described an air-promoted photoredox cyclization of substituted hydrazines with activated alkene (Michael addition reaction acceptors) to produce the corresponding Pyrazoles with good to outstanding yields [12]. Harigae *et al.* reported synthesizing 3,5-disubstituted pyrazoles in one pot with good yields using a regioselective method [13]. 3,5-disubstituted 1H-pyrazoles were also produced using propargylic alcohols, the reduced form of ynones [14]. According to Guo *et al.*, β-amino vinyl ketone may cyclize with tosyl hydrazine in water to produce completely substituted pyrazoles when iodine and tert-butyl hydroperoxide (TBHP) were introduced [15].

1,3-dipolar cycloaddition plays a significant role in creating substituted Pyrazoles due to its inherent high regioselectivity and efficiency [16]. As a departing group, bromine works well. In order to create 3,5-diaryl-4-bromopyrazoles, Sha *et al.* used gemdibromoalkene as the substrate and devised a straightforward, highly effective, and regioselective approach [17]. Li and colleagues described a cycloaddition of dicarboxylic alkynes and hydrazines that was catalyzed by rhodium [18]. Kobayashi et al.

proposed a one-pot, multicomponent method for creating multisubstituted Pyrazoles starting with primary alcohols [19]. In order to assemble monosubstituted Pyrazoles, Yi et al. reported a brand-new silver-mediated [3 + 2] cycloaddition of alkynes and N-isocyanoiminotriphenylphosphorane (NIITP) [20]. Aldehyde hydrazones and maleimides were combined in a moderate reaction by Zhu et al. that used CuCl as a catalyst to produce dihydropyrazoles [21].

All of the [NN] pieces in each of the pyrazole synthesis methods discussed above were derived from azo compounds or hydrazine derivatives. Recent research by Pearce and colleagues describes an unique fragment combination mode [NC] + [CC] + [N] that produces multi-substituted Pyrazoles [22].

The numerous pharmaceutical uses of Pyrazoles have sped up the methodological advancement of pyrazole synthesis. Many general and practical methods, such as the use of transition-metal catalysts, photoredox reactions, one-pot multicomponent processes, new reactants, and novel reaction types, have resulted in fruitful advancements in the fields of the synthesis and functionalization of pyrazole derivatives over the past ten years [7]. A number of noteworthy biological properties of this molecule include those that are antibacterial, anti-inflammatory, anti-cancer, analgesic, anticonvulsant, anthelmintic, antioxidant, and herbicidal. Considering that Pyrazoles are heterocyclic planar five-membered rings, the research suggests that they have a variety of pharmacological effects [4].

1.1 Strategies for pyrazole synthesis

Pyrazoles are the five-membered heterocycles that constitutes several derivatives or compounds which are useful in various fields like drugs, dyes and in organic synthesis. In this section we represents description and discussion on most of the synthetic methods or strategies of pyrazole heterocyclic system.

There are various routes for pyrazole nucleus synthesis which is described as below:

1. Multicomponent approach
2. Dipolar cycloadditions
3. Cyclocondensation of hydrazine with carbonyl system
4. Heterocyclic system

1.2 Multicomponent approach

The multicomponent approach is used for synthesis of pyrazole nucleus by performing one pot synthesis reaction to get high yield of product.

1.2.1 In situ *formation of carbonyl derivatives*

The 3,5-substituted pyrazole derivatives 4 can be synthesized in good yield by the treatment of terminal alkynes 1 with aromatic aldehyde, molecular iodine and hydrazines. It is a very simple and practical method for the preparation of 3,5-substituted pyrazole [13].

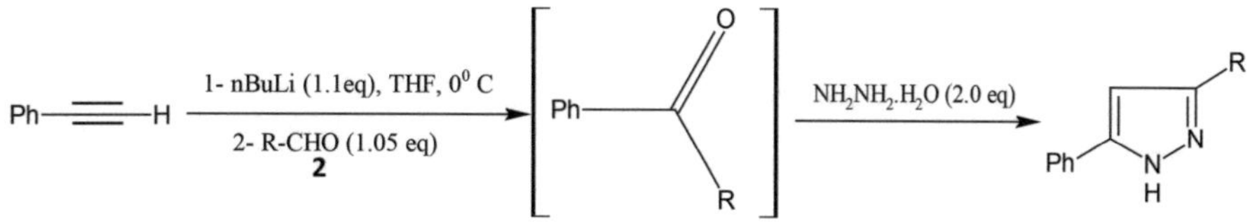

1 **3** **4**

The 1,3,5-substituted pyrazoles 6 was prepared by palladocatalyzed carbonylation of acetylenic acids on aryl iodides 5 in the presence of hexacarbonyl molybdenum with excellent yield [23].

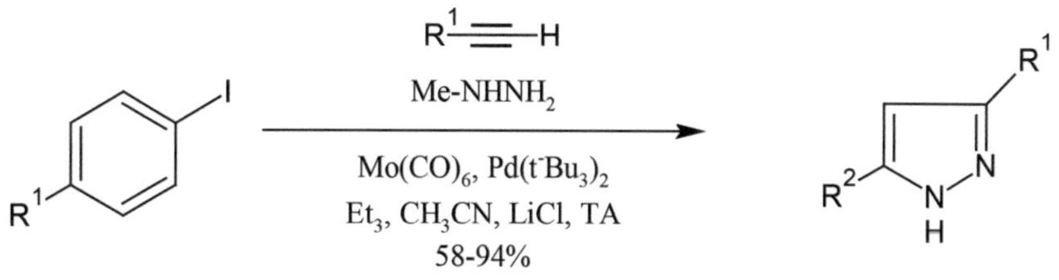

R^1 =NO_2, CO_2ET, OME, Ac
R^2 = Ph, N-Bu

5 **6**

1.2.2 In situ *formation of β-Aminoenones*

The β Aminoenones synthesized by via coupling between alkyne 8 and an oxime 7 in dimethylformamide which was transformed into pyrazoles 10 with the addition of hydrazine in one pot procedure in good yield [24].

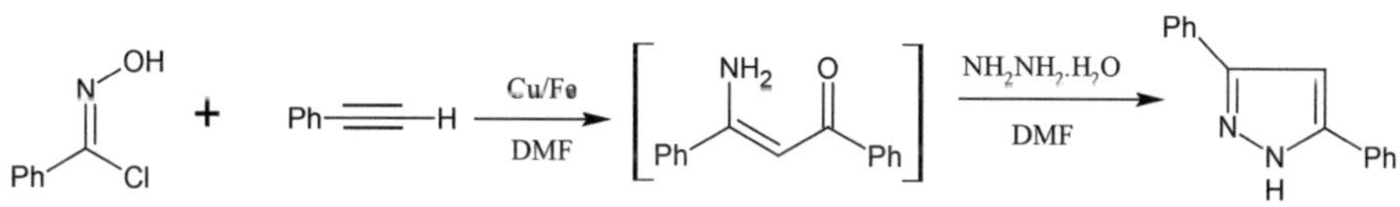

7 **8** **9** **10**

1.2.3 In situ *formation of a Hydrazone*

It is a novel reaction in which the cyclization of diethyl oxalate 12 with the dianions of hydrazones 11 afforded the pyrazole-3-carboxylates 13 in good yields [25].

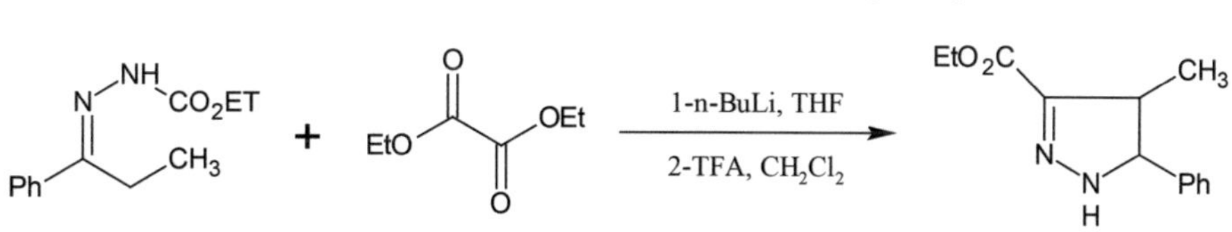

11 **12** **13**

The condensation of hydrazine in the presence of phosphorus oxychloride gives the 4-formyl pyrazole 15 which is called as Vilsmeier-Haack reaction [26].

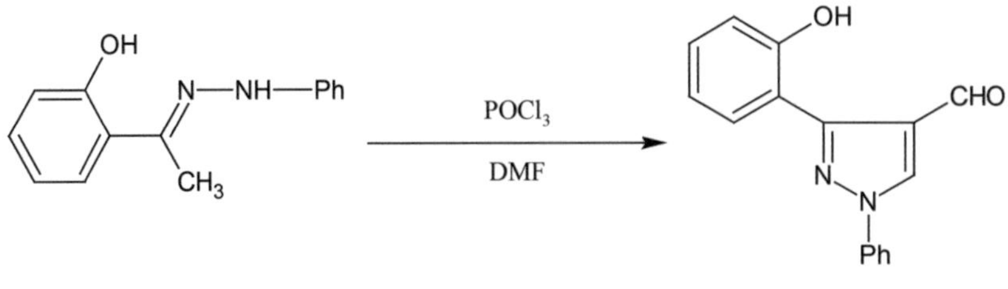

14 **15**

1.2.4 In situ *formation of diazo compounds*

The Aggarwal team has developed a multicomponent process in which diazo 17 derivatives are generated in situ from various aldehydes 16 and tosylhydrazines, thus limiting the risks associated with the isolation of these compounds. These are then used in a 1,3-dipolar cycloaddition reaction to give corresponding pyrazoles 18 and 19 Diazo compounds derived from aldehydes were reacted with terminal alkynes to furnish regioselectively 3,5-disubstituted pyrazoles in 24–67% yields [27].

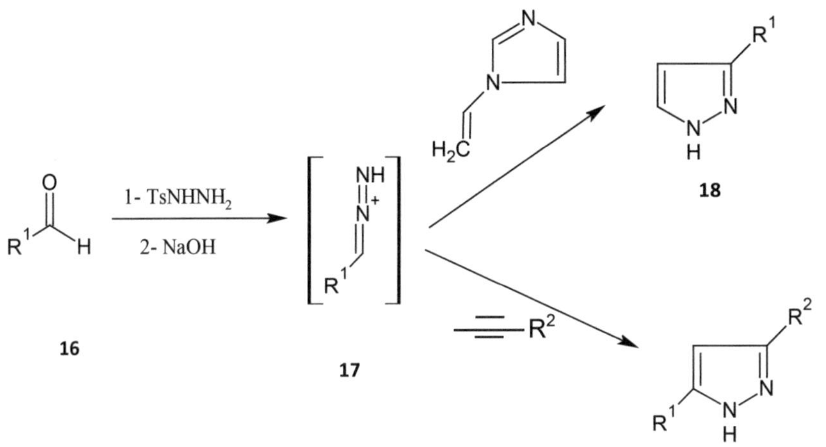

16 **17** **18** **19**

1.2.5 Ring opening reaction

A palladium-catalyzed ring opening reaction of 2H-azirines with hydrazones provides polysubstituted pyrazoles 23 with a wide substrate scope [28].

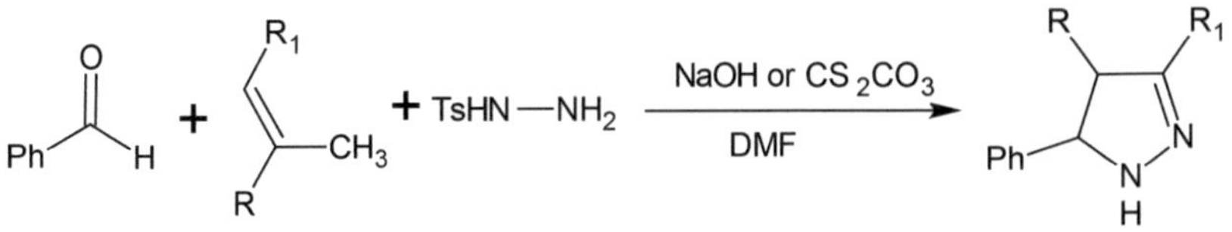

20 **21** **22** **23**

1.2.6 Multicomponent reaction

Pyrazole or isoxazole derivatives 27 are prepared by a palladium-catalyzed four-component coupling of a terminal alkyne 24, hydrazine 25 (hydroxylamine), carbon monoxide under ambient pressure, and an aryl iodide 26 [29].

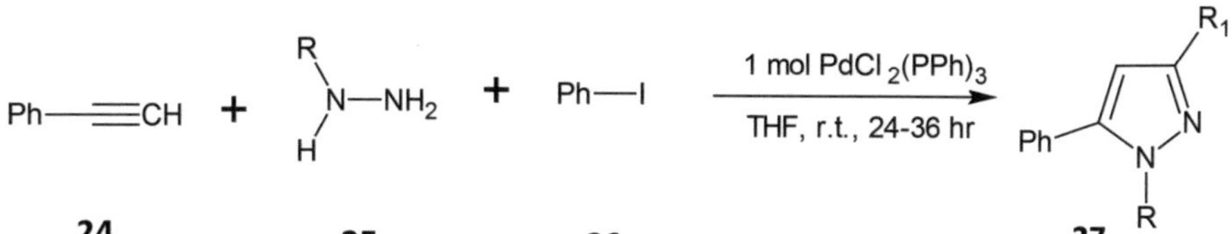

24 25 26 27

1.3 Dipolar cycloadditions

In this method the pyrazole nucleus was synthesized by the cycloaddition between an alkyne and 1,3-dipolar compounds such as diazo compounds.

1.3.1 Cycloaddition of Diazocarbonyl compounds

The action of ethyl diazoacetate 29 on phenylpropargyl 28 in triethylamine and in the presence of zinc triflate as a catalyst; the 1,3-dipolar cycloaddition reaction, leads to the corresponding pyrazole 30 in good yield (89%) [30].

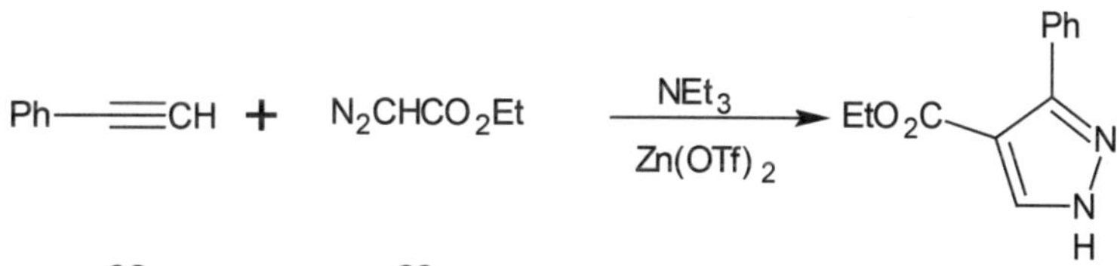

28 29 30

1.3.2 Cycloaddition of ethyl diazoacetate

A facile one-pot procedure for the synthesis of pyrazole-5-carboxylates 31 by 1,3-dipolar cycloaddition of ethyl diazoacetate 32 with methylene carbonyl compounds utilizing 1,8-diazabicyclo[5.4.0]undec-7-ene (DBU) as base and acetonitrile as solvent [31].

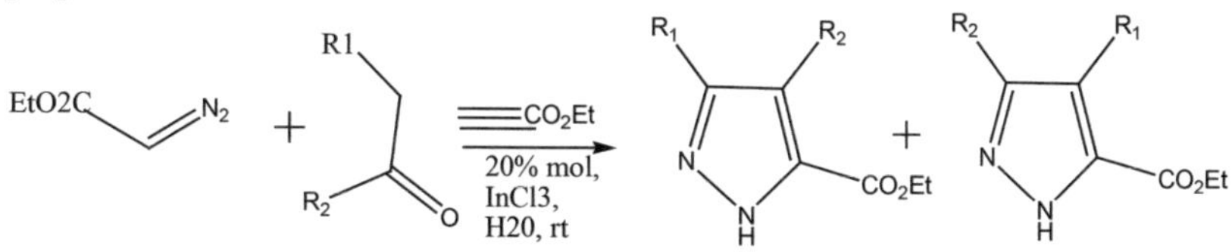

31 32 33 34

1.3.3 Cycloaddition of acetylides with diazocarbonyl compounds

A direct and efficient access towards 3-acylpyrazoles 36 that involves the copper-promoted cycloaddition of acetylides 35 with diazocarbonyl compounds under mild conditions. A wide variety of substituents is tolerated at both the acetylide and the diazo compound [32].

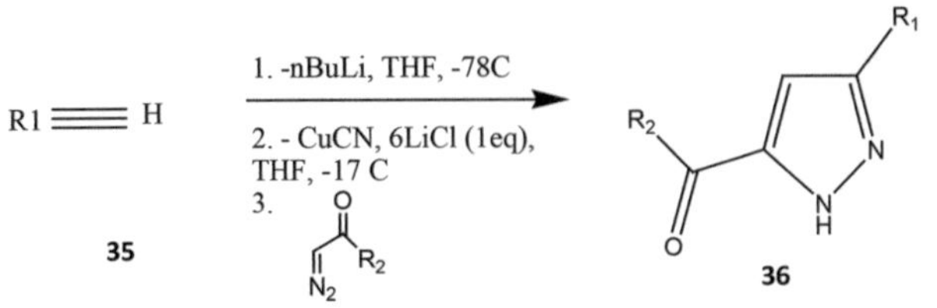

35 36

1.3.4 The syndones

The pyrazoles can be obtained by a cycloaddition reaction of sydnones. The synthesis of a trisubstituted pyrazole 39, by 1,3-dipolar cycloaddition of arylsydnones and unsaturated ketone in dry xylene [33].

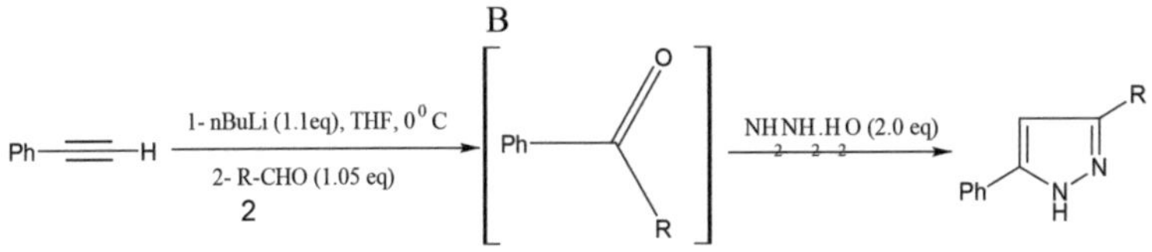

37 38 39

1.3.5 Cycloaddition of N-isocyanoiminotriphenylphosphorane

A silver-mediated [3 + 2] cycloaddition of *N*-isocyanoiminotriphenylphosphorane as "CNN" building block to terminal alkynes provides pyrazoles 42. *N* isocyanoiminotriphenylphosphorane is a stable, safe, easy-to-handle, and odorless solid isocyanide 41. The reaction offers mild conditions, broad substrate scope, and excellent functional group tolerance [20].

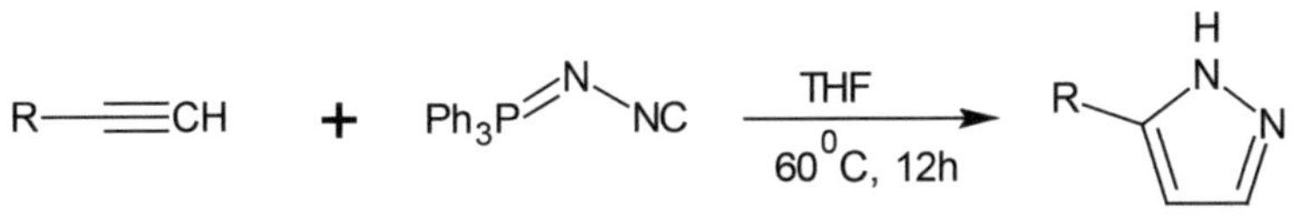

40 41 42

1.3.6 Cycloaddition reaction of dialkyl azodicarboxylates

A phosphine-free [3 + 2] cycloaddition reaction of dialkyl azodicarboxylates 44 with substituted propargylamines 43 provides functionalized pyrazoles 45 in good yields and high selectivity at room temperature [34].

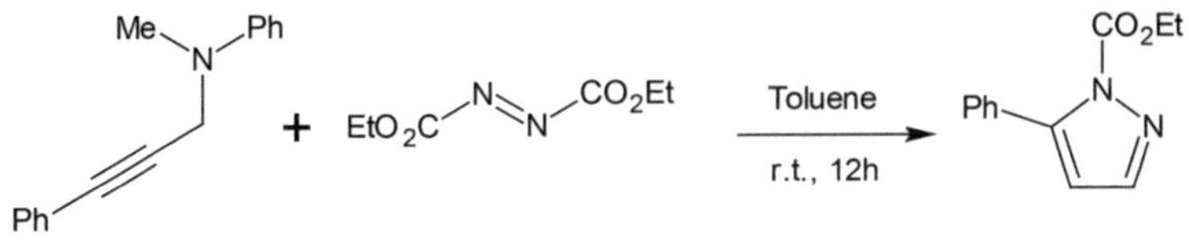

43 44 45

1.3.7 Cycloaddition of diazo compounds and alkynyl bromides

A simple, highly efficient, 1,3-dipolar cycloaddition of diazo compounds 46 and alkynyl bromides 47 gives 3,5-diaryl-4-bromo-3*H*-pyrazoles 48 or the isomerization products 3,5-diaryl-4-bromo-1*H*-pyrazoles 49 in good yields. The diazo compounds and alkynyl bromides were generated in situ from tosylhydrazones and *gem*-dibromoalkenes, respectively. The reaction system exhibited high regioselectivity and good functional group tolerance [35].

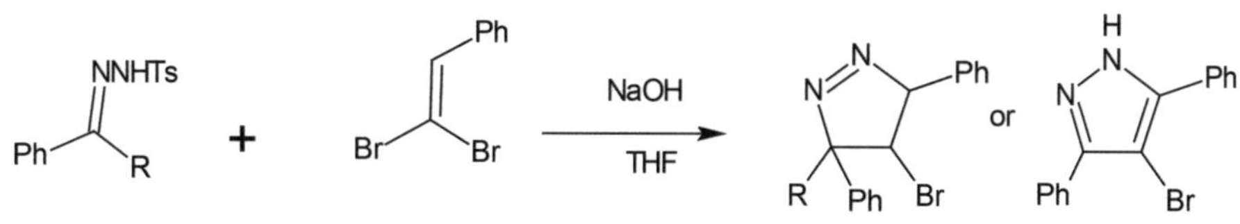

46 **47** **48** **49**

1.3.8 Cycloaddition of diazoacetonitrile and nitroolefins

A transition-metal-free [3 + 2] cycloaddition reaction between diazoacetonitrile 51 and nitroolefins 50 provides multisubstituted cyanopyrazoles 52. This protocol offers mild reaction conditions, broad substrate scope, good yields, and regioselectivities. A one-pot three component reaction of nitroolefins with diazoacetonitrile and alkyl halides also provides multisubstituted cyanopyrazoles in good to high yields [36].

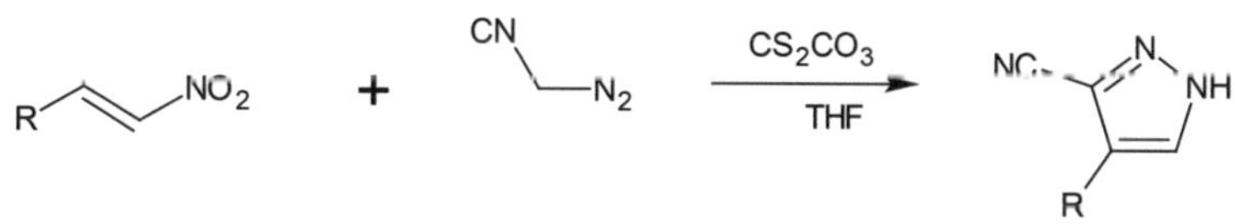

50 **51** **52**

1.3.9 Cyclocondensation of hydrazine with carbonyl system

This is a leading method used for obtaining substituted pyrazoles is a cyclocondensation reaction between an appropriate hydrazine acting as a bidentate nucleophile and a carbon unit like a 1,3-dicarbonyl compound, a 1,3-dicarbonyl derivatives or an unsaturated ketone.

1.3.10 From 1,3-diketones

The cyclocondensation of the 1,3-dicarbonyl compounds 53 with the hydrazine derivatives is a simple and rapid approach to obtain polysubstituted pyrazoles 54 and 55. The first synthesis of the substituted pyrazoles was carried out in 1883 by Knorr et al. [11] who reacted diketone with hydrazine derivatives to give two regioisomers [37].

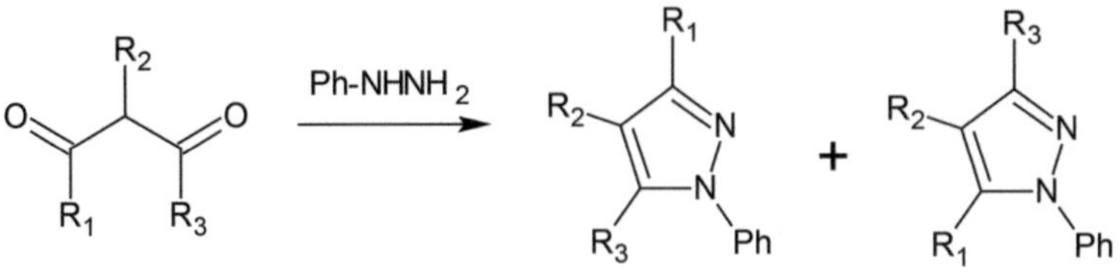

53 **54** **55**

The condensation of phenylhydrazine 57 with the trifluoromethyl)-1,3-diketone 56 in ethanol, affording 1,3,4,5-substituted pyrazole in good yield (63%) [38].

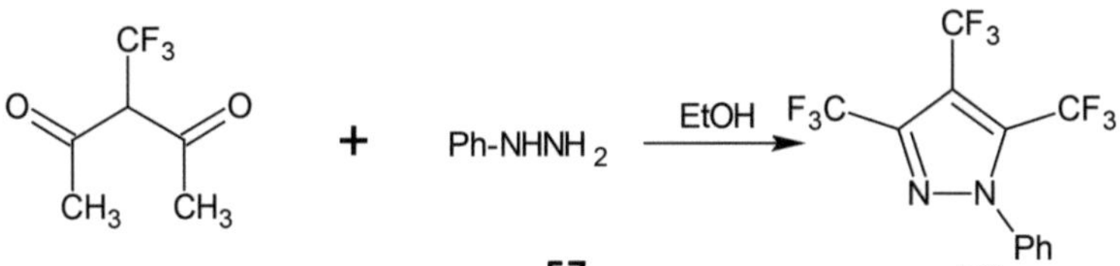

56 **57** **58**

1.3.11 From Acetylenic ketones

The cyclocondensation reaction of hydrazine derivatives 60 on acetylenic ketones 59 to form pyrazoles. The reaction between hydrazine derivatives and acetylenic ketones forms pyrazoles and the reaction again results in a mixture of two regioisomers 61 and 62 [39].

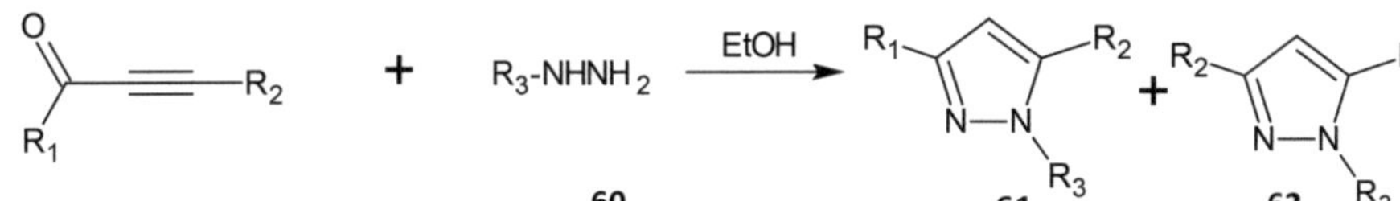

59 **60** **61** **62**

The cyclocondensation of acetylenic ketones 63 on methylhydrazine or aryl hydrazines 64 in ethanol, which provides two difficultly separable regioisomeric pyrazoles 65 and 66 [40].

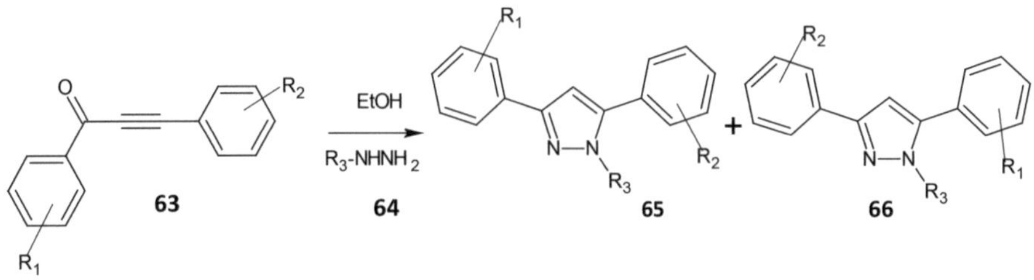

63 **64** **65** **66**

1.3.12 From vinyl ketones

The cyclocondensation reaction between an ethylenic ketone and a hydrazine derivative results in the synthesis of pyrazolines which, after oxidation, provide the pyrazole ring.

The condensation of an ethylenic ketone 67 with p-(4-(tert-butyl)phenyl) hydrazine 68 in the presence of copper triflate and 1-butyl-3-methylimidazolium hexafluorophosphate bmim] (PF6) as catalysts, to access pyrazoline 69. The corresponding 1,3,5-trisubstituted pyrazole was obtained after oxidation in situ of this pyrazoline [41].

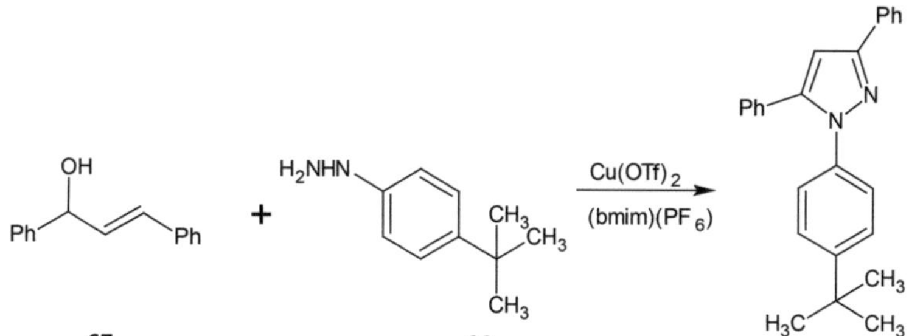

67 68 69

Cyclocondensation of the ethylenic ketone 70 with phenylhydrazine (1.2 eq.) in acetic acid and in the presence of iodine (1.0 eq.) afforded the corresponding pyrazole 71 in good yield (70%) [42].

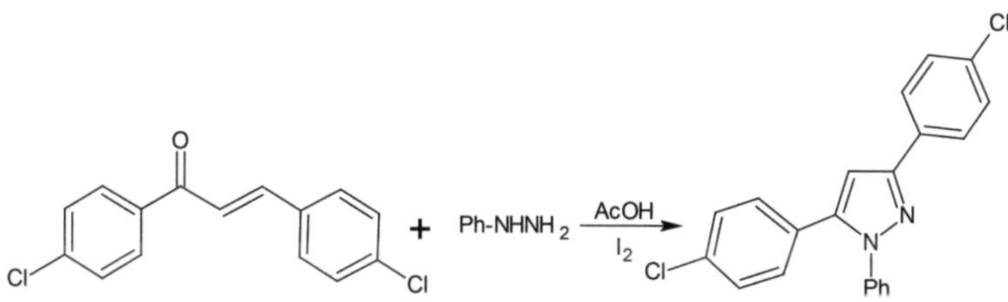

70 71

1.4 From heterocyclic system

1.4.1 From imidazole

Cycloaddition of (5Z)-1-acyl-5-(cyanomethylidene)-3-methylimidazolidine-2,4-diones 72 with arylhydrazonyl chloride under basic conditions to give pyrazole-5-carboxamides 73 in moderate 27–40% yields [43].

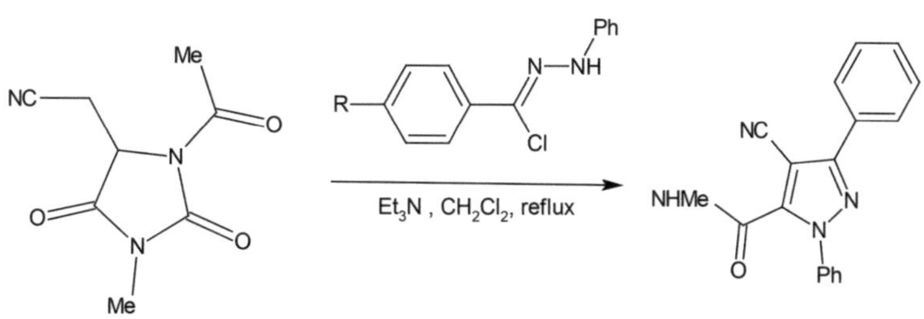

72 73

1.4.2 From oxazole

5-Trifluoromethyl-3-hydroxypyrazoles 75 were obtained in good yield (46–95%) by heating phenylhydrazine and 4-trifluoroacetyl-1,3-oxazolium-5-olates 74 under reflux of benzene [44].

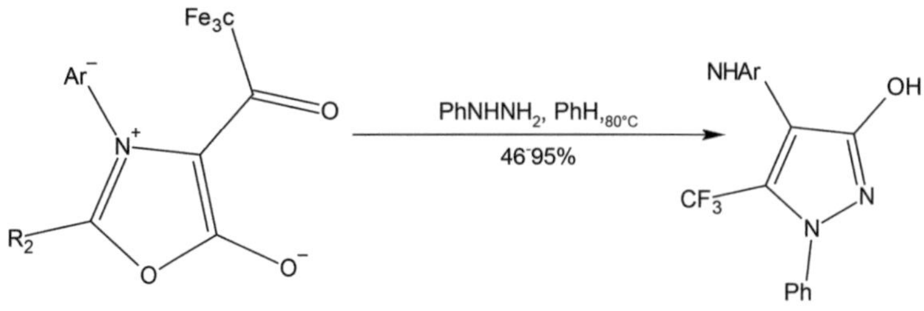

74 75

1.4.3 From pyrimidine

The reaction of nitropyrimidine 76 with arylhydrazines in methanol at room temperature, to afford 4-nitro-3,5-diaminopyrazoles 77 in yields of 21–61% [45].

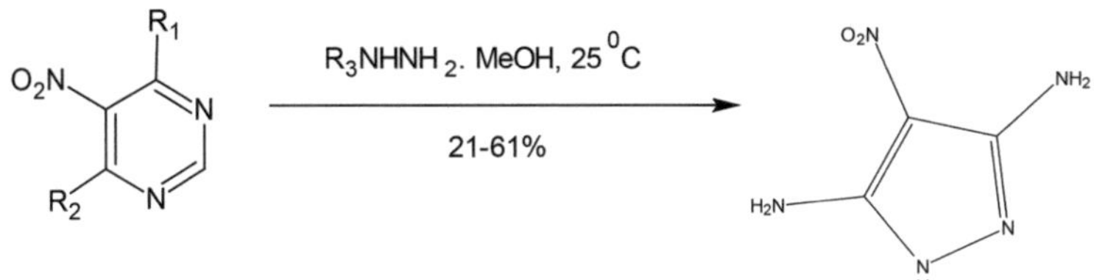

76 77

1.4.4 From Tetrazole

Tetrazolyl acroleins 78 reacts with fumaronitrile in xylene at 140°C to give the corresponding pyrazole formation 79 [46].

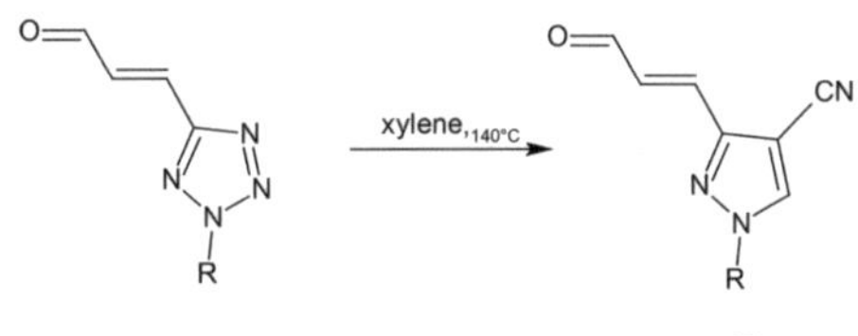

78 79

2. Applications of pyrazole scaffold

Pyrazole moiety have wide applications and are effective therapeutic scaffolds that display a wide range of biological actions as listed in **Figure 3**.

The synthesis of a novel, powerful family of 5-reductase and aromatase inhibitors derived from 1, 2, 3-triazole derivative uses pyrazole-4-carbaldehyde as the starting material. The appropriate Schiff bases were created by condensation of the starting material with active methylene and various amino pyrazoles. In contrast, starting material was treated in a single pot with ethyl cyanoacetate and thiourea to produce pyrazolo-6-thioxopyridin-2-[3H]-one. Additionally, beginning chemical was reacted with p-methoxy acetophenone, which then reacted with each of the ethyl cyanoacetates to create an unsaturated chalcone derivative. The following derivatives showed 5-α reductase inhibitor and aromatase inhibitor activity prominently

Figure 3.
Biological activity of pyrazole moiety.

compared to reference drug. It is due to the pyridine moiety was present and it was connected to the pyrazoline and 1, 2, 3-trizole moieties (**Figure 4**) [47].

Sequence of pyrazolinyl and pyrazolyl pregnenolones were produced and their ability for both series to inhibit 5- α reductase was examined by multiple step synthesis and pregnenolone used as starting material. Cyclization reaction was found to be main step in this synthesis. Derivatives 4b, 4c and 6b were found to be more active for this activity as it contains fluoro group and para position chloro group. 4b and 4c contain Ar ring as follows (**Figures 5** and **6**) [48].

Khalillulha H et al. synthesized pyrazole derivatives covering 1, 4-dioxane ring which have low-molecular-weight molecules that are simple to manufacture. On the other hand, silybin is a complicated, highly molecular substance that is difficult to manufacture. In addition, the substances are anticipated to be easily metabolizable, in contrast to silybin, which is straightforward and has a low molecular

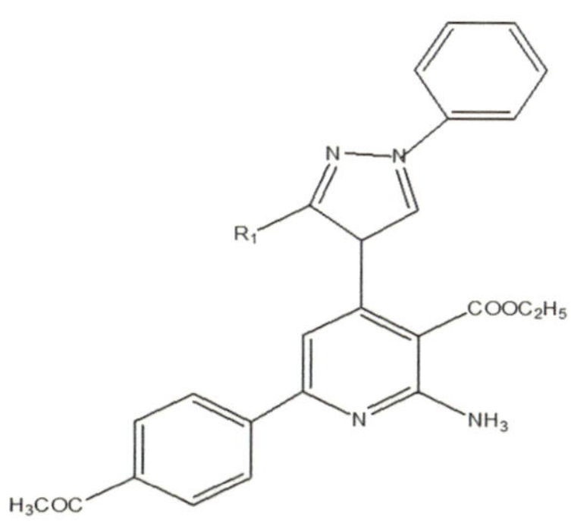

Figure 4.
Potent derivative showing 5- α reductase inhibitor activity.

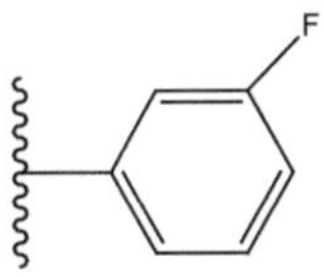

4a

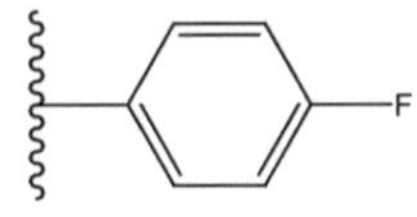

4b

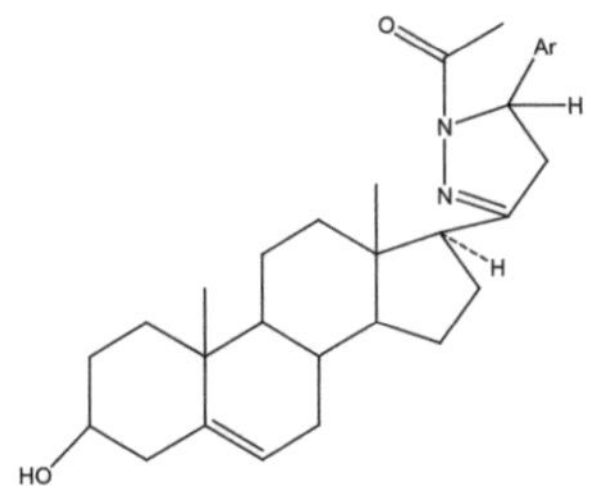

4c

Figure 5.
Potent compound of pyrazolinyl pregnenolone.

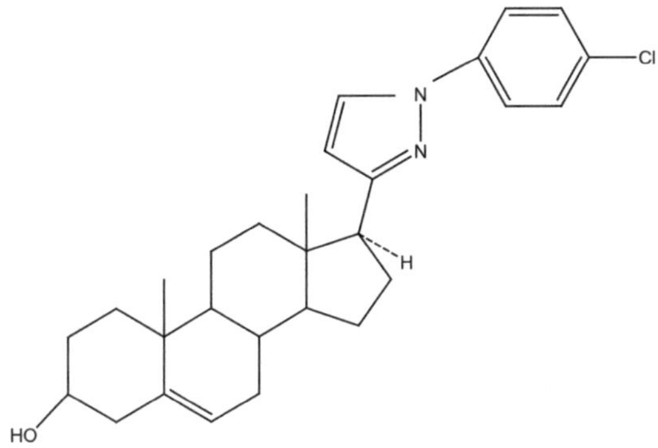

Figure 6.
6b derivatives of pyrazolyl pregnenolone.

weight. These derivatives were prepared by Claisen–Schmidt condensation reaction using substituted acetophenone chalcones. Rats' liver damage caused by CCl_4 was used in a hepatotoxicity investigation (**Figure** 7) [49].

MAO enzyme having EC number 1.4.3.4 that contain flavin Hence inhibitory activity was done by different researchers like Chimenti F et al. synthesized 1-Thiocarbamoyl-3, 5-diaryl-4, 5-dihydro-(1H) - pyrazole Derivatives by using

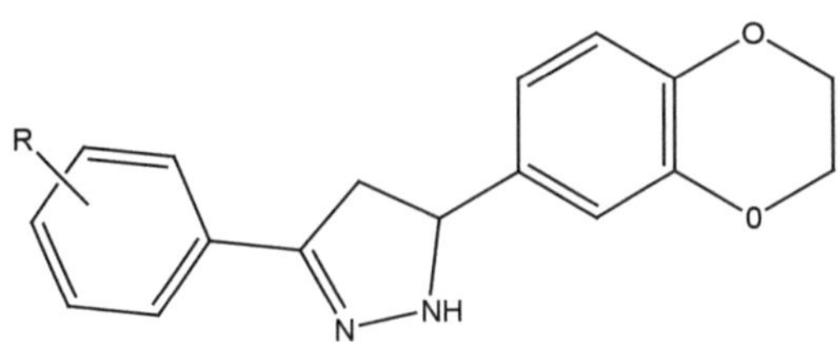

Figure 7.
Pyrazoline derivative.

chalcones, thiosemicarbazide with potassium hydroxide in ethanol as solvent. Its isomers were also tested for MAO inhibitory action. Substrate used was kynuramine. Derivative containing following group those are highly active for MAO inhibition. To check bonding affinity for MAO computational methods were also carried out. Following derivative shows potent activity (**Figure 8**) [50].

Secci D prepared pyrazole derivatives are produced either by intermolecular [3 + 2] cycloadditions of 1, 3-dipoles to alkynes excellent inhibitory effect primarily against MAO-B contain halo group when placed in the para position of the aryl group (**Figure 9**) [51].

Alam MS et al. used schiff base ligand 4-amino-1, 5-dimethyl-2-phenylpyrazole-3-one with benzaldehyde to form single crystal which was checked by X-ray diffraction analysis. For detection of antioxidant activity they used DPPH Radical Scavenging Activity assayed by Blois method (**Figure 10**) [52].

Gressleri, A et al. prepared derivatives by refluxing for 24 hours with the use of 1, 5-diarylpenta-1, 4-dien-3-ones, aminoguanidine hydrochloride, triethylamine, and ethanol. For antioxidant activity DPPH was used and the color of the DPPH shifts from violet to yellow. From all the synthesized derivatives 1-carboxyamidino 1*H*-pyrazole derivatives showed potent activity. [53] Hanam A et al. made an effort to produce new heterocycles, 2-cyano-3-(1,3-diphenyl-1H-pyrazol-4-yl) acryloyl chloride was subjected to reactions with various mono-, 1,2-, 1,3-,1,4-, and 1,5-binucleophiles. Assay was performed by using ABTS [2, 20 -azinobis-(3-

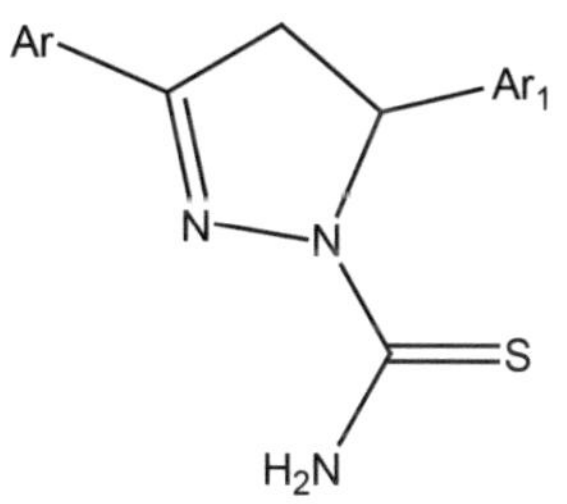

Figure 8.
Pyrazole derivatives.

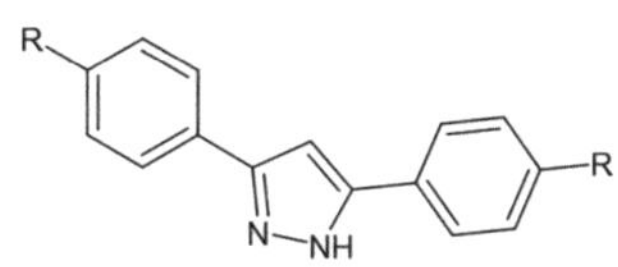

Figure 9.
Cycloaddition derivative.

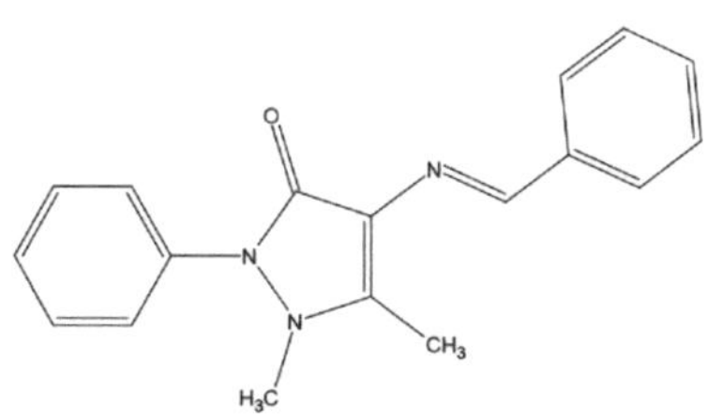

Figure 10.
Pyrazole Schiff base derivative.

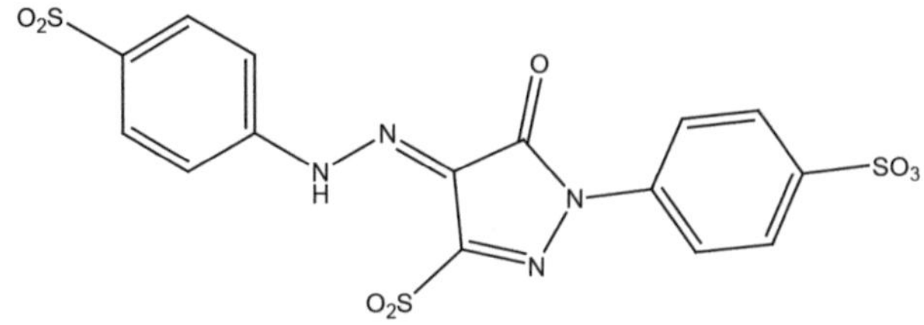

Figure 11.
Structure of tetrazine.

ethylbenzthiazoline-6-sulphonic acid)] method. In that comparison was done by ascorbic acid as standard (**Figure 11**) [54, 55].

Alsayari A, et al. prepared pyrazole derivatives by A sulforhodamine B assay method which was used to assess the antiproliferative effects on cancer cell lines were identified: hepatocellular carcinoma (HepG2), colorectal carcinoma (HCT-116), and breast cancer (MCF-7). These are effective xanthine oxidase inhibitory action (IC50 0.83 M) and a high IC50 against the human colon cancer cell line (**Figure 12**) [56].

Different anticancer activity showing in pyrazole moiety were listed here, the preclinical or early-phase clinical trials for these described drugs were passed (**Figure 13**) [57].

Three different breast cancer cell lines, such as MDA-MB-231, MCF-7, and 4 T1, all were used by authors for breast cancer cell line study as well as HepG2 liver cancer cells, were used to test the cytotoxicity of pyrazole. Synthesized pyrazole 13 derivative (5-oxo-N′-(2-oxoindolin-3-ylidene)-3-phenyl-4,5-dihydro-1H-pyrazole-1-carbothiohydrazide), mechanism was discussed as caused 4 T1 cells to die by preventing wound healing and colony formation, delaying the G0/G1 phase of the cell cycle, activating p27 levels, and most likely inducing apoptosis through DNA intercalation. IC50 value of this synthesized derivative was found to be 25 ± 0.4 μm (**Figure 14**) [58].

Human lung carcinoma A549 cells, murine P388 leukemia cells, and human ovarian adenocarcinoma A2780 cells were all tested with pyrazole derivatives by synthesizing this derivatives 2-pyridinyl moiety containing compound 12 is mostly active as shown below (**Figure 15**) [59].

Cyclin-dependent kinases (CDKs), a subfamily of the protein kinase which control the cell cycle. Given that cyclin E is selective for CDK2 and the dysregulation of particular cancer types, CDK2 is an alluring target for malignancies with particular genotypes. According to the ongoing clinical trials, CDKs inhibitor, specifically CDK2/cyclin A-E, has the potential to be a reliable cancer target. The majority of the pyrazole scaffolds have demonstrated CDK2 inhibitor selectivity and potency [60, 61]. The antibacterial activity of a series of 1H-pyrazole-3-carboxylic acid derivatives against Bacillus cereus, Staphylococcus aureus, Escherichia coli, and Pseudomonas putida were assessed by Akbas et al. Having antibacterial action against both Gram-positive

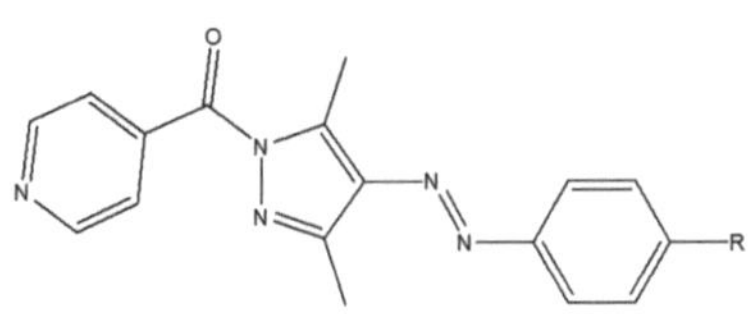

Figure 12.
Pyrazole derivative as anticancer.

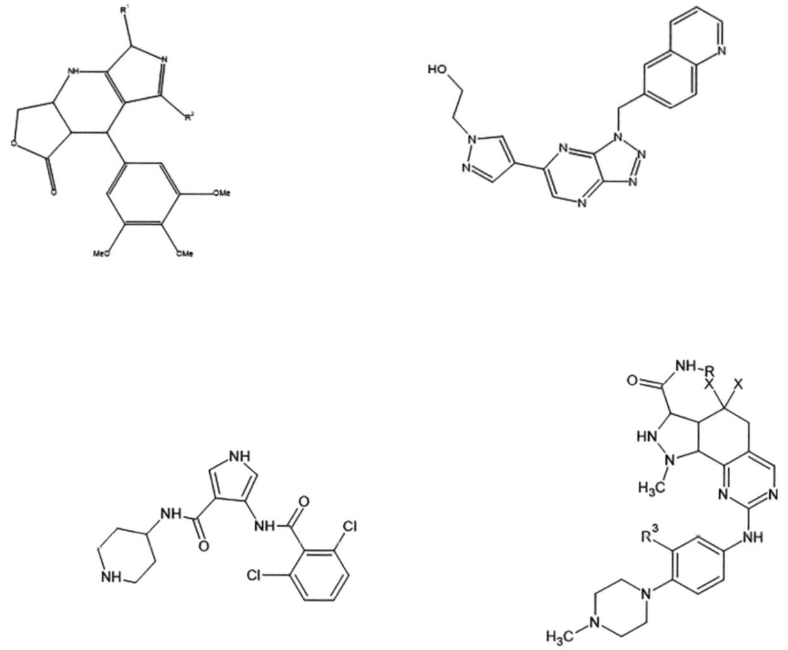

Figure 13.
Pyrazole derivative in preclinical study.

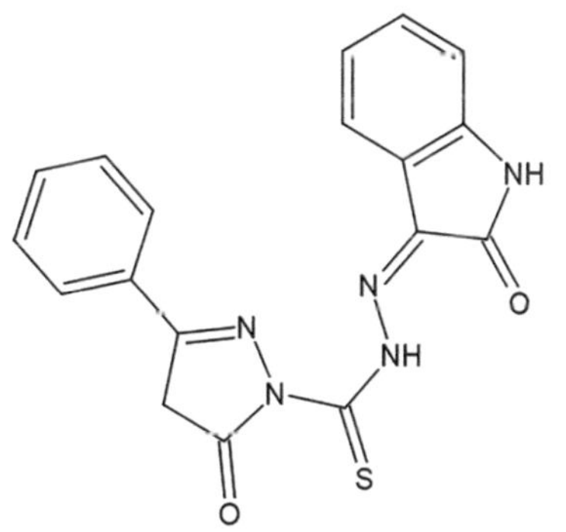

Figure 14.
Pyrazole 13 derivative.

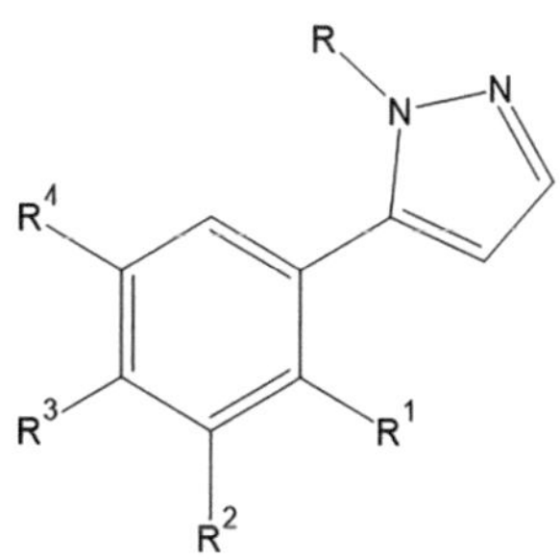

Figure 15.
Pyrazole 12 derivative.

and Gram-negative bacteria, the results revealed that compound 5c was the best compound in the series (**Figure 16**) [62].

For the detection of pyrazole pesticides such as fibronil in water samples of environment, method solid-phase extraction (SPE) approach combining with high-performance liquid chromatography as adsorbent multi-walled carbon nanotubes was developed by Ma Jiping et al. [63] (**Figure 17**).

An essential scaffold is 3-bromo-1-(3-chloropyridin-2-yl)-1H-pyrazole-5-carbohydrazide, which is produced by Wang Y et al. as an insecticidal compounds. By using species *Helicoverpa armigera* and *Plutella xylostella* as standard tebufenozide. Fluoro group containing compound showed potent activity at fourth position and decreased by iodo group [64] (**Figure 18**).

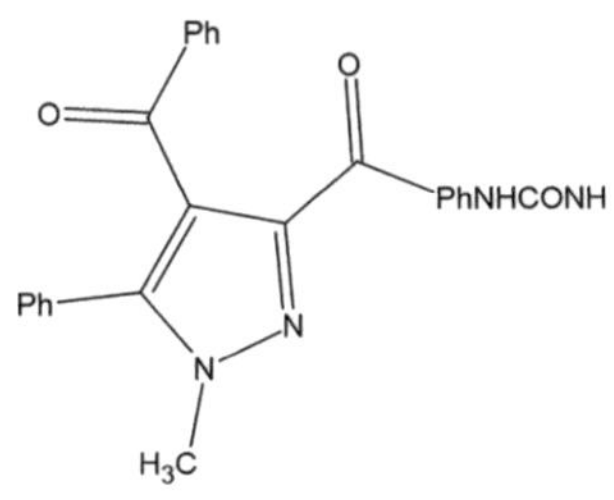

Figure 16.
5C (Pyrazole carboxylate derivative).

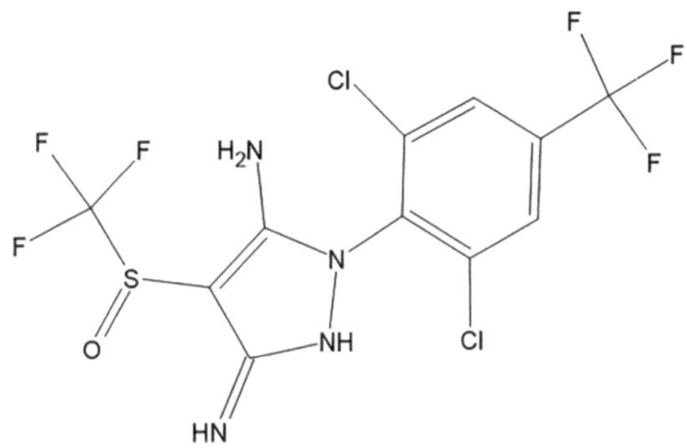

Figure 17.
Fipronil.

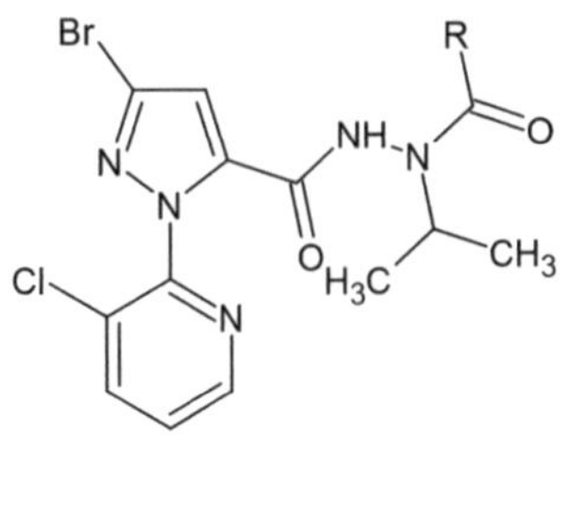

Figure 18.
Pyrazole derivative.

3. Conclusion

Pyrazole nucleus and its various derivatives have been studied extensively in the past and found to be effective in a variety of pharmacological and pathological conditions. Its structure allows for numerous applications in fields such as technology, medicine, and agriculture. In the agrochemical industry, pyrazole derivatives, in particular, have a long history of use as herbicides, insecticides, fungicides, and acaricides. We have covered a variety of pyrazole-related synthetic strategies and biological applications in this chapter. Even though a wide variety of pyrazole synthesis techniques have been developed by organic chemists, and new techniques are continually being developed, the creation of novel, regioselective pyrazole-forming processes remains an exciting area of study. This chapter serves as a foundation and helps researchers to create novel synthetic approaches and potent new compounds.

Author details

Deweshri Nandurkar*, Kishor Danao, Vijayshri Rokde, Ruchi Shivhare and Ujwala Mahajan
Department of Pharmaceutical Chemistry, Dadasaheb Balpande College of Pharmacy, Nagpur, Maharashtra, India

*Address all correspondence to: kerzarepritee@gmail.com

References

[1] Eicher T, Hauptmann S, Speicher A. The Chemistry of Heterocycles: Structure, Reactions Synthesis and Applications. 2nd ed. Weinheim: Wiley-VCH; 2003

[2] Alam MJ, Alam O, Alam P, Naim MJ. A review on pyrazole chemical entity and biological activity. International Journal of Pharmaceutical Sciences and Research. 2015;**6**:1433-1442

[3] Dash B, Karim S. Pyrazoline heterocyclic: A review. International Journal of Pharmaceutical Sciences and Research. 2021;**12**:2570-2588

[4] Jamwal A, Javed A, Bhardwaj V. A review on Pyrazole derivatives of pharmacological potential. Journal of Pharmaceutical and Biological Sciences. 2013;**3**:114-123

[5] Dewangan D, Kumar T, Alexander A, Nagori K, Tripathi DK. Pyrazole: Their chemistry and pharmacological potentials: A review. Journal of Current Pharma Research. 2011;**1**(4):369-374

[6] National Center for Biotechnology Information. PubChem Compound Summary for CID 1048, Pyrazole. 2022. Available from: https://pubchem.ncbi.nlm.nih.gov/compound/Pyrazole [Retrieved November 16, 2022]

[7] Li X, Yu Y, Tu Z. Pyrazole scaffold synthesis, functionalization, and applications in Alzheimer's disease and Parkinson's disease treatment (2011–2020). Molecules. 2021;**26**(5): 1202

[8] Karrouchi K, Radi S, Ramli Y, Taoufik J, Mabkhot YN, Al-Aizari FA, et al. Synthesis and pharmacological activities of pyrazole derivatives: A review. Molecules. 2018;**23**(1):134

[9] Turkan F, Cetin A, Taslimi P, Gulçin İ. Some pyrazoles derivatives: Potent carbonic anhydrase, α-glycosidase, and cholinesterase enzymes inhibitors. Archiv der Pharmazie. 2018;**351**(10):1800200

[10] Wang H, Sun X, Zhang S, Liu G, Wang C, Zhu L, et al. Efficient copper-catalyzed synthesis of substituted pyrazoles at room temperature. Synlett. 2018;**29**(20):2689-2692

[11] Zhang X, Kang J, Niu P, Wu J, Yu W, Chang J. I2 -mediated oxidative C-N bond formation for metal-free one-pot synthesis of di-, tri-, and tetrasubstituted pyrazoles from α,β-unsaturated aldehydes/ketones and hydrazines. The Journal of Organic Chemistry. 2014;**79**:10170-10178

[12] Ding Y, Zhang T, Chen QY, Zhu C. Visible-light photocatalytic aerobic annulation for the green synthesis of pyrazoles. Organic Letters. 2016;**18**: 4206-4209

[13] Harigae R, Moriyama K, Togo H. Preparation of 3,5-disubstituted pyrazoles and isoxazoles from terminal alkynes, aldehydes, hydrazines, and hydroxylamine. The Journal of Organic Chemistry. 2014;**79**:2049-2058

[14] Raji Reddy C, Vijaykumar J, Grée R. Facile one-pot synthesis of 3,5-disubstituted 1H-pyrazoles from propargylic alcohols via propargyl hydrazides. Synthesis. 2013;**45**:830-836

[15] Guo Y, Wang G, Wei L, Wan JP. Domino C-H sulfonylation and pyrazole annulation for fully substituted pyrazole synthesis in water using hydrophilic enaminones. The Journal of Organic Chemistry. 2019;**84**:2984-2990

[16] Fustero S, Sanchez-Rosello M, Barrio P, Simon-Fuentes A. From 2000

to mid-2010: A fruitful decade for the synthesis of pyrazoles. Chemical Reviews. 2011;**111**:6984-7034

[17] Sha Q, Wei YY. An efficient one-pot synthesis of 3,5-diaryl-4-bromopyrazoles by 1,3-dipolar cycloaddition of in situ generated diazo compounds and 1-bromoalk-1-ynes. Synthesis. 2013;**45**:413-420

[18] Li DY, Mao XF, Chen HJ, Chen GR, Liu PN. Rhodium-catalyzed addition-cyclization of hydrazines with alkynes: Pyrazole synthesis via unexpected C-N bond cleavage. Organic Letters. 2014;**16**: 3476-3479

[19] Kobayashi E, Togo H. Facile one-pot transformation of primary alcohols into 3-aryl- and 3-alkyl-isoxazoles and -pyrazoles. Synthesis. 2019;**51**:3723-3735

[20] Yi F, Zhao W, Wang Z, Bi X. Silver-mediated [3+2] cycloaddition of alkynes and N-isocyanoiminotriphenylphosphorane: Access to monosubstituted pyrazoles. Organic Letters. 2019;**21**:3158-3161

[21] Zhu JN, Wang WK, Jin ZH, Wang QK, Zhao SY. Pyrrolo[3,4-c] pyrazole synthesis via copper(I) chloride-catalyzed oxidative coupling of hydrazones to maleimides. Organic Letters. 2019;**21**:5046-5050

[22] Pearce AJ, Harkins RP, Reiner BR, Wotal AC, Dunscomb RJ, Tonks IA. Multicomponent pyrazole synthesis from alkynes, nitriles, and titanium imido complexes via oxidatively induced N-N bond coupling. Journal of the American Chemical Society. 2020;**142**:4390-4399

[23] Iizuka M, Kondo Y. Palladium-catalyzed Alkynylcarbonylation of aryl iodides with the use of Mo (CO) 6 in the presence of tBu3P ligand. European Journal of Organic Chemistry. 2007; **2007**:5180-5182

[24] Kovacs S, Novak Z. Copper on iron promoted one-pot synthesis of -aminoenones and 3,5-disubstituted pyrazoles. Tetrahedron. 2013;**69**: 8987-9893

[25] Dang TT, Dang TT, Fischer C, Görls H, Langer P. Synthesis of pyrazole-3-carboxylates and pyrazole-1,5-dicarboxylates by one-pot cyclization of hydrazone dianions with diethyl oxalate. Tetrahedron. 2008;**64**:2207-2215

[26] Lokhande P, Hasanzadeh K, Konda SG. A novel and efficient approach for the synthesis of new halo substituted 2-arylpyrazolo[4,3-c] coumarin derivatives. European Journal of Chemistry. 2011;**2**:223-228

[27] Aggarwal VK, de Vicente J, Bonnert RV. A novel one-pot method for the preparation of pyrazoles by 1,3-dipolar cycloadditions of diazo compounds generated in situ. The Journal of Organic Chemistry. 2003;**68**: 5381-5383

[28] Zhang G, Ni H, Chen W, Shao J, Liu H, Chen B, et al. Nickel-catalyzed direct amination of Arenes with Alkylamines. Organic Letters. 2013;**15**: 5967-5969

[29] Ahmed MSM, Kobayashi K, Mori A. One-pot construction of Pyrazoles and Isoxazoles with palladium-catalyzed four-component coupling. Organic Letters. 2005;**7**:4487-4489

[30] He S, Chen L, Niu Y-N, Wu L-Y, Liang Y-M. 1,3-dipolar cycloaddition of diazoacetate compounds to terminal alkynes promoted by Zn (OTf)2: An efficient way to the preparation of pyrazoles. Tetrahedron Letters. 2009;**50**: 2443-2445

[31] Gioiello A, Khamidullina A, Fulco MC, Venturoni F, Zlotsky S, Pellicciari R. New one-pot synthesis of pyrazole-5-carboxylates by 1,3-dipole cycloadditions of ethyl diazoacetate with _-methylene carbonyl compounds. Tetrahedron Letters. 2009;**50**:5978-5980

[32] Qi X, Ready JM. Copper-promoted cycloaddition of Diazocarbonyl compounds and Acetylides. Angewandte Chemie, International Edition. 2007;**46**: 3242-3244

[33] Chen F, Liu F-M, Shi H, Chen S-L. A facile access to 1,3,4-trisubstituted pyrazoles via 1,3-dipolar cycloaddition of 3-arylsydnones with ,-unsaturated ketones. Monatshefte für Chemie/ Chemical Monthly. 2013;**144**:879-884

[34] Zhang Y, Liu J, Jia X. Phosphine-free [3+2] cycloaddition of Propargylamines with Dialkyl Azodicarboxylates: An efficient access to Pyrazole backbone. Synthesis. 2018;**50**:3499-3505

[35] Sha Q, Wei Y. An efficient one-pot synthesis of 3,5-Diaryl-4-bromopyrazoles by 1,3-dipolar cycloaddition of In situ generated diazo compounds and 1-Bromoalk-1-ynes synthesis. Synthesis. 2013;**45**:413-420

[36] Chen Z, Zhang Y, Nie J, Ma J-A. Transition-metal-free [3 + 2] cycloaddition of Nitroolefins and Diazoacetonitrile: A facile access to multisubstituted Cyanopyrazoles. Organic Letters. 2018;**20**:2024-2027

[37] Knorr L. Einwirkung von acetessigester auf phenylhydrazin. European Journal of Inorganic Chemistry. 1883;**16**:2597-2599

[38] Ohtsuka Y, Uraguchi D, Yamamoto K, Tokuhisa K, Yamakawa T. Syntheses of 2-(trifluoromethyl)-1, 3-dicarbonyl compounds through direct trifluoromethylation with CF3I and their application to fluorinated pyrazoles syntheses. Tetrahedron. 2012;**68**: 2636-2649

[39] Moureu C, Delange R. Over some Acetylenketone and over a new method to the synthesis of -Diketones. Bulletin de la Société Chimique de France. 1901; **25**:302-313

[40] Bishop BC, Brands KM, Gibb AD, Kennedy DJ. Regioselective synthesis of 1,3,5-substituted pyrazoles from acetylenic ketones and hydrazines. Synthesis. 2004;**2004**:43-52

[41] Rao VK, Tiwari R, Chhikara BS, Shirazi AN, Parang K, Kumar A. Copper triflate-mediated synthesis 1,3,5-triarylpyrazoles in [bmim][PF6] ionic liquid and evaluation of their anticancer activities. RSC Advances. 2013;**3**: 15396-15403

[42] Ponnala S, Prasad SD. Iodine-mediated synthesis of 2-Arylbenzoxazoles, 2-Arylbenzimidazoles, and 1,3,5-Trisubstituted Pyrazoles. Synthetic Communications. 2006;**36**:2189-2194

[43] Grošelj U, Drobniˇc A, Reˇcnik S, Svete J, Stanovnik B, Golobiˇc A, et al. 1,3-dipolar cycloadditions to (5Z)-1-Acyl-5-(cyanomethylidene)-imidazolidine-2,4-diones: Synthesis and transformations of Spirohydantoin derivatives. Helvetica Chimica Acta. 2001;**84**:3403-3417

[44] Kawase M, Koiwai H, Yamano A, Miyamae H. Regioselective reaction of mesoionic 4-trifluoroacetyl- 1,3-oxazolium-5-olates and phenylhydrazine: Synthesis of trifluoromethyl substituted pyrazole and 1,2,4-triazine derivatives. Tetrahedron Letters. 1998;**39**:663-666

[45] Guillard J, Goujon F, Badol P, Poullain D. New synthetic route to diaminonitropyrazoles as precursors of energetic materials. Tetrahedron Letters. 2003;**44**:5943-5945

[46] Simoni D, Rondanin R, Furnò G, Aiello E, Invidiata FP. Facile synthesis of pyrazoles and pyrroles via thermolysis of tetrazolo[1,5-b]pyridazines, tetrazolo [1,5-a]pyrimidines and tetrazolo [1,5-a] pyridines. Tetrahedron Letters. 2000;**41**: 2699-2703

[47] El-Naggar M, El-All ASA, El-Naem SIA, Abdalla MM, Rashdan HRM. New potent 5α- reductase and aromatase inhibitors derived from 1, 2, 3-Triazole derivative. Molecules. 2020;**25**(3):672

[48] Banday AH, Shameem SA, Jeelani S. Steroidal pyrazolines and pyrazoles as potential 5α-reductase inhibitors: Synthesis and biological evaluation. Steroids. 2014;**92**:13-19

[49] Khalilullah H, Khan S, Ahsan MJ, Ahmed B. Synthesis and antihepatotoxic activity of 5-(2, 3-dihydro-1, 4-benzodioxane-6-yl)-3-substituted-phenyl-4, 5-dihydro-1H-pyrazole derivatives. Bioorganic & Medical Chemistry Letters. 2011;**21**(24): 7251-7254

[50] Chimenti F, Maccioni E, Secci D, Bolasco A, Chimenti P. Synthesis, molecular modeling studies, and selective inhibitory activity against monoamine oxidase of 1-Thiocarbamoyl-3, 5-diaryl-4,5-dihydro-(1H)- pyrazole derivatives. Journal of Medicinal Chemistry. 2005;**48**(23): 7113-7122

[51] Secci D, Bolasco A, Chimenti P, Carradori S. The state of the art of Pyrazole derivatives as monoamine oxidase inhibitors and antidepressant/ anticonvulsant agents. Current Medicinal Chemistry. 2011;**18**(33): 5114-5144

[52] Alam MS, Lee DU. Synthesis, molecular structure and antioxidant activity of (E)-4-[Benzylideneamino]-1, 5-dimethyl-2-phenyl-1H-pyrazol-3(2H)-one, a Schiff Base ligand of 4-Aminoantipyrine. Journal of Chemical Crystallography. 2012;**42**:93-102

[53] GresslerI A, Moura S, Flores AFC, Flores DC, Colepicolo P. Antioxidant and antimicrobial properties of 2-(4,5-dihydro-1H-pyrazol-1-yl)-pyrimidine and 1-carboxamidino-1H-pyrazole derivatives. Journal of the Brazilian Chemical Society. 2010;**21**:1477-1483

[54] Sallam HA, Elgubbi AS, El-Helw EAE. Synthesis and antioxidant screening of new 2-cyano-3-(1, 3-diphenyl-1H-pyrazol-4-yl) acryloyl amide derivatives and some pyrazole-based heterocycles. Synthetic Communications. 2020;**2020**:1-12

[55] Mor S, Khatri M, Punia R, Sindhu S. Recent Progress in anticancer agents incorporating Pyrazole scaffold. Mini Reviews in Medicinal Chemistry. 2022; **22**(1):115-163

[56] Alsayari A, Asiri YI, Muhsinah AB, Hassan MZ. Anticolon cancer properties of Pyrazole derivatives acting through xanthine oxidase inhibition. Journal of Oncology. 2021;**2021**:5691982

[57] Simpal K, Paliwal S. Chauhan R synthesis of Pyrazole derivatives possessing anticancer activity: Current status. Synthetic Communications. 2014; **44**(11):1521-1578

[58] Abdel-Motaal M, El-Senduny FF, Shaaban S. One-pot synthesis and anticancer activity of novel Pyrazole hybrids. Chemistry Select. 2021;**6**: 7306-7316

[59] Balbi A, Anzaldi M, Maccio C, Aiello C, Mazzei M, Gangemi R, et al. Synthesis and biological evaluation of novel pyrazole derivatives with anticancer activity. European Journal of Medicinal Chemistry. 2011;**2011**(46): 5293-5309

[60] Shaikh J, Patel K, Khan T. Advances in Pyrazole based scaffold as cyclin-dependent kinase 2 inhibitors for the treatment of cancer. Mini Reviews in Medicinal Chemistry. 2022;**22**(8): 1197-1215

[61] Hafner M, Mills CE, Subramanian K, Chen C, Chung M, Boswell SA, et al. Multiomics profiling establishes the Polypharmacology of FDA-approved CDK4/6 inhibitors and the potential for differential clinical activity. Cell Chemical Biology. 2019;**26**(8): 1067-1080

[62] Akbas E, Berber I, Sener A, Hasanov B. Synthesis and antibacterial activity of 4-benzoyl-1-methyl-5-phenyl-1*H*-pyrazole-3-carboxylic acid and derivatives. Farmaco. 2005;**60**:23-26

[63] Ma J, Lu X, Xia Y, Yan F. Determination of pyrazole and pyrrole pesticides in environmental water samples by solid-phase extraction using multi-walled carbon nanotubes as adsorbent coupled with high-performance liquid chromatography. Journal of Chromatographic Science. 2015;**53**(2):380-384

[64] Wang Y, Xu F, Yu G, et al. Synthesis and insecticidal activity of diacylhydrazine derivatives containing a 3-bromo-1-(3-chloropyridin-2-yl)-1H-pyrazole scaffold. Chemistry Central Journal. 2017;**11**:50

Chapter 3

Synthesis and Biological Applications of Thiazolidinone

Ramarajan Rajalakshmi and Subramaniyan Ramkumar

Abstract

Thiazolidinone scaffold has become a highly powerful scaffold in the current era when it comes to its clinical importance. Its wide variety of biological functions have piqued the researchers' intense curiosity. The 1,3-thiazolidin-4-ones have numerous pharmacological properties, including anti-cancer, anti-diabetic, anti-microbial, antiviral, anti-inflammatory and anticonvulsant properties because of these wide spectrum biological properties. Thiazolidinones are called as molecule of magic. In the recent years, a number of innovative synthetic techniques have been developed to create a variety of scaffolds to investigate a range of biological activities. Numerous researchers have been drawn to this skeleton by the variety in the biological response profile to investigate its potential against various activities.

Keywords: thiazolidinones, synthesis, biological activities, thiazolidinone scaffold, pharmacological properties

1. Introduction

Heterocyclic compounds play an important role in organic chemistry and of supreme practical and theoretical consequence. Therefore, more research being done on heterocyclic compound in chemistry. Heterocyclic compounds have an important role in therapeutic agents, drugs, dye stuffs etc. Therefore heterocyclic compound has taken a prominent place in the field of chemistry. Heterocyclic compounds retain the key to improve the quality of human life for example; more than 70% of the medications (or) drug used today is heterocyclic compounds. These are widely available in nature and act as key intermediates in biological processes. Over the years, chemists have paid close attention to a variety of physiologically active compounds that contain heteroatoms including nitrogen, sulphur, and oxygen because of their significance to biology. Thiazolidinone is considered as indispensable anchor for development of new therapeutic agents because this five member magic moiety possesses all types biological activities. Thus the thiazolidinone nucleus has been studied in the field of organic, medicinal and photochemistry. There are many examples of biologically active such as antibiofilm [1, 2], hypoglycemic [3], antimicrobial [4], analgesic [5], antipyretic [6] and anti-inflammatory activities [7, 8] anticonvulsant [9], antihistaminic [10], anti HIV [11] cardio protective [12], antinociceptive [13] With a sulphur atom at position 1, a nitrogen atom at position 3, and a carbonyl group at position 2, 4, or 5, thiazolidinones are thiazolidine derivatives. However, its derivatives are among the most researched moieties, and the discovery of its existence in penicillin was the first indication that it existed in nature. Similar to 1,3-thiazolidin-4-ones, which are heterocyclic nuclei with a sulphur atom at position 1, a nitrogen atom at position 3, and a carbonyl group at position 4, they have been the focus of substantial research recently. Since it is so adaptable that the 4- thiazolidinone scaffold has been used in a number of clinically effective medications. They have been used anti-HIV, anti-tubercular, anti-microbial, anti-inflammatory and anti-viral medicines. The presence of arylazo, sulfamoylphenyl or phenylhydrazone moiety at various postions of the thiazolidone ring has been extensively reported to enhance anti-microbial activity. Its antibacterial activity may be caused by its inhibitory activity of enzyme Mur B, which is a precursor acting during the biosynthesis of peptidoglycan. Numerous publications highlighting their chemistry and pharmacological uses have been published in the literature. Thiazolidinone possess wide range of biological actions from antibacterial to anticancer. Various recent new drug developments of thiazolidinone derivatives show better effect and less toxicity. Moreover the possible improvements in the activity can be achieved by slight modifications in the substituent on the thiazolidinone nucleus. This has been noticed so far, that the introduction of another heterocyclic moiety into the thiazolidinone nuclei will enhance the biological activities.

Recently, thiazole and thiazolidinone derivative have been used as a potent antitrypanosomal agents [14]. Therfore in this chapter it is planned to discuss the synthesis of various thiazolidinone derivatives in brief and their widespectrum bilogical activities will be discussed elaborately.

2. Synthesis

In the literature, a variety of techniques for creating 4-thiazolidinones have been extensively described. An amine, a carbonyl molecule, and a mercapto-acid are the three elements that are often used in the main synthesis pathways for 1,3-thiazolidin-4. The disclosed classical synthesis can be either a two-step method or a one-pot, three-component condensation (**Figure 1**). The first step in the reactions is the creation of an imine (the nitrogen of the amine attacks the carbonyl of the aldehyde or ketone). This is followed by intramolecular cyclization on the removal of water.

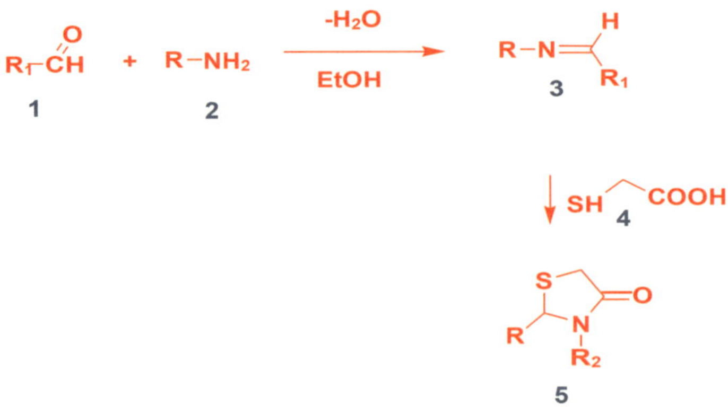

Figure 1.
Synthesis of thiazolidinone by condensation.

The most common method for making 4-thiazolidinone derivatives is syntheic [15]. On being treated with mercaptoacetic acid while having silica chloride present as a heterogeneous catalyst to speed up the intramolecular cyclocondensation in a solvent-free environment, a variety of quinazolinyl azomethines produce 4-thiazolidinones (**Figure 2**).

By using benzylidene-anilines and mercaptoacetic acid in benzene at 30 C for 10 min, Bolognese et al. [16] produced a variety of 1,3-thiazolidin-4-one derivatives. Following chromatographic purification, the 1,3-thiazolidin-4-ones are extracted at a yield of 65–90% (**Figure 3**).

A solvent-free, nano-titania-supported sulfonic acid [nano-TiO2- SO3H (n-TSA)] catalysed method has been developed by Ruby Singh et al. [17]. for the synthesis of novel hybrids of isoxazolyl-spiro-thiazolidinones that are promising for use in pharmaceuticals (**Figure 4**).

To create a few thiazolidinone derivatives using the recommended techniques to look into their potential antiamoebic effect as described in a three step reaction process for obtaining target chemicals (**Figure 5**). In the first stage,

Figure 2.
Synthesis of thiazolidinone by intramolecular cyclocondensatio.

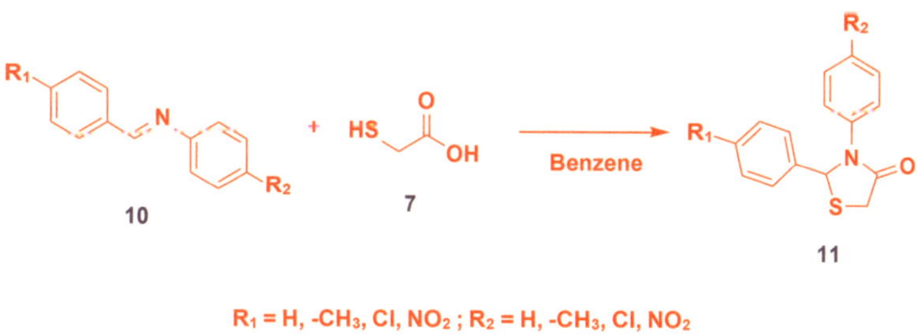

Figure 3.
Synthesis of thiazolidinone by Bolognese et al.

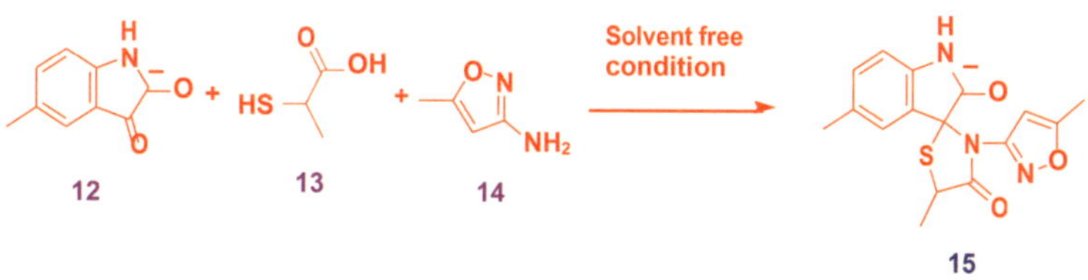

Figure 4.
Synthesis of thiazolidinone by Ruby Singh et al.

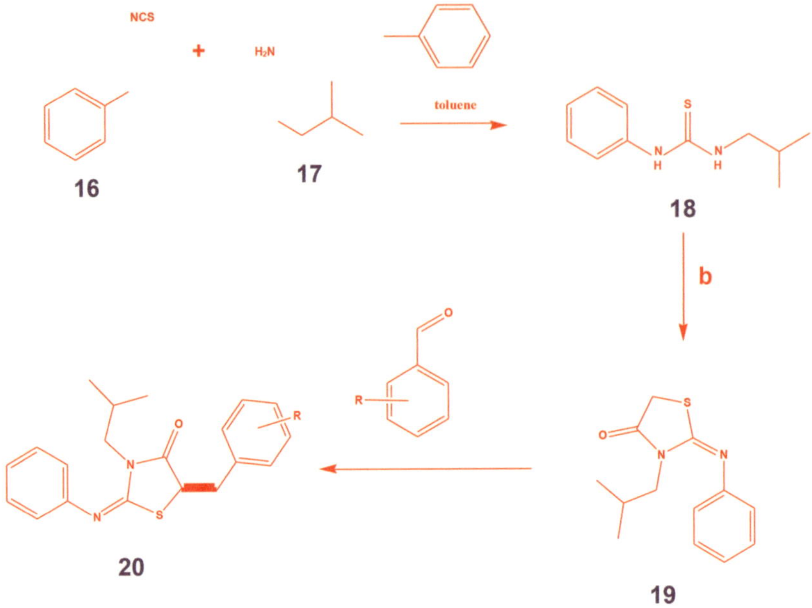

Figure 5.
Synthesis of thiazolidinone by Knoevenagel condensation.

2-methylpropan-1-amine and phenylisothiocyanate were combined with toluene to create 1-(2-methylpropyl)-3 phenylthiourea. The second stage was the cyclization of 3-(2-methylpropyl)-2-(phenylimino)- 1,3-thiazolidin-4-one with the use of sodium acetate and chloroacetic acid. In the third stage, 3-(2-methylpropyl)-2-(phenylimino)-1,3-thiazolidin-4-one was Knoevenagel condensation with various substituted aldehydes in ethanol to produce the target compounds [18].

Another method involves employing dialkyl substituted bromoacetic acid or bromacetyl chloride as the starting material to create 5-substituted dialkyl thiazolidinones. In order to produce the intermediate, which was then hydrolyzed to produce 5,5-dialkyl- thiazolidin-2,4-dione, the reaction begins with the refluxing of dialkyl substituted bromoacetic acid or bromoacetyl chloride with thiourea in the presence of sodium acetate in ethanol (**Figure 6**) [19].

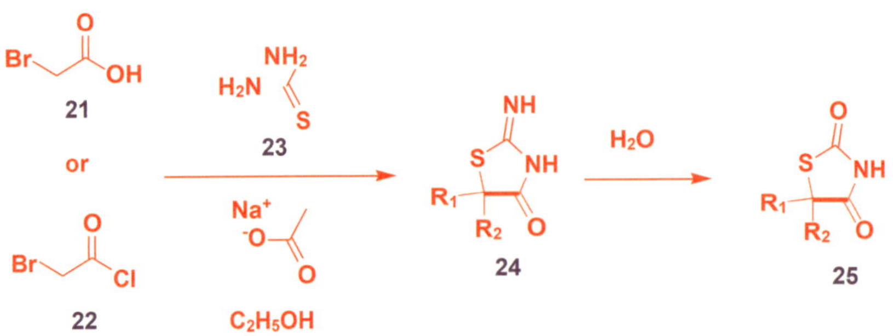

Figure 6.
Synthesis of thiazolidinone by Knoevenagel condensation.

By refluxing the mixture of substituted benzaldehyde **35** with L-cysteine **36** in the presence of ethanol for 1.5–48 hours, 2-substituted phenyl thiazolidine-4-carboxylic acid derivatives have been created (**Figure 7**) [20].

By reacting 4-hydrazinobenzenesulfonamide hydrochloride with 4-substituted aldehydes in the presence of sodium acetate in ethanol, followed by the reaction of the resulting phenyl hydrazones with excess thiolactic acid at 60°C for 3 h, Abdellatif et al. [21], were able to produce 2,4,5-trisubstituted thiazolidinone derivatives, as shown in (**Figure 8**).

Velmurugan. V et al. [22] have been synthesized Thiazolidinone derivatives from thiourea (**Figure 9**).

Novel oxazinyl thiazolidinone compounds have been created by Rajalakshmi et al. [23], as powerful antidiabetic agents (**Figure 10**).

Novel Thiazinyl-thiazolidinone compounds have been created by Rajalakshmi et al. [24], as potential in vitro anti-diabetic and antioxidant agents (**Figure 11**).

26 27 28

Figure 7.
Synthesis of thiazolidinone.

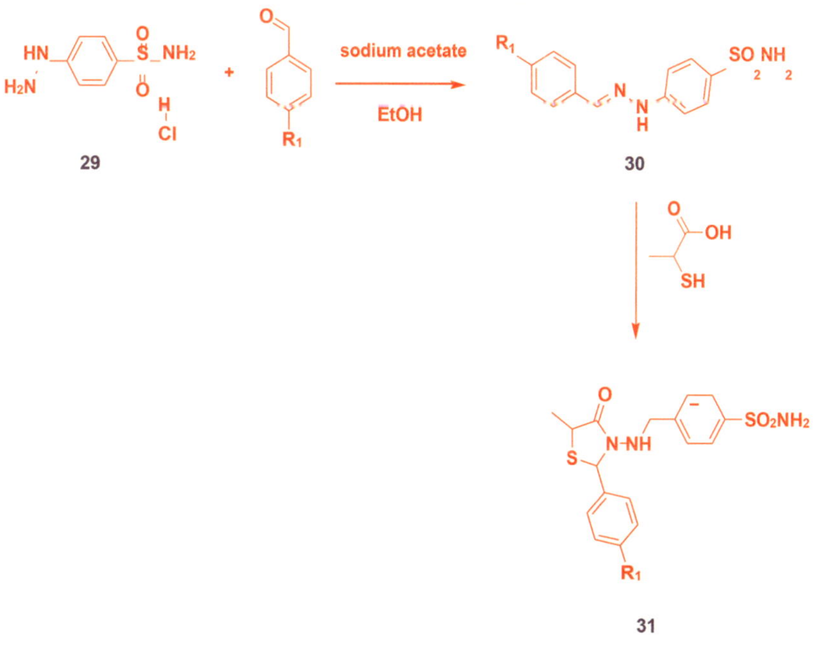

Figure 8.
Synthesis of thiazolidinone by Abdellatif et al.

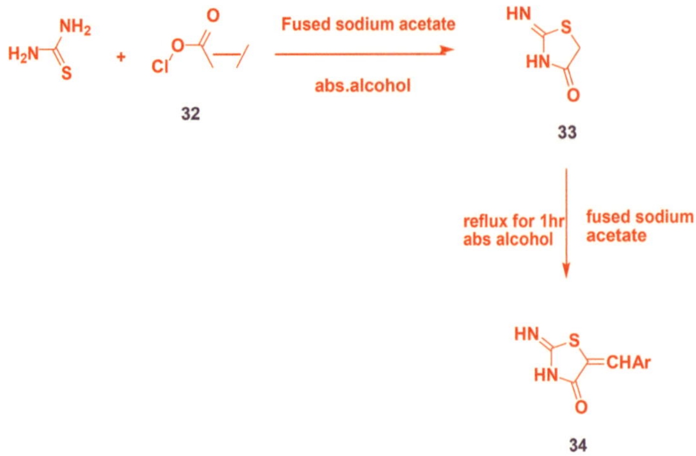

Figure 9.
Synthesis of thiazolidinone by Velmurugan. V et al.

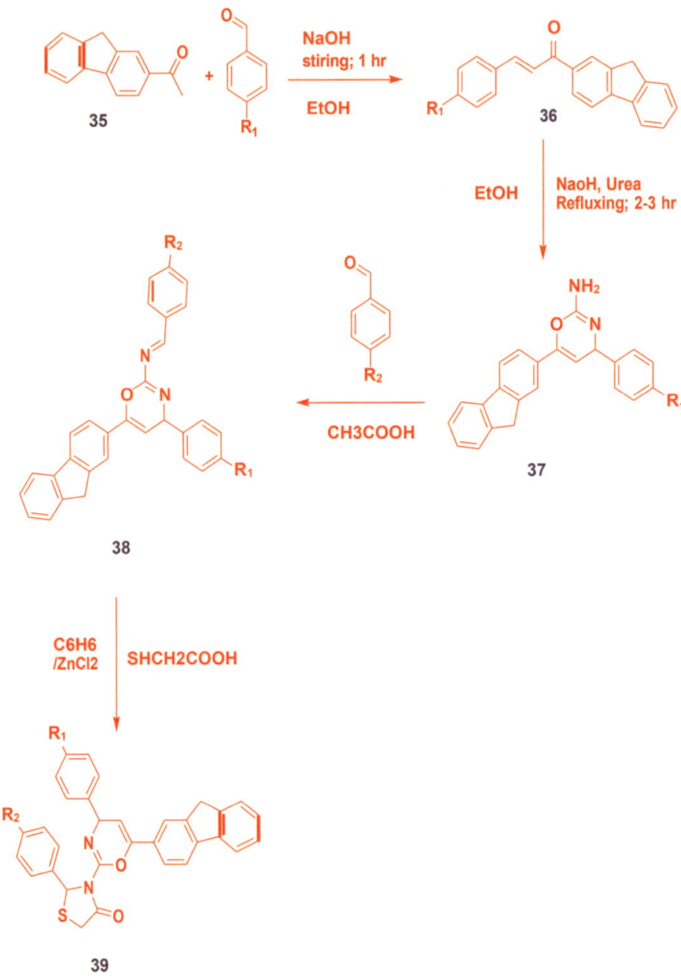

Figure 10.
Synthesis of thiazolidinone by Rajalakshmi et al.

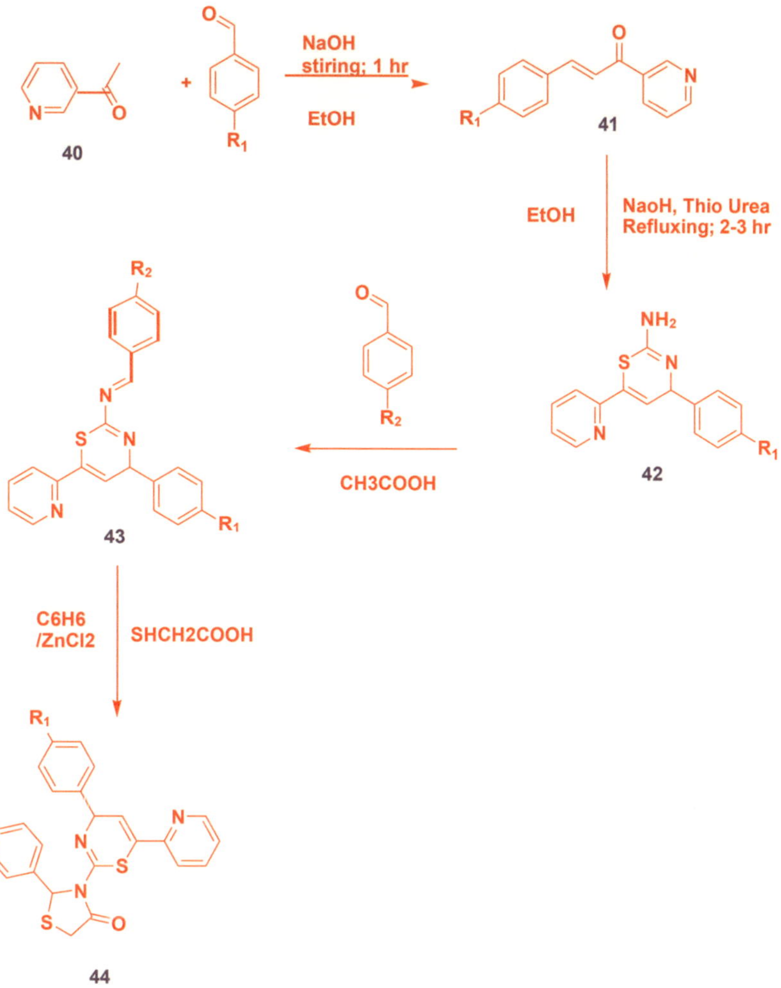

Figure 11.
Synthesis of thiazolidinone by Rajalakshmi et al.

Substituted benzaldehyde/acetophenone were reacted with thiosemicarbazide in HCl to get the intermediate schiffbase. Schiff base intermediate react with chloro acetic acid and sodium acetate in acetic acid to obtain 4-thiazolidinone analogs by Rahim et al. [25]. Saiz et al. [26] have been synthesized 2-hydrazolyl-4-thiazolidinones by the reaction involving aldehydes, thiosemicarbazides, and maleic anhydride, effectively assisted by microwave irradiation (**Figure 12**).

Benmohammed et al. [27] a series of reaction involving thiosemicarbazones react with ethyl 2-bromoacetate in anhydrous sodium acetate afforded to the thiazolidin-4-one derivatives (**Figure 13**).

Ottana et al. [28] has been reported the series of new thiazolidinone derivatives by reacting *N*-hydroxypropyl-*N*′–phenylthiourea of methyl bromoacetate in the presence of triethylamine (**Figure 14**).

One pot three component synthesis containing aldehyde, thiourea and chloroform to give 2-amino-4-thiazolidinone derivatives (**Figure 15**) reported. Various imino thiazolidinones were developed by using different reagents with different reaction conditions by Jieping et al. [29].

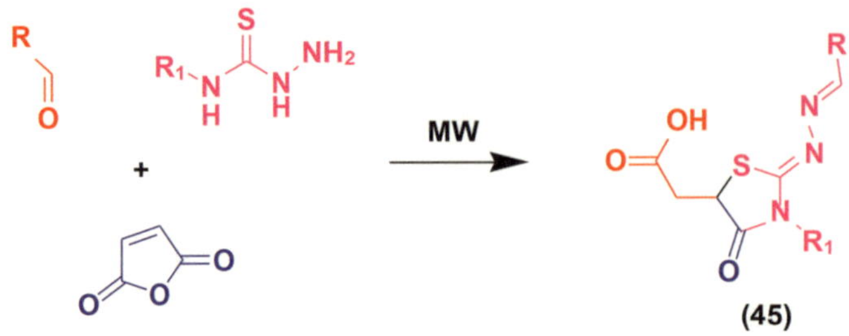

(45)

Figure 12.
Synthesis of thiazolidinone by Saiz et al.

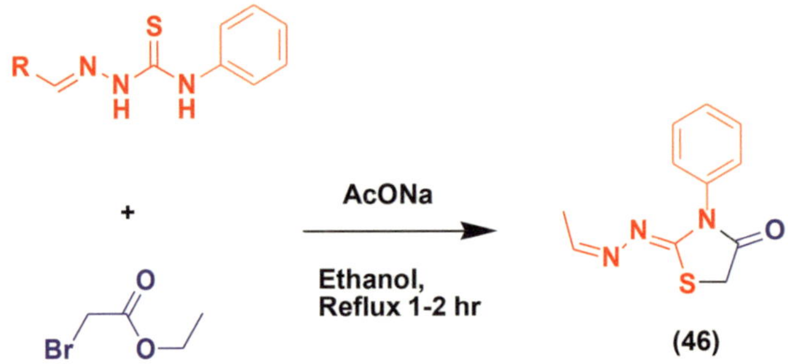

(46)

Figure 13.
Synthesis of thiazolidinone by Benmohammed et al.

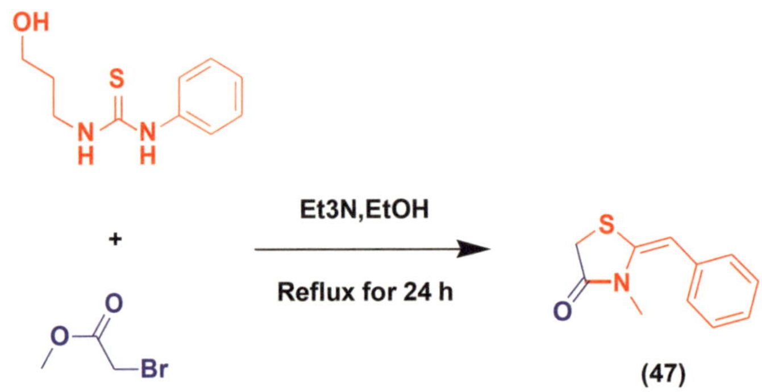

(47)

Figure 14.
Synthesis of thiazolidinone by Ottana et al.

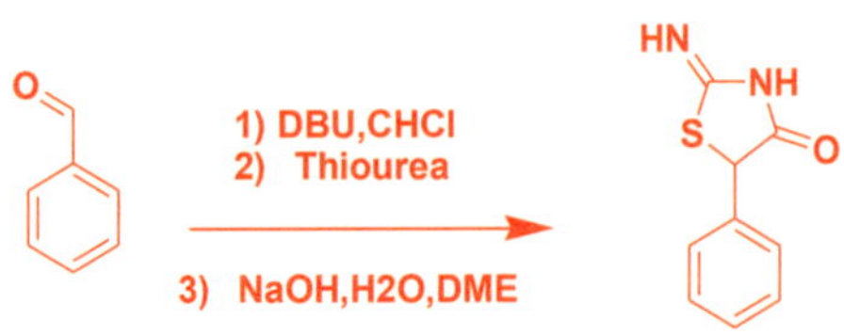

(48)

Figure 15.
Synthesis of thiazolidinone by Jieping et al.

Pang et al. [30] were synthesized by the Mesoporous MCM-41 supported Schiff base and CuSO4.5H2O mediated in the cyclocondensation of mercaptoacetic acid with imines (or aldehydes and amines) to afford thiazolidinone derivatives (**Figure 16**).

2.1 The Bioimplication of Thiazolidinones

Thiazolidinones are the primary compounds with a diverse range of biological activity. This ring handles several pharmacological processes. Thiazolidinones are being studied biologically using a variety of mechanisms, including receptor-mediated mechanisms and enzymatic activity. The biological study of thiazolidinones has shown that substitution at positions 2, 3, and 5 imparts various activity. As seen in **Figure 17**, many commercial medications that contain the thiazolidinone scaffold exhibit a variety of biological actions.

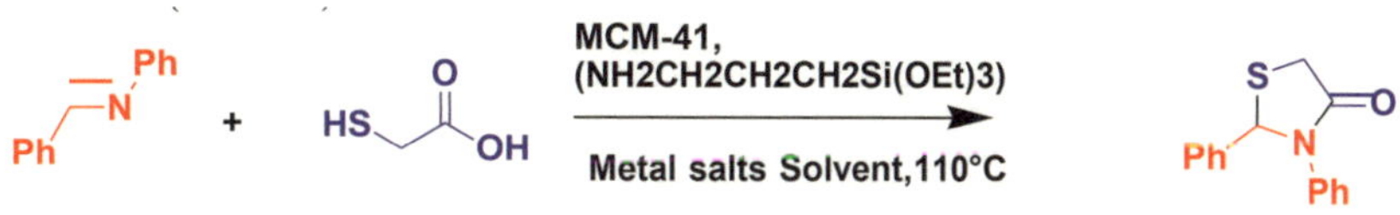

(49)

Figure 16.
Synthesis of thiazolidinone by Pang et al.

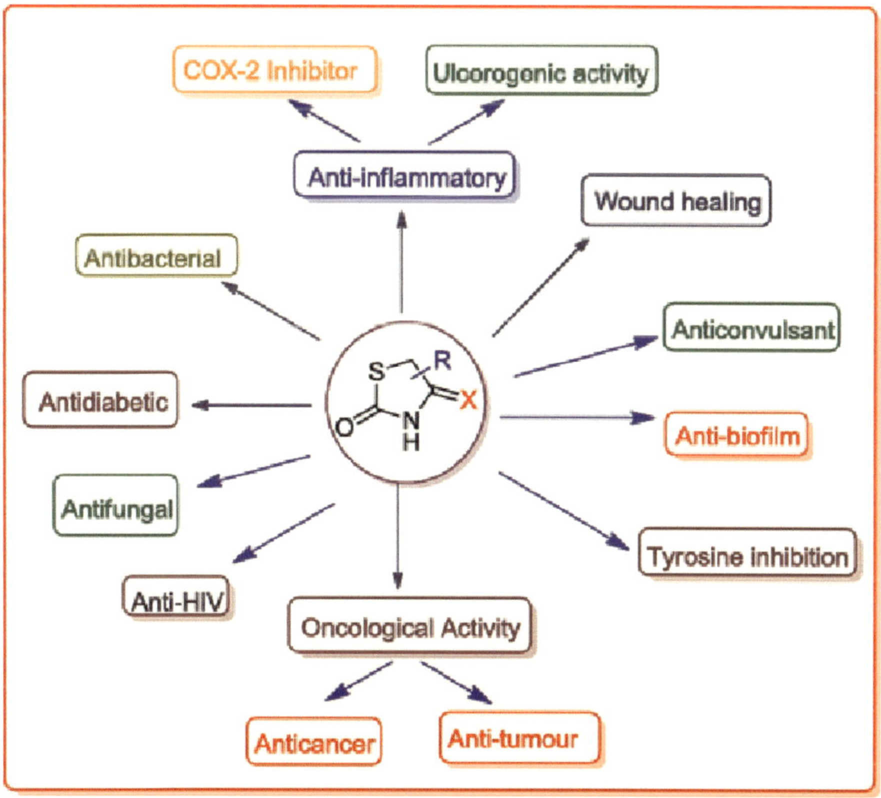

Figure 17.
Pharmacological action of thiazolidinones and its derivatives.

2.2 Antibacterial and antifungal activity

Veerasamy et al. [31] have prepared a series of 1,3-thiazolidin-4-one derivatives (**50**).These compounds were also evaluated for *in-vitro* antibacterial, antifungal and anti-viral activities. Preliminary results indicated that most of the compounds demonstrated moderate to good antimicrobial activity, comparable to standard drugs.

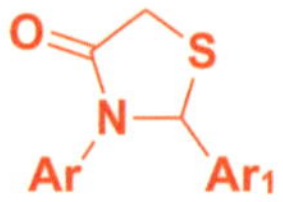

(50)

Different levels of inhibition against bacteria and fungus are present in thiazolidinones with substituted positions at C-2 and N-3. Multi-drug resistance microbial infections have rapidly increased in frequency during the past few decades, posing a serious health risk. Nearly every location of the 4-thiazolidinone has been investigated in an effort to increase its antibacterial and antifungal activities. Thiazolidinone derivatives' SAR analyses revealed that they are more efficient against gram-negative bacteria than gram-positive bacteria. Therefore, finding novel antimicrobial drugs will continue to be a difficult and vital work for medicinal chemists. According to Liesen et al., 4-thiazolidinone compounds made from ethyl (5-methyl-1-H-imidazole-4-carboxylate). The entire produced chemicals were tested for their antibacterial and antifungal activities against a variety of diseases. The findings demonstrated that, in comparison to common antibacterial and antifungal medications like *chloramphenicol*, *rifampicin*, and *ketoconazole*, the examined compounds had modest antibacterial and antifungal properties. Against *Bacillus subtilis*, compound (**51**) displayed a MIC of 270 lg mL1 [32].

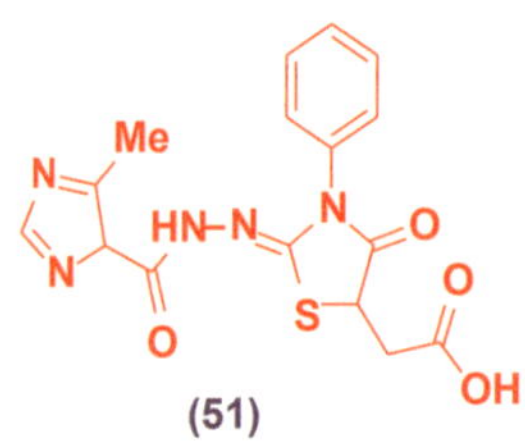

(51)

The kind of the substituents at the thiazolidinone ring's C-2 and N-3 substantially influences antibacterial activity. Compound 51–3-(1,5-dimethyl-3-oxo-2-phenyl-2,3-dihydro-1H-pyrazol-4-yl) -2- \s(2-hydroxy-3,5-diiodophenyl) -thiazolidin-4-one (**52**) had a zone of inhibition of 27, 24, and 25 mm against *Escherichia coli, B. subtilis*, and *Salmonella typhi*, respectively, and had antipyrine at N-3 and a 3-iodo substituted phenyl ring at C-2.

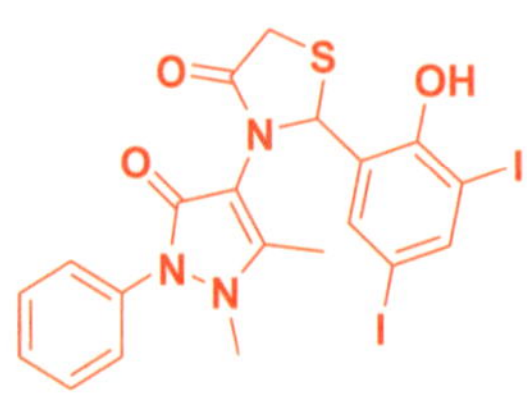

(52)

Recently, a group of 2-thioxo-4-thiazolidinones and 4,40-bis(2-thioxo-4-thiazolidinone-3-yl)diphenylsulfone derivatives were synthesised by El-Gaby et al.

The majority of the compounds were found to have moderate efficacy against the tested bacterial strain. *Bacillus cereus* was found to be the target of the highest antibacterial activity in thiazolidinones (**53**) with sulfamoyl and thioxo moieties, while *Staphylococcus* aureus was the target of the highest antibacterial activity in thiazolidinone derivative (**54**) with pyrimidine nucleus, sulfamoylphenyl, and thioxo moieties [33].

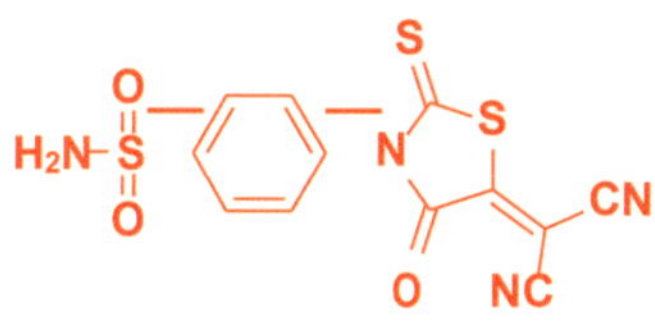

(53)

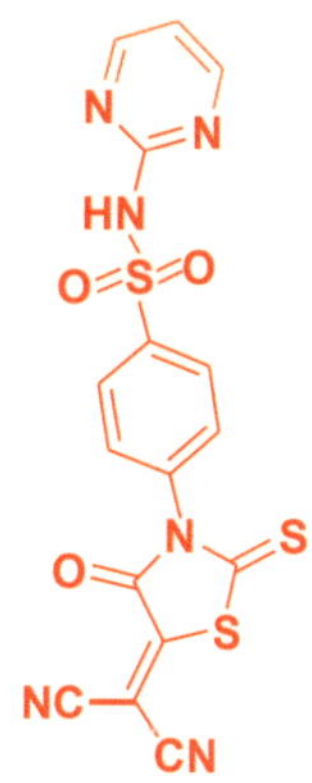

(54)

Bondock et al. [34]. created thirteen compounds and tested them against *B. subtilis, Bacillus megaterium,* and *E. coli* for antibacterial activity. By using *ampicillin* and *chloramphenicol* in a concentration of 25 mg/mL as a reference medicine, the majority of the produced thiazolidinone derivatives (**55, 56**) revealed comparable action against tested bacteria.

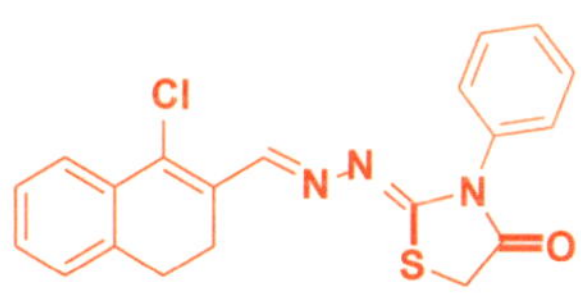

(55)

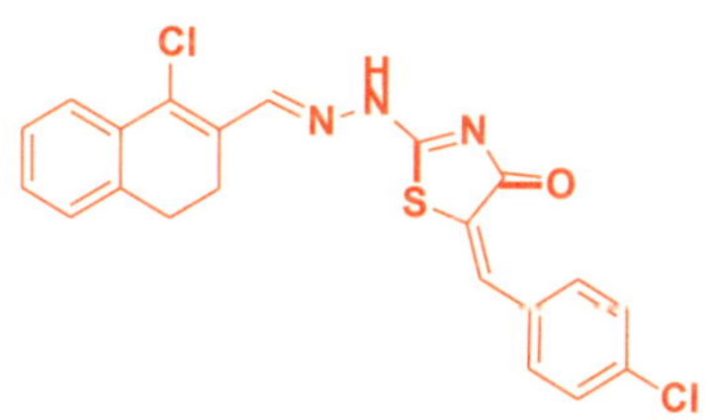

(56)

Bonde et al. [35] reported the synthesis of *N*- [(2Z)-3-(4- bromophenyl)-4- oxo-1,3-thiazolidin–2-ylidene]–2-(pyrazin-2-yloxy)acetohydrazide (**57**) exhibiting for antibacterial activity against two different strains of Gram-negative (E. coli and S. typhi), Gram-positive (S.aureus and B.subtilis) bacteria and the antimycobacterial activity against H37Rv strain of Mycobacterium tuberculosis. The minimum inhibitory concentration (MIC) was determined for test compounds and for reference standards.

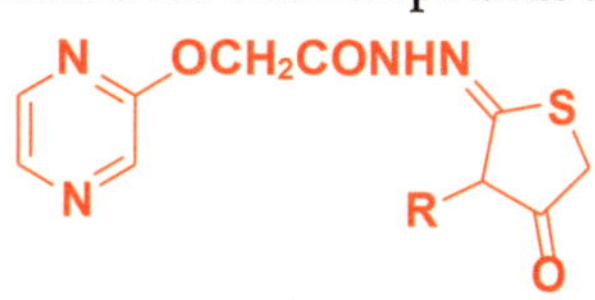

(57)

2.3 Antitubercular activity

Kucukguzel et al. [36]. reported substituted 4-thiazolidinones have antimycobacterial action, although only compounds (**58** and **59**) demonstrated 90 and 98% inhibitions at 6.25 lg mL1, respectively.

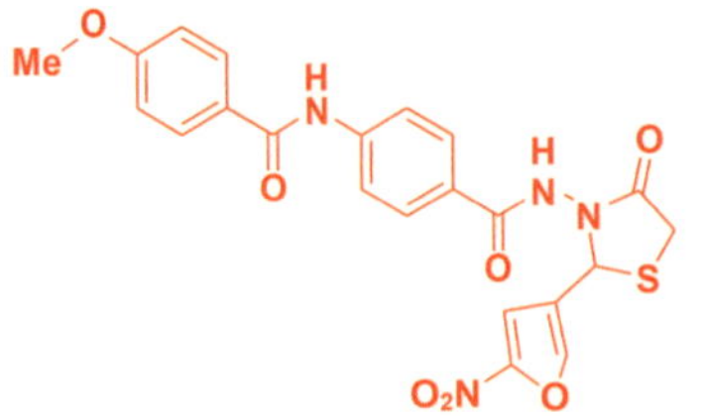

(58)

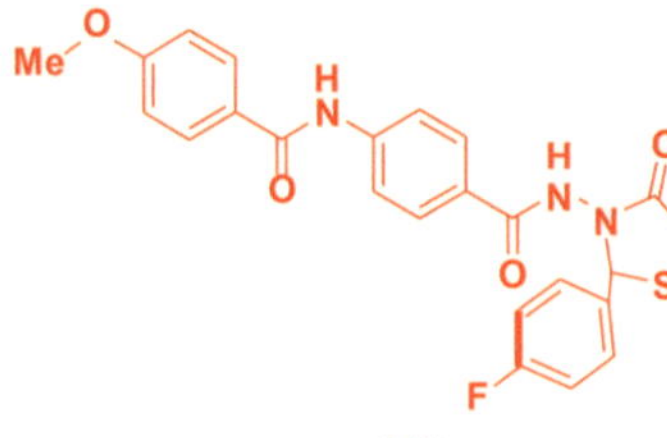

(59)

The synthesis of N-pyridyl-N0 -thiazolylhydrazine derivatives was described by Zitouni et al. [37]. High antituberculosis activity was demonstrated by compound **60** (IC50: 6.22 lg/mL and IC90: 6.78 lg/mL). Analysis of compound (60) structural details revealed that 2-pyridyl and 2-hydroxy-5-methoxyphenyl groups are necessary for antimycobacterial activity while 3-pyridyl and 4-pyridyl groups are unfavourable for activity.

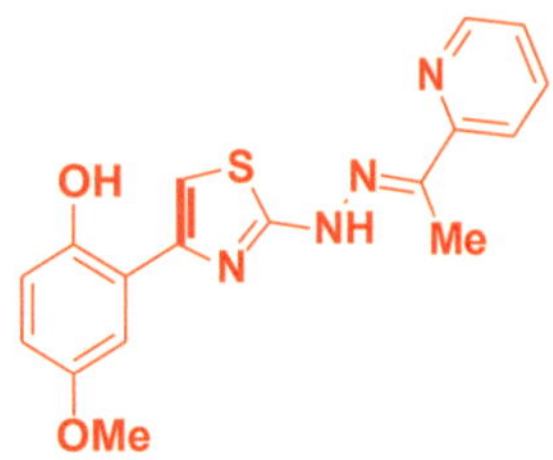

(60)

2.4 Anticancer activity

A series of 2-arylthiazolidine-4-carboxylic acid amides were examined by Gududuru et al. for potential cytotoxic action against prostate cancer. With an IC50 of 0.55 lM and a 38- fold selectivity in *PPC-1* cells, compound **61** was discovered to be the most powerful and selective cytotoxic agent. The SAR study demonstrated that as the chain length increased from C7 to C18, the potency also increased. However, continuing to lengthen the alkyl chain by one carbon unit resulted in a significant loss of activity, making the alkyl chain with a C18 unit the ideal length for the efficacy of thiazolidine analogues. The potency was decreased when the phenyl ring was replaced with an alkyl or cyclohexyl group, but the cytotoxicity was maintained when the furanyl ring derivative was used in its place. The same research team created a novel series of 2-aryl-4-oxo-thiazolidin-3-yl amides (**62**), and each chemical was tested against five different types of human prostate cancer cell lines. According to their findings, replacing the alkyl chain with an aryl group decreased biological activity while increasing the alkyl chain increased the antiproliferative activity.

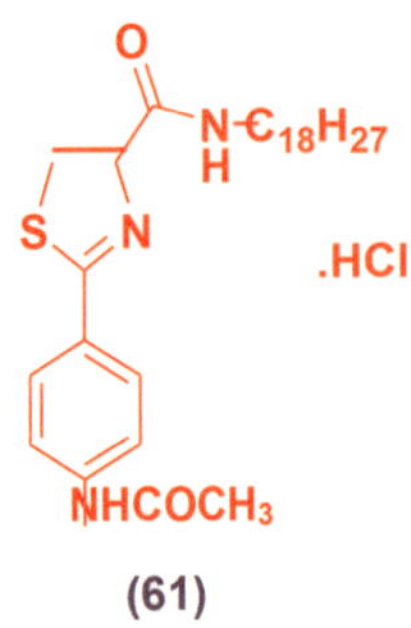

(61)

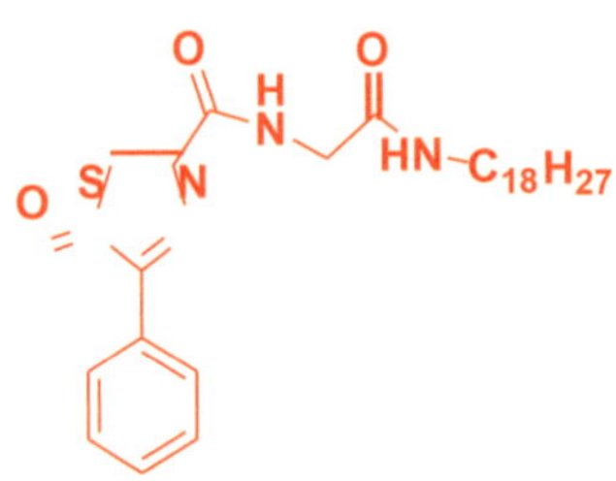

(62)

The primary cytotoxic activity of several 5-bromo-3-[(3-substituted-5-methyl-4-thiazolidinone-2-ylidene)hydrazono]-1H-2-indolinones (**63**) against a panel of three cell lines, NCI-H460 (Lung), MCF7 (Breast), and SF-268, was examined (CNS). Thiosemicarbazone derivatives of indolinones were shown to have potential cytotoxic properties [38].

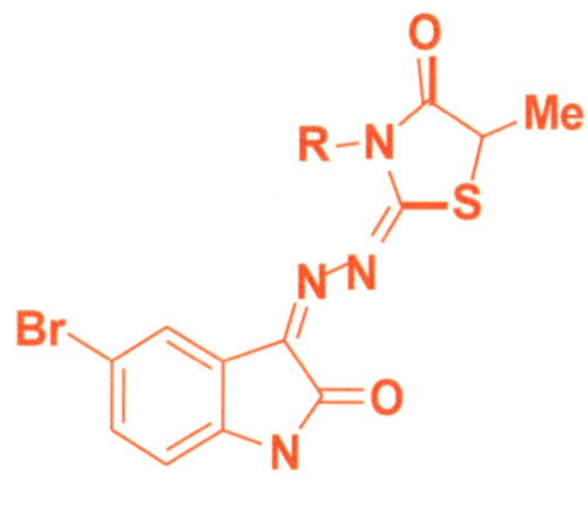

(63)

Moorkoth et al. [39] have synthesized thiazolidinone (**64**) derivatives and evaluated in vivo anti-cancer activity performed in albino mice bearing Dalton's ascites carcinoma showed that the new analogs enhanced life span and prevented increases in body weight owing to tumor volumes. Moreover, cell-cycle analysis and Hoechst staining analysis proved the apoptotic potential of these analogs. Preliminary pharmacokinetic evaluation was carried out on the synthesized compounds to determine the lipophilicity and pKa. Lipophilicity was determined usinghigh-performance liquid chromatography and the results showed a direct correlation between the observed anti- cancer activity and log P value, while pKa values indicated the ionizing range which is a prediction tool for solubility and permeability.

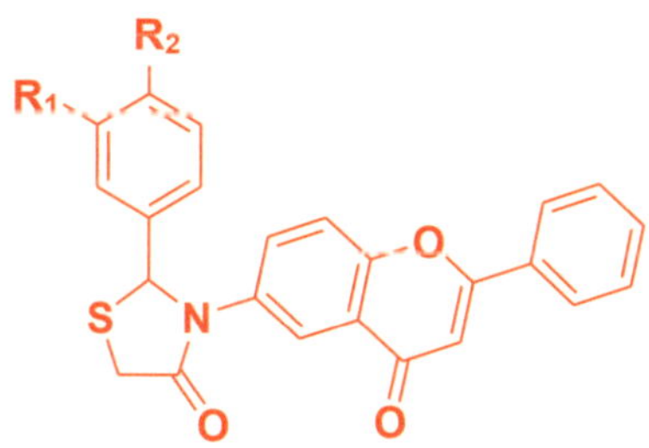

(64)

Kaminskyy et al. [40] have synthesized thiazolidinone derivatives these compounds were evaluated for their anticancer activity. Among the tested compounds, 3-(2,4-thiazolidinedione-5-ylidene)-carboxyimino]olean-12-en-28-oic acid methyl ester was superior to other related compounds with mean values of pGI50 = 5.51/5.57, pTGI = 5.09/5.13, and pLC50 = 4.62/4.64,low toxicity and moderate activity level in vivo hollowfiberassay. Zhou et al. have prepared a series of 2- thioxo-4-thiazolidinone derivatives (**65**) and evaluated them on *peroxisomeproliferator* activated receptor g (PPARg) binding activities. Through thebiological assays, compounds5-(2-(allyloxy)- 5-bromobenzylidene)-2-thioxothiazolidin-4-one and 5-(5-bromo-2-iodobenzylidene)-2-thioxothiazolidin-4- one were highlightedwith Ki values of 12.15 nmol/L and 14.46 nmol/L, respectively.

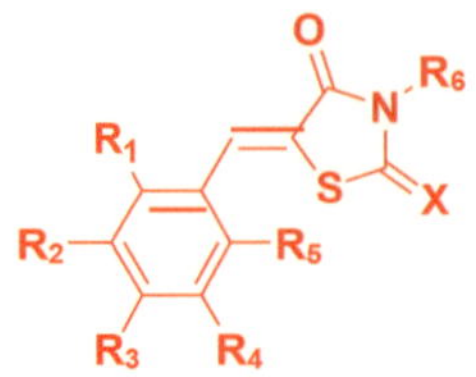

(65)

Sala et al. [41] have synthesized 2,3-thiazolidin-4-one **(66)** possessing strong inhibitory effects on breast cancer cell growth. Our results indicate that some of thiazolidin-based resveratrol derivatives may become a new potent alternative tool for the treatment of human breast cancer.

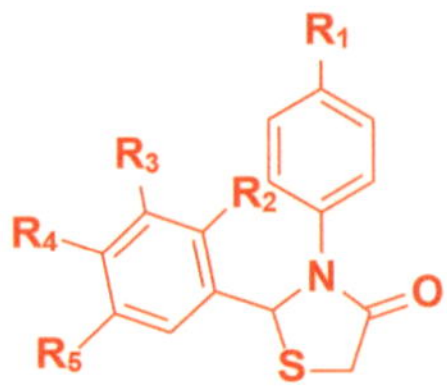

(66)

2.5 Anti-inflammtory and analgesic activity

Inflammation is a biological reaction to damage stimuli that is complicated and associated with numerous pathophysiological diseases. The short-lived free radical nitric oxide is one of the pro-inflammatory chemicals that are released by macrophages in response to inflammatory stimuli (NO). The commonly used nonsteroidal anti-inflammatory medicines naproxen and ibuprofen, which inhibit the COX enzyme that catalyses the manufacture of prostaglandins and tromboxane from arachidonic acid, are derived from arylalkanoic acids. These medications' modes of action are linked to unpleasant side effects include renal and gastrointestinal toxicity. The anti-inflammatory and analgesic effectiveness of a new series of quinazolinone compounds with thiazolidinone at the second position was reported by Kumar et al. to combat the aforementioned side effects. Interestingly, compound **67**, which replaced the phenyl ring at the second position with a chloro group, shown nearly identical anti- inflammatory effect to phenylbutazone at 50 mg/kg. Another study found that 5-(arylidene)- 2-(aryl)-4-oxothiazolidin-3-yl amide derivatives of biphenyl-4-carboxylic acid were significantly anti-inflammatory. Bromine substitutions on both aromatic rings in compound 58 resulted in percentage inhibition values of 44.59 and 55.73 at 2 and 4 hours, respectively [42, 43].

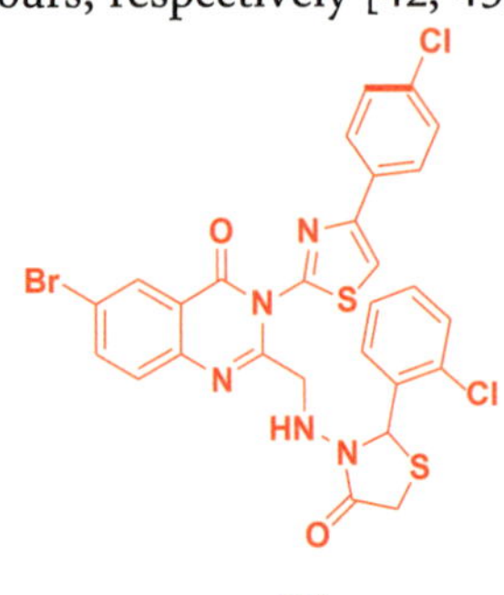

(67)

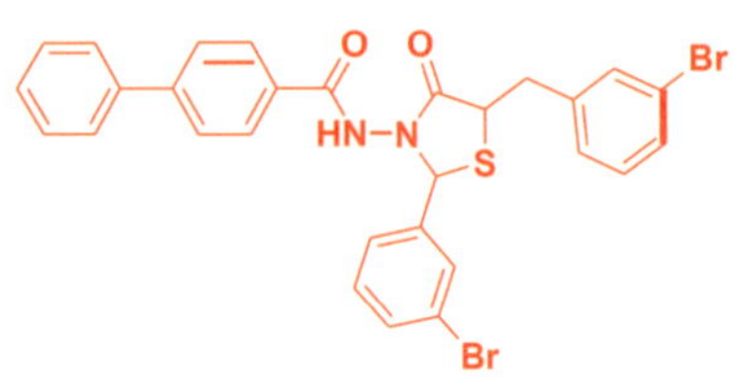

(68)

In their investigation of 3,30 -(1,2-ethanediyl)-bis[2-aryl-4- thiazolidinone] derivatives (**69**), Ottana et al. hypothesised that these derivatives would interact preferentially with inducible COX-2 isoform due to their remarkable stereo selective anti-inflammatory/analgesic actions. The anti-inflammatory activity of 3-[2- (4 methylphenyl)-2- oxo-l-phenylethyl]-2,4-thiazolidinedione (70) was increased and the analgesic activity was lowered in its absence of the 5-arylmethylidene moiety. The 2,4-thiazolidinedione ring's bulkiness at the NH group either reduced or eliminated the anti-inflammatory action [44, 45].

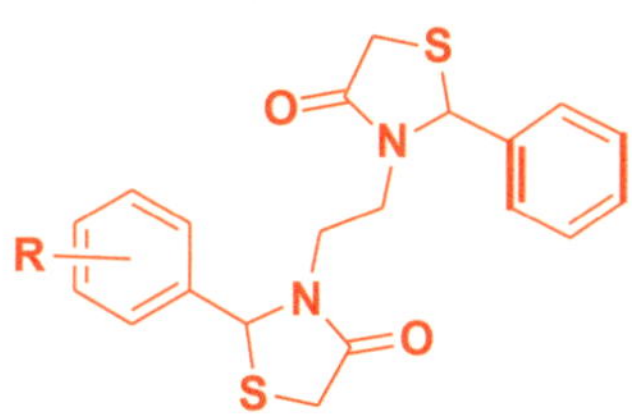

(69)

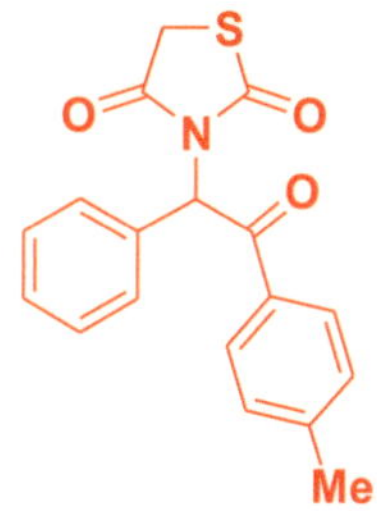

(70)

The ability of 2-aryl-3-[([1, 3, 4] thiadiazino [6,5-b]indol-3-ylamino]methyl] to reduce inflammation In order to study 1,3,4-thiadiazol-2-yl, 1,3-thiazolidin-4-one (**71**), rats' paw edoema caused by carrageenan was used. Azetidinones were shown to have stronger anti- inflammatory and analgesic properties than their comparable thiazolidinone molecules, according to Bhati and Kumar [46].

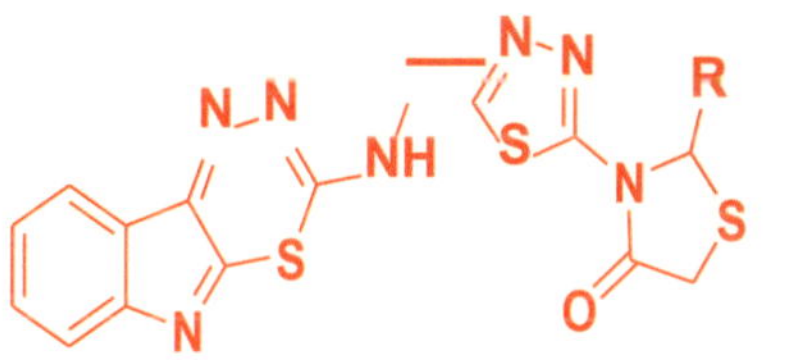

(71)

Vigorita et al. [47] have synthesized 3,3′-(1,2-Ethanediyl)-bis[2-aryl-4-thiazolidinone] (**72**) exhibiting for Antiinflammatory activity was investigated by the carrageen in-inducedpawedema test and analgesic activity by acetic acid writhing and hot plate tests in rats.

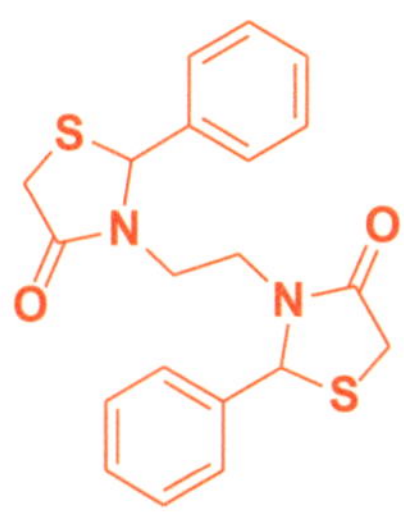

(72)

2.6 Anticonvulsant and antidepressant activity

Shiradkar et al. have created a brand-new series of clubbed thiazolidinone-barbituric acid (**73**) and thiazolidinone-triazole derivatives (**74**) in order to research the effects of a hydrophobic unit, hydrogen bonding domain, and electron-donor group on the compounds' anticonvulsant action. They discovered that the -OH function at the 4-position of the phenyl ring is necessary for anticonvulsant activity and that its removal or substitution by the moieties -Cl, CH3, or -NO2 results in a loss of activity. The activity was eliminated when the hydroxyl group responsible for hydrogen bonding was replaced since there was no longer an HBD (hydrogen bonding domain). It was discovered that compounds with p-methoxyphenyl substitution at the C-2 of the thiazolidinone ring were more active than standard medication sodium phenytoin in a series of thiazolidinonyl 2-oxo/thiobarbituric acid derivatives [48, 49].

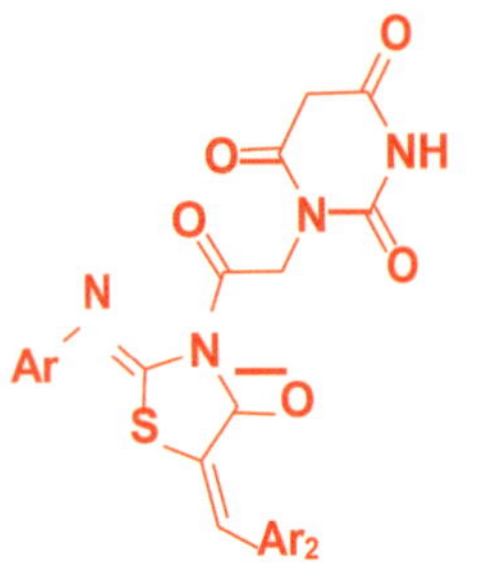

(73)

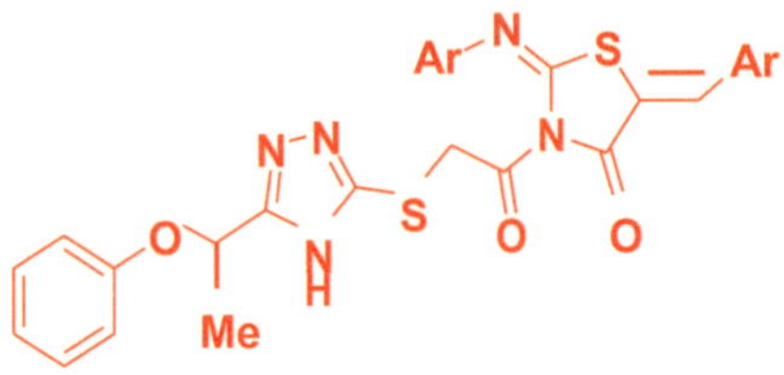

(74)

2.7 Antiviral/anti-HIV activity

Different 2-(2,6-dihalophenyl)-3-(4,6-dimethyl-5-(un)substituted-pyrimidin-2-yl)- thiazolidin-4-ones were created by Chen et al. The analytical and spectral data of these newly synthesised chemicals supported their structures. The ability of these substances to inhibit HIV-RT was also tested. It was claimed that a high hydrophobicity value will have a significant impact on HIV-RT inhibitory efficacy. Compounds **75** and **76**, which had an ethyl group at the 5-position on the N-3 position of the pyrimidine ring, were found to be the most effective ones, with IC50 values of 0.26 and 0.23 lM, respectively. According to their research, the analogues' overall hydrophobicity as well as the steric and electronic properties of the 3-hetero-aryl moiety's meta- and parasubstituents on thiazolidin-4-one resulted in a significant increase in antiviral activity [50–52].

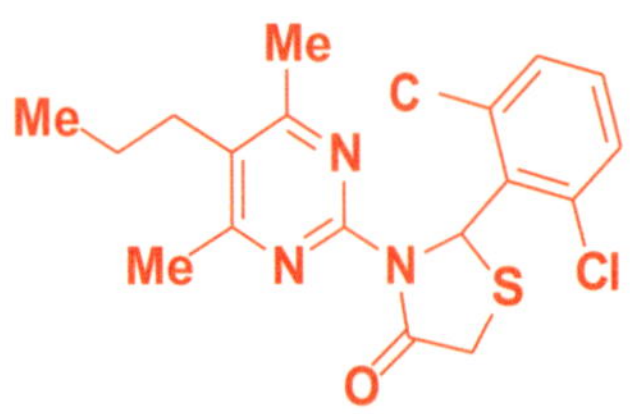

(75)

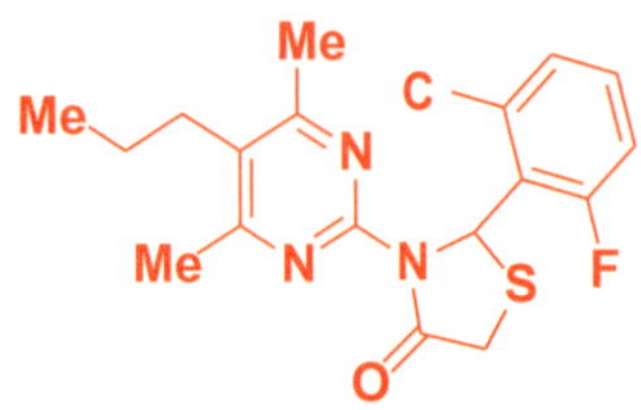

(76)

HIV-RT inhibitory action in a group of 2-aryl-3-(4,5,6-trimethylpyrimidin-2-yl) thiazolidin-4-ones was negatively influenced by substitution at C-5 and positively affected by the addition of a chlorine atom on the phenyl at C-2. The anti-HIV-RT activity of compound 66 was remarkably strong (IC50 = 2.95 lM) [53].

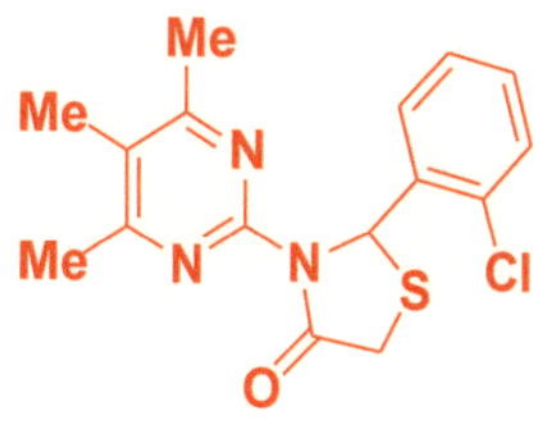

(77)

Ravichandran et al. [54] have synthesized 1,3,4-thiazolidinone derivatives. The present 3D-QSAR study attempts to explore the structural requirements of thiazolidinone derivatives for anti-HIV activity. Based on the structures and biodata of previous thiazolidinone analogs,3D-QSAR studies have been performed with a training set consisting of 96 molecules, which resulted in two reliable computational models, *CoMFA* and *CoMSIA* with r^2 values of 0.931 and 0.972, standard error of estimation (SEE) of 0.173 and 0.089, andq2 values of 0.663 and 0.784, respectively, with the number of partial least-squares (PLS) components being six. Rao et al. [55] have synthesized 2,3-diaryl-1,3 thiazolidin-4-ones **(78)** bearing a methyl group at C- 5 position and tested as anti-HIV agents. The results of the *in vitro* tests showed that some of them proved to be effective inhibitors of HIV-1 replication.

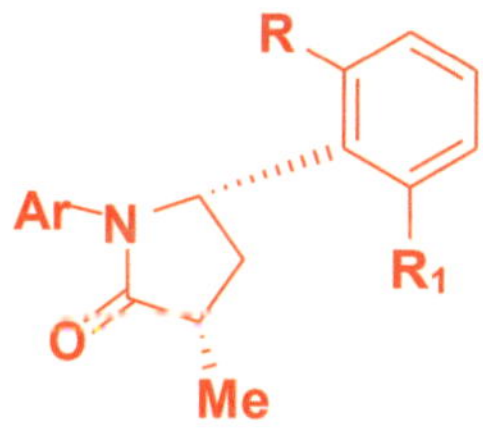

(78)

2.8 Antidiabetic activity

Compound **79**, which had the benzylidene-2,4-thiazolidinedione ring at position C-5 replaced with 5-(2-(3,5-bis(trifluoromethyl) benzyloxy)-5bromobenzylidene), was the most effective at inhibiting PTP1B. The anti-hyperglycemic action of a number of thiazolidine-2,4- dione derivatives with carboxylic ester moieties at N-3 and benzyl and heteroaryl substituents at C-5 has also been investigated. The most promising anti-hyperglycemic activity was found to be compound **80** [56].

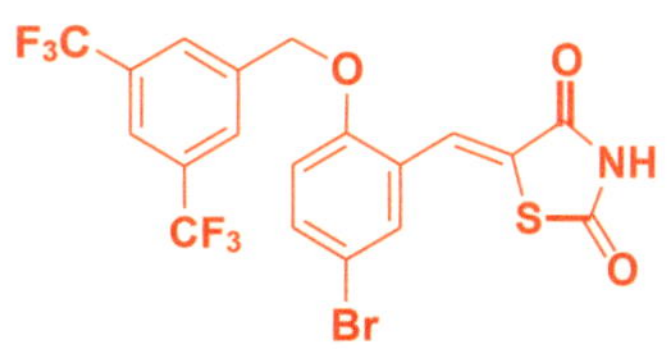

(79)

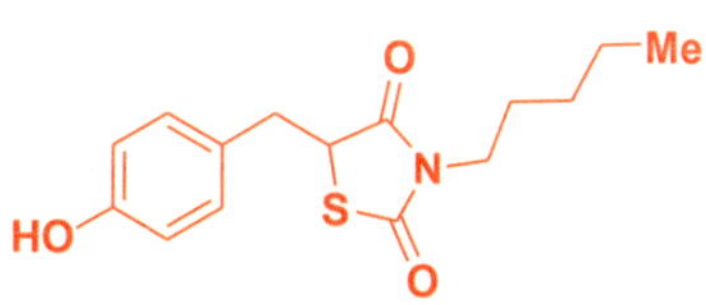

(80)

Liu et al. [57] studied a series of thiazolidinone-substituted biphenyl scaffold**(81)** as PTP1B inhibitors and reported that introduction of the 4- oxothiazolidine-2-thione moiety showed better inhibitory activity against PTP1B.

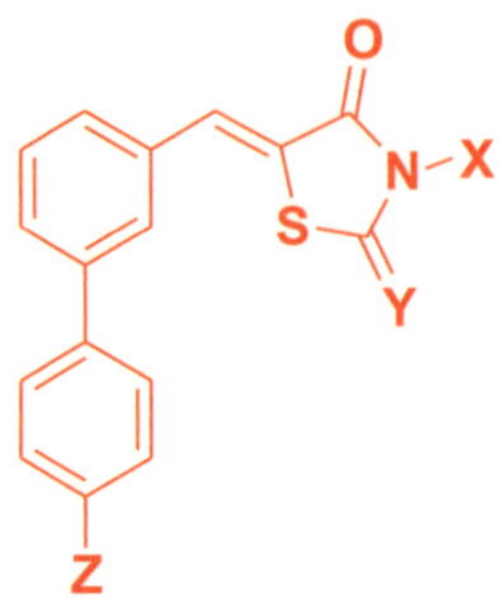

(81)

Bhosle et al. [58, 59] have synthesized 2-hydrazolyl-4-thiazolidinone-5- carboxylic acids with pyrazolyl pharmacophore and evaluated for the antihyperglycemic activity in sucrose loaded rat model. Maccari et al. [59] have synthesized 5-arylidene-2-thioxo-4- thiazolidinones **(82)** and evaluated as aldose reductase inhibitors (ARIs) and most of them exhibited good or excellent in vitro efficacy. Out of the tested compounds, most N-unsubstituted analogues were found to possess inhibitory effects at low micromolar doses and two of them exhibited higher potency than sorbinil, used as a reference drug. The insertion of an acetic chain on N-3 of the thiazolidinone scaffold led to analogues with submicromolar affinity for ALR2 and IC50values very similar to that of epalrestat, the only ARI currently used in therapy.

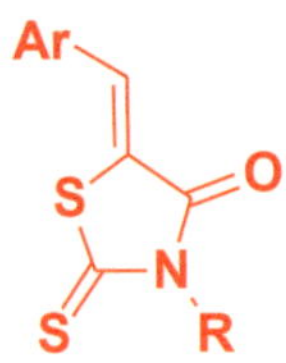

(82)

3. Conclusion

In this chapter the various work up for the synthesis of biologically potent Thiazolidinone derifvatives are discussed and the wide spectrum biological activities are elaborated. The clinically utilised medications no longer have any of the 4-thiazolidinone nucleus' efficacy. Although the four main clinical uses of title compounds are antibacterial, antitubercular, antiviral, and antidiabetic, other potential targets need to be investigated. The majority of locations were investigated to enhance the 4-thiazolidinone's antibacterial and antitubercular profile, however none of the derivatives shown encouraging antitubercular activity. However, few derivatives of 4-thiazolinone with C-2 and N-3 substituted positions and the presence of electron-withdrawing substituent on the aromatic ring at C-2 position exibit varying degrees pharmacological activity. The clinically utilised medications no longer have any of the 4-thiazolidinone nucleus' efficacy.

Author details

Ramarajan Rajalakshmi* and Subramaniyan Ramkumar
Department of Chemistry, Annamalai University, Tamil Nadu, India

*Address all correspondence to: chemrajalaksmi@gmail.com

References

[1] Tripathi AC, Gupta SJ, Gupta GN, Sonar PK, Verma A, Saraf SK. 4-Thiazolidinones: The advances continue. Journal of Medicinal Chemistry. 2014;**72**: 52-77

[2] Chadha N, Bahia MS, Kaur M, Silakari O. Thiazolidine-2, 4-dione derivatives: Programmed chemical weapons for key protein targets of various pathological conditions. Bioorganic & Medicinal Chemistry. 2015;**23**(13):2953-2974

[3] Jain AK, Vaidya A, Ravichandran V, Kashaw SK, Agrawal RK. Recent developments and biological activities of thiazolidinone derivatives: A review. Bioorganic & Medicinal Chemistry. 2012;**20**:3378-3395

[4] Verma A, Saraf SK. 4-Thiazolidinone-A biologically active scaffold. European Journal of Medicinal Chemistry. 2008; **43**:897-905

[5] Doran WJ, Shonle HA. Dialkyl thiazolidiones. The Journal of Organic Chemistry. 1938;**3**(3):193-197

[6] Halimehjani AZ, Marjani K, Ashouri A. A one-pot, three-component synthesis of thiazolidine-2-thiones. Tetrahedron Letters. 2012;**53**(27): 3490-3492

[7] Nitsche C, Klein CD. Aqueous microwave-assisted one-pot synthesis of Nsubstituted rhodanines. Tetrahedron Letters. 2012;**53**(39):5197-5201

[8] Srivastava T, Haq W, Katti SB. Carbodiimide mediated synthesis of 4-thiazolidinones by one-pot three-component condensation. Tetrahedron Letters. 2002;**58**(38): 7619-7624

[9] Rawal RK, Srivastava T, Haq W, Katti SB. An expeditious synthesis of thiazolidinones and tetathiazanones. Journal of Chemical Research. 2004;**5**: 368-369

[10] Pratap UR, Jawale DV, Bhosle MR, Mane RA, et al. Saccharomyces cerevisiae catalyzed one-pot three component synthesis of 2, 3-diaryl-4-thiazolidinones. Tetrahedron Letters. 2011;**52**(14):1689-1691

[11] Rawal RK, Tripathi R, Katti SB, Pannecouque C, De Clercq E. Design, synthesis, and evaluation of 2-aryl-3-heteroaryl-1, 3-thiazolidin-4-ones as anti-HIV agents. Bioorganic & Medicinal Chemistry. 2007;**15**(4):1725-1731

[12] Markovic R, Stodanovic M. Stereocontrolled synthesis of new tetrahydrofuro [2, 3-d] thiazole derivatives via activated vinylogous iminium ions. Heterocycles. 2005; **56**:2635

[13] Pawar RB, Mulwad VV. Synthesis of some biologically active pyrazole, thiazolidinone, and azetidinone derivatives. Chemistry of Heterocyclic Compounds. 2004;**40**:219

[14] Ocal N, Aydogan F, Yolacan C, Turgut Z. Synthesis of some furo-thiazolidine derivatives starting from aldimines. Journal of Heterocyclic Chemistry. 2003;**40**:721

[15] Eltsov OS, Mokrushin VS, Belskaya NP, Kozlova NM. Synthesis and transformations of (imidazolylimino) thiazolidinones. Russian Chemical Buletin, International Edition. 2003; **52**:461

[16] Bolognese A, Correale G, Manfra M, Lavecchia A, Novellino E, Barone V.

Thiazolidin-4-one formation. Mechanistic and synthetic aspects of the reaction of imines and mercaptoacetic acid under microwave and conventional heating. Organic & Biomolecular Chemistry. 2004;**2**:2809

[17] Dandia A, Singh R, Khaturia S, Mérienne C, Morgant G, Loupy A. Regioselective Synthesis of Unsymmetrical 3-(Quinolin-3-yl) Pentane-1, 5-Diones in the Aqueous Medium through Montmorillonite KSF Catalysis. Bioorganic & Medicinal Chemistry. 2006;**14**(7):2409-2417

[18] Liu X, Zheng C, Sun L, Liu X, Piao H. Synthesis of new chalcone derivatives bearing 2, 4-thiazolidinedione and benzoic acid moieties as potential anti-bacterial agents. European Journal of Medicinal Chemistry. 2011;**46**:3469

[19] Gaba V, Bariwal J, Sharma V, Gupta MK, Rawal RK. Synthetic and medicinal perspective of thiazolidinones: A review. Chemistry and Biology Interface. 2014;**4**(5):2249-4820

[20] Patel KH, Mehta AG. Synthesis and Antifungal Activity of Azetidinone and Thiazolidinones Derivatives of 2-Amino-6-(2-naphthalenyl) thiazolo [3, 2-d] thiadiazole. Journal of Chemistry. 2006; **3**(4):267-273

[21] Abdellatif KRA et al. Design, synthesis and biological screening of new 4-thiazolidinone derivatives with promising COX-2 selectivity, anti-inflammatory activity and gastric safety profile. Bioorganic Chemistry. 2016;**64**:1-12

[22] Shanmuga Priya S, Aanandhi V. Synthesis and anticonvulsant activity of thiazolidinone derivatives. Synthesis. 2012;**4**(1):01-04

[23] Rajalakshmi R, Santhi R, Elakkiya T. Synthesis and Evaluation of novel oxazinyl thiazolidinone derivatives antioxidant agent as potent antidiabetic. International Journal of Advanced Science and Technology. 2020;**29**(5): 1724-1733

[24] Rajalakshmi R, Santhi R, Elakkiya T. Synthesis, Characterization, Docking studies, Evaluation of Thiazinyl-thiazolidinone derivatives as potential in vitro antidiabetic and antioxidant agents. Synthesis. 2020;**29**(08):6091-6103

[25] Rahim F, Zaman K, Ullah H, Taha M, Wadood A, Javed MT, et al. Synthesis of alpha amylase inhibitors based on privileged indole scaffold. Bioorganic Chemistry. 2015;**63**:123-131

[26] Saiz C, Pizzo C, Manta E, Wipf P, Mahler SG. Microwave-assisted tandem reactions for the synthesis of 2-hydrazolyl-4-thiazolidinones. Tetrahedron Letters. 2009;**50**:901-904

[27] Benmohammed A, Khoumeri O, Djafri A, Terme T, Vanelle P. Synthesis of novel highly functionalized 4-thiazolidinone derivatives from 4-phenyl-3-thiosemicarbazones. Molecules. 2014;**19**:3068-3083

[28] Ottana R, Maccari R, Ciurleo R, Vigorita MG, Panico AM, Cardile V, et al. Design, synthesis, evaluation of thiazolidinone derivatives incorporated with oxazine pharmacophore as potential in vitro antidiabetic and antioxidant agents-Docking studies. Bioorganic & Medicinal Chemistry. 2007;**15**:7618-7625

[29] Jieping Z, Blanchet J. Reeves, Reeve's synthesis of 2-imino-4-thiazolidinone from alkyl (aryl) trichloromethylcarbinol revisited, a three-component process from aldehyde, chloroform and thiourea. Tetrahedron Letters. 2004;**45**: 4449-4452

[30] Pang H-X, Hui Y-H, Fan K, Xing X-J, Wua Y, Yang J-H, et al. Discovery of novel double pyrazole Schiff base derivatives as anti-tobacco mosaic virus (TMV) agents [J]. Chinese Chemical Letters. 2015;**27**:335-339

[31] Veerasamy R, Rajak H, Kumar KS, Agrawal RK. Validation of QSAR models-strategies and importance. Letters in Drug Design & Discovery. 2011;**8**:82-86

[32] Liesen AP, Aquino TM, Carvalho CS, Lima VT, Araujo JM, Lima JG, et al. Synthesis and evaluation of anti-Toxoplasma gondii and antimicrobial activities of thiosemicarbazides, 4-thiazolidinones and 1, 3, 4-thiadiazoles. European Journal of Medicinal Chemistry. 2010;**45**:3685

[33] Sayed M, Mokle S, Bokhare M, Mankar A, Surwase S, Bhusare S, et al. Synthesis of some new 2, 3-diaryl-1, 3-thiazolidin-4-ones as antibacterial agents. ARKIVOC. 2006;**II**:187

[34] Abdou MM, El-Saeed RA, Bondock S. Recent advances in 4-hydroxycoumarin chemistry. Part 1: Synthesis and reactions. Arabian Journal of Chemistry. 2019;**12**(1):88-121

[35] Bonde CG, Gaikwad NJ. Synthesis and preliminary evaluation of some pyrazine containing thiazolines and thiazolidinones as antimicrobial agents. Bioorganic & Medicinal Chemistry. 2004;**12**:2151-2161

[36] Kucukguzel SG, Oruc EE, Rollas S, Sahin F, Ozbek A. Synthesis, characterisation and biological activity of novel 4-thiazolidinones, 1, 3, 4-oxadiazoles and some related compounds. European Journal of Medicinal Chemistry. 2002;**37**:197

[37] Turan-Zitouni G, Kaplancikli ZA, Ozdemir A. Synthesis and antituberculosis activity of some N-pyridyl-N′-thiazolylhydrazine derivatives. European Journal of Medicinal Chemistry. 2010;**45**:2085

[38] Zhou H, Wu S, Zhai S, Liu A, Sun Y, Li R, et al. Design, synthesis, cytoselective toxicity, structure-activity relationships, and pharmacophore of thiazolidinone derivatives targeting drug-resistant lung cancer cells. Journal of Medicinal Chemistry. 2008;**51**:1242

[39] Moorkoth S. Synthesis and anti-cancer activity of novel thiazolidinone analogs of 6-aminoflavone. Chemical & Pharmaceutical Bulletin. 2015;**63**:974-985

[40] Kaminskyy D, Bednarczyk-Cwynar B, Vasylenko O, Kazakova O, Zimenkovsky B, Zaprutko L, et al. Synthesis of new potential anticancer agents based on 4-thiazolidinone and oleanane scaffolds. Medicinal Chemistry Research. 2012;**21**:3568-3580

[41] Sala M, Chimento A, Saturnino C, Gomez-Monterrey IM, Musella S, Bertamino A, et al. Synthesis and cytotoxic activity evaluation of 2, 3-thiazolidin-4-one derivatives on human breast cancer cell lines. Bioorganic & Medicinal Chemistry Letters. 2013;**23**:4990-4995

[42] Gududuru V, Hurh H, Dalton JT, Miller DD. Discovery of 2-arylthiazolidine-4-carboxylic acid amides as a new class of cytotoxic agents for prostate cancer. Journal of Medicinal Chemistry. 2005;**48**:2584

[43] Gududuru V, Hurh H, Dalton JT, Miller DD. Synthesis and antiproliferative activity of 2-aryl-4-oxo-thiazolidin-3-yl-amides for prostate cancer. Bioorganic & Medicinal Chemistry Letters. 2004;**14**:5289

[44] Kamel MM, Ali HI, Anwar MM, Mohamed NA, Soliman AM. Synthesis,

antitumor activity and molecular docking study of novel Sulfonamide-Schiff's bases, thiazolidinones, benzothiazinones and their C-nucleoside derivatives. European Journal of Medicinal Chemistry. 2010;**45**:572

[45] Havrylyuk D, Mosula L, Zimenkovsky B, Vasylenko O, Gzella A, Lesyk R. Synthesis and anticancer activity evaluation of 4-thiazolidinones containing benzothiazole moiety. European Journal of Medicinal Chemistry. 2010;**45**:5012

[46] Karali N, Terzioglu N, Gursoy A. Synthesis and primary cytotoxicity evaluation of new 5-bromo-3-substituted-hydrazono-1H-2-indolinones. Archiv der Pharmazie Pharmaceutical and Medicinal Chemistry. 2002;**8**:374

[47] Vigorita MG, Ottana R, Monforte F, Maccari R, Trovato A, Monforte MT, et al. Synthesis and antiinflammatory, analgesic activity of 3, 3′-(1, 2-Ethanediyl)-bis [2-aryl-4-thiazolidinone] chiral compounds. Bioorganic & Medicinal Chemistry Letters. 2001;**11**:2791-2794

[48] Kumar A, Rajput CS, Bhati SK. Synthesis of 3-[4′-(p-chlorophenyl)-thiazol-2′-yl]-2-[(substituted azetidinone/thiazolidinone)-aminomethyl]-6-bromoquinazolin-4-ones as anti-inflammatory agent. Bioorganic & Medicinal Chemistry. 2007;**15**:3089

[49] Deep A, Jain S, Sharma PC. Synthesis and anti-inflammatory activity of some novel biphenyl-4-carboxylic acid 5-(arylidene)-2-(aryl)-4-oxothiazolidin-3-ylamides. Acta Poloniae Pharmaceutica - Drug Research. 2010;**67**:63

[50] Ottana R, Mazzon E, Dugo L, Monforte F, Maccari R, Sautebin L, et al. Modeling and biological evaluation of 3, 3′-(1, 2-ethanediyl) bis [2-(4-methoxyphenyl)-thiazolidin-4-one], a new synthetic cyclooxygenase-2 inhibitor. European Journal of Pharmacology. 2002;**448**:71

[51] Ali AM, Saber GE, Mahfouz NM, El-Gendy, MA, Radwan AA, Hamid MA. Synthesis and three-dimensional qualitative structure selectivity relationship of 3, 5-disubstituted-2, 4-thiazolidinedione derivatives as COX2 inhibitors. Archives of Pharmacal Research. 2007;**30**:1186

[52] Bhati SK, Kumar A. Synthesis of new substituted azetidinoyl and thiazolidinoyl-1, 3, 4-thiadiazino (6, 5-b) indoles as promising anti-inflammatory agents. European Journal of Medicinal Chemistry. 2008;**43**:2323

[53] Shiradkar MR, Ghodake M, Bothara KG, Bhandari SB, Nikalje A, Akula KC, et al. Synthesis and anticonvulsant activity of clubbed thiazolidinone-barbituric acid and thiazolidinone-triazole derivatives. ARKIVOC. 2007;**XIV**:58

[54] Ravichandran V, Prashantha Kumar BR, Sankar S, Agrawal RK. Predicting anti-HIV activity of 1, 3, 4-thiazolidinone derivatives: 3D-QSAR approach. European Journal of Medicinal Chemistry. 2009;**44**:1180-1187

[55] Rao A, Carbone A, Chimirri A, De Clercq E, Monfortea AM, Monforte P, et al. Il Farmaco. 2002;**57**:747-751

[56] Liu Z, Chai Q, Yuan-yuan LI, Shen Q, Ma L-p, Zhang L-n, et al. Discovery of novel PTP1B inhibitors with antihyperglycemic activity. Acta Pharmacologica Sinica. 2010;**31**:1005-1012

[57] Bhosle MR, Mali JR, Pal S, Srivastava AK, Mane RA. Synthesis and

antihyperglycemic evaluation of new 2-hydrazolyl-4-thiazolidinone-5-carboxylic acids having pyrazolyl pharmacophores. Bioorganic & Medicinal Chemistry Letters. 2014;**24**: 2651-2654

[58] Maccari R, Del Corso A, Giglio M, Moschini R, Mura U, Ottanà R. In vitro evaluation of 5-arylidene-2-thioxo-4-thiazolidinones active as aldose reductase inhibitors. Bioorganic & Medicinal Chemistry Letters. 2011;**21**: 200-203

[59] Arulmani R, Ramkumar S, Rajalakshmi R. Substituent Effect on the Infrared Spectra of Thiazolyl Styryl Ketone. Asian Journal of Applied Chemistry Research. 2021:25:1-5

Chapter 4

Synthetic Approaches for Pharmacologically Active Decorated Six-Membered Diazines

Yousef Najajreh and Maha Awwad Khoury

Abstract

Diazine alkaloid (pyridazine, pyrimidine and pyrazine) scaffold, a widespread two-nitrogen containing compounds in nature (DNA, RNA, flavors, and fragrances), constitutes a central building block for wide range of pharmacological applications. Diazines are reported to exhibit antimetabolite (antifolate and), anticancer, antibacterial, antiallergic, tyrosine kinase, antimicrobial, calcium channel antagonistic, anti-inflammatory, analgesic, antihypertensive, antileishmanial, antituberculostatic, anticonvulsant, diuretic and potassium-sparing, to antiaggressive activities. Pyridazine (1,2-diazine), pyrimidine (1,3-diazine) and pyrazine (1,4-diazine) are found as mono-systems, fused or annulated in pharmaceutical, agrochemical or materials. These six-membered heterocyclic aromatic moieties defined as privileged scaffolds constitute diverse chemical structures and as such hold substantial interest for organic, medicinal and biological chemists. This chapter will focus on elaboration of the different synthetic approaches applied in preparing pharmacologically active decorated diazines with special care on pyrimidines (non-fused substituted forms) that are endowed with clinical applications. Synthetic approaches applied in preparing selected FDA approved drugs with pyrimidine as a central unit bearing different substituents will be intensively explored. Special attention will be given to novel synthetic methodologies that served molecules with improved druglikeness and ADME-Tox properties.

Keywords: diazine, pyrimidine, chemistry, synthesis, methods, substituted, medicinal, bioactive, anticancer, ADME-Tox

1. Introduction

Pyrimidine derivatives, broadly applied in therapeutic disciplines, owing to their high degree of structural diversity. This “privileged scaffold” and “derivatives” either as substituted or as fused systems gain wide interest due to plentiful biological activities reported over the years. Those versatile biological activities include (and not limited to) modulation of myeloid leukemia (for example imatinib, Dasatinib and nilotinib are pyrimidine-based drugs and well established treatments for leukemia) [1, 2], breast cancer and idiopathic pulmonary fibrosis [3], antimicrobial [4], antifungal [5], antiparasitic [6], diuretic [7], antitumor [8–13], antifilarial [14], DNA topoisomerase II inhibitors [15], antitubercular agents [16], antimalarial [17] and antiplasmodial [18], anti-inflammatory and analgesic activities [19–21], anti-HIV [5, 22, 23], cardiovascular agents and antihypertensive [24, 25], antiulcer agents [18], hair disorders activities [26], calcium-sensing receptor antagonists, DPP-IV inhibitors [27, 28], antidiabetic [29], potent adenosine A2a receptor agonistic or antagonist action [30], TLR8 or [15] or interferon beta (IFN-β) modulators [31], vascular relaxation for ocular ciliary artery and neuroprotection on retinal ganglion cell [32], 3-hydroxy-3-methylglutaryl-coenzyme A (HMG-CoA) reductase inhibitors and coronary heart disease therapeutics [33], anticancer [8, 34–36], key intermediate for Vitamin B1 [37], pyruvate dehydrogenase kinase inhibitors [38]. In addition, many of pyrimidine derivatives are reported to possess potential central nervous system (CNS) and antidepressant properties [39], antihypnotic [40], anti-Alzheimer's Disease (AD) agents [41, 42], anticonvulsant [43], antiallergic [44] and for treatment of hypoglycemic and hypolipidemic activities [45, 46].

Over the years a large interest in fused pyrimidines compiled and exceeded in certain extent substituted pyrimidine derivatives. Though highly appealing the synthesis of fused pyrimidines- is beyond the scope of this chapter. Additionally, this chapter will not address in depth the diverse bioactivities neither in structure activity relationship (SAR), nor in detailed mechanism of actions (MOA). The main focus of this manuscript is to explore various synthetic methods employed to produce this diverse group of compounds.

Synthesis of pyrimidine and its derivatives gains a great deal of interest due to wide applications in medical and therapeutic.

The current chapter focuses on chemical process and methods to derivatize the pyrimidine heterocyclic core with di-, tri- and tetrasubstituted pyrimidines.

2. Challenges facing pyrimidine synthesis

Although widely used, the existing protocols for the synthesis of pyrimidines were labor intensive, oftentimes impractical, inefficient or incompatible with a number of functional groups, harsh reaction conditions, involvement of transition metal catalysis, multistep synthesis of the starting material, or need for inert gas protection, remain as shortcomings in known pyrimidine synthesis (**Figure 1**).

3. Synthetic methods in decorated pyrimidine core

A diverse group of drugs or drug candidates comprise pyrimidine core. The instances elaborated below belong to recently FDA approved ones or with special medical interest.

3.1 Synthesis of N-2,4-Disubstitited pyrimidines

Polyamino-pyrimidines play a very important role in biological and pharmaceutical chemistry. 2,4- diamino-pyrimidine derivatives as potential antimalarial agents [17], SNSR4 antagonists [47], antitumor agents [48] and caspase-1 inhibitors [49]. This report focused on the method development for the synthesis of the latter class of compounds.

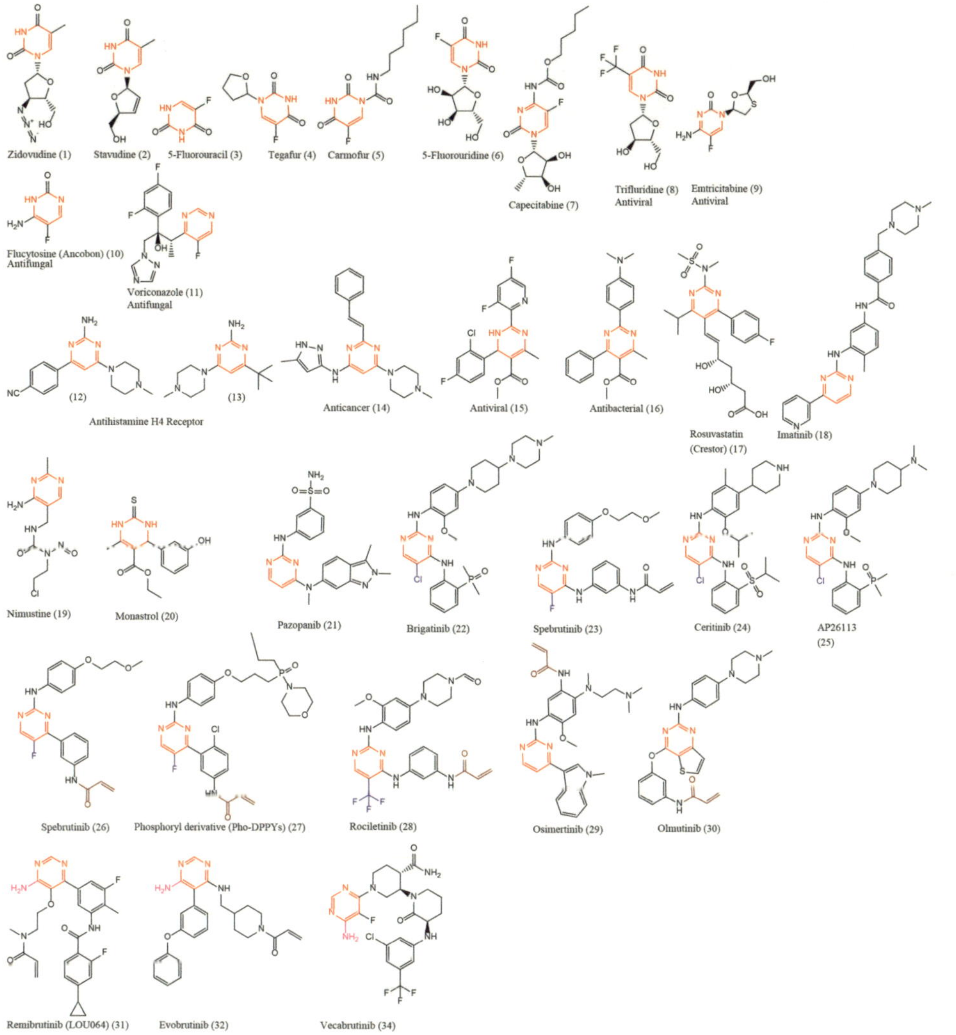

Figure 1.
Examples of chemical structures of bioactive pyrimidine-based derivatives. Pyrimidine heterocyclic rings found in potent anti-HIV agents like zidovudine (1), Stavudine (2); antiviral like Trifluridine (8), Emtricitabine (9, 15); antifungal like Voriconazole (11) Flucytosine (Ancobon) (10); antihistamine H4 receptor (12 and 13), antibacterial (16), cholesterol lowering agent Rosuvastatin (Crestor) (17); antitumor agents like fluorouracil (3), Tegafur (4), Carmofur (5), 5-Fluorouridine (6), Capecitabine (7), ENMD-2076 (14), Imatinib (18), Nimustine (19), Monastrol (20), Pazopanib (21), Brigatinib (22), Spebrutinib (23), Ceritinib (24), AP26113 (25), Spebrutinib (26), phosphoryl derivative (pho-DPPYs) (27), Rociletinib (28), Osimertinib (29), Olmutinib (30), Remibrutinib (LOU064) (31), Evobrutinib (32), Vecabrutinib (34).

Polysubstituted pyrimidines received sub- special attention due to their pronounced physiological activity.

3.1.1 Synthesis of N-2,4-Disubstituted N-2,4,5-trisubstituted pyrimidines

3.1.1.1 Approaches for synthesis of N-2,4-disubstituted pyrimidines

2,4-disubstited pyrimidines are classifies according to the type of bond linking the substituent to the core heterocycle: i) 2,4-diamminosubstituted (2,4-diN), ii) 2,4-

monoaminomonocarbon (2 N, 4C), iii) 2,4-dicarbon (2,4-diC), iv) 2,4-monoaminomonooxo (2 N, 4O), v) 2,4-monoaminomonothio (2 N, 4S), vi) 2,4-monothiomonoamino (2S, 4 N), vii) 2,4-monothiomonocarbo (2S, 4C) and so on. Substituents reported so far can be aromatic, aliphatic, heteroaromatic and heteroaliphatic.

3.1.1.2 2,4-Disubstitited pyrimidines: a heterocyclic core with privileged regioisomers

An increased number of 2,4-disubstituted pyrimidine derivatives that are endowed with bioactivities indicated high number of research articles and patent application shown in the literature (**Figure 2**). Though most of the reported activities of 2,4-disubstituted pyrimidines were related to antitumor due to ability to exert antiproliferative effect like inhibitors of KDR [50] and Aurora kinase [51], ERα/VEGFR-2 [52], T790M-EGFR [53] and L858R/T790M-EGFR double mutant [54] inhibitors, bioactivities of high significance of some 2,4-disubstituted pyrimidines such as antibacterial [55], anti-neuronal nitric oxide synthase [56], anticholinesterases [57] actions were also disclosed [58].

In the following section synthetic methods of four FDA approved drugs Pazopanib (21), Spebrutinib (51), Dabrafenib (60) and Rilpivirine (62) will be presented and discussed in details.

3.1.1.2.1 Case study—1: synthesis of Pazopanib hydrochloride [Votrient®, GW786034, 5-[[4-[(2,3-dimethylindazol-6-yl)-methylamino]pyrimidin-2-yl]amino]-2-methylbenzenesulfonamide. HCl, (21)]

Pazopanib [Votrient®, GW786034, 5-[[4-[(2,3-dimethylindazol-6-yl)-methylamino]pyrimidin-2-yl]amino]-2-methylbenzenesulfonamide (21)] is a potent and selective multi-targeted receptor tyrosine kinase inhibitor of VEGFR-1, VEGFR-2, VEGFR-3, PDGFR-a/b, and c-kit. The drug blocks tumor growth and inhibits angiogenesis. It was approved for renal cell carcinoma by the U.S. Food in 2009 and advanced soft tissue sarcoma [59]. The drug is marketed under the trade name Votrient by the drug's manufacturer, GlaxoSmithKline [59].

3.1.1.2.1.1 Pazopanib One drug different Synthetic Routes

In the procedure reported by Qi et al. the methylation of 3-methyl-6-nitroindazole (36) was carried in the presence of dimethyl carbonate (DMC) and 1,4-diazabicyclo [2.2.2] octane (DABCO) ending in the N2,3-dimethyl-6-nitroindazole (36) in high yield. In another report the methylation of 3-methyl-6-nitroindazole (36) was performed using trimethyl orthoformate in the presence of BF_3OEt to give rise to the N2,3-dimethyl-6-nitroindazole (37) in 65% yield.

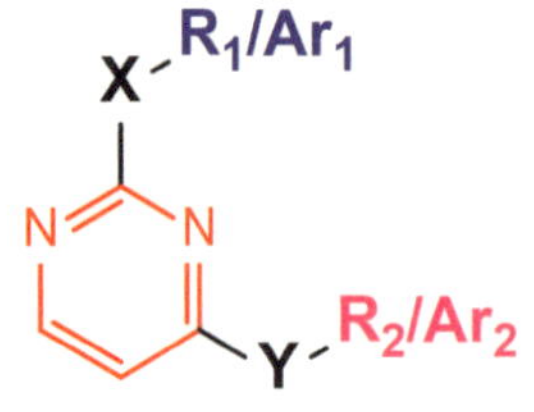

Figure 2.
Types of 2,4-disubstitited pyrimidines (X = O, S, C, NH; Y = X = O, S, C, NH).

The reduction step was carried out under hydrogenation of (36) in the presence of Pd/C and H_2 to afford the aminoindazole derivative (37) in 97% yield [60]. The subsequent condensation of aminoindazole derivative (37) with 2,4-dichloro-pyrimidine (38) to yield the pyrimidinylaminoindazole (39) and (40). A second methylation at the secondary aniline nitrogen of (40) with CH_3I and Cs_2CO_3 resulted gave (41) in 83% yield, which was afterwards condensed with aniline derivative 5-amino-2-methylbenzenesulfonamide (42) in acidic alcohol solution (HCl/ isopropanol) to furnish the target compound (21) as hydrochloride salt in 81% yield.

Allowing 6-amino-2,3-dimethyl-2H-indazole (37) to react with 2,4-dichloro-pyrimidine (38) gave rise exclusively to the C4-aminated product (40) indicating regioselectivity of the reaction with 2,4-dichloropyrimidine (38). The intermediate (40) was transformed into (41) following the reaction with the sulfonamide aniline derivative i.e. 5-amino-2-methylbenzenesulfonamide (42).

The coupling of the aniline derivative like 5-amino-2-methylbenzenesulfonamide (42) to (41) was accomplished in good to high yields in refluxing ethanol.

To overcome the reduced regioselectivity in the methylation reaction of N-H indazoles, usually yielded a mixture of N-alkyl 1*H*-indazoles and N-alkyl 2*H*-indazoles regioisomers (**Figures 3** and **4**).

YiCheng Mei *et al* reported a synthetic process where the 2,3-dimethyl-6-nitro-2*H*-indazol (36) was prepared in a regioselective manner (**Figure 5**) [61].

3-Methyl-6-nitro-1*H*-indazole (35) was prepared 93.9% yield by allowing 2-ethyl-5-nitroaniline (45) to react with sodium nitrite in glacial acetic acid. The regioselective conversion of 3-methyl-6-nitro-1*H*-indazole (35) to prepare N2,3-dimethyl-6-nitro-2H-indazole (37) was accomplished using trimethyl orthoformate in toluene and DMF (10, 1 mL) at room temperature (should be kept below 35°C). These conditions

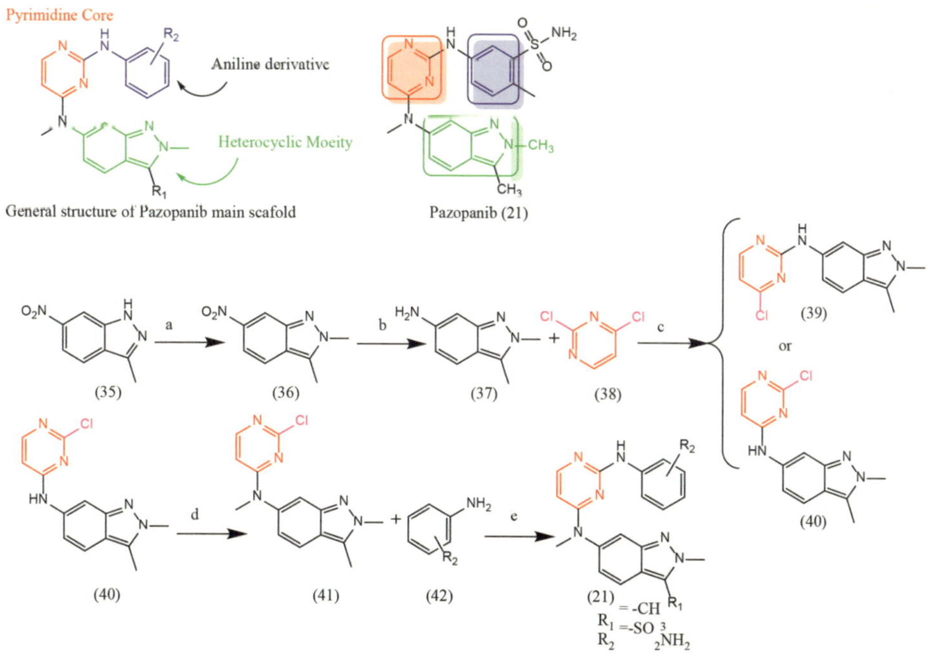

Figure 3.
Synthesis of Pazopanib (21). Reagents and conditions: a) DMC, DABCO, DMB; b) Pd/C, H_2, EtOH; c) $NaHCO_3$, EtOH; d) CH_3I, Cs_2CO_3, DMF; e) EtOH [60].

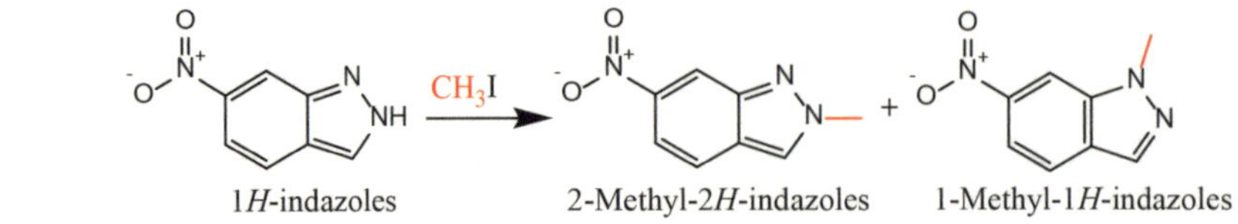

Figure 4.
Regioselective methylation of 1H-indazoles.

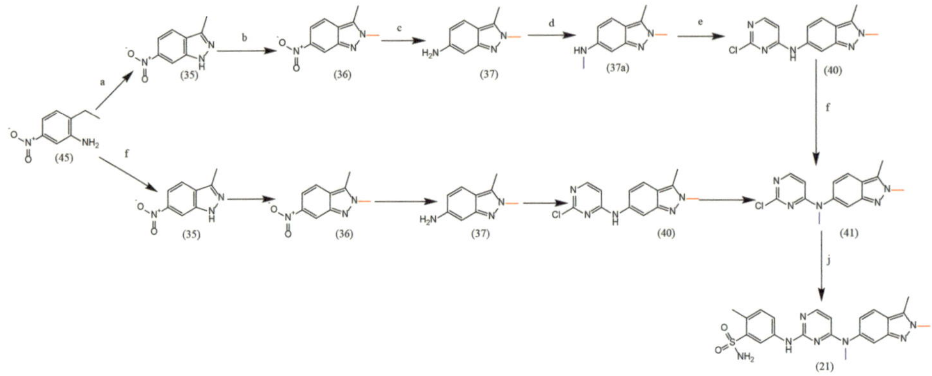

Figure 5.
Synthesis of Pazopanib (21) following YiCheng Mei et al *report. Reagents and conditions: a)* t-*BuONO, acetic acid; b) trimethyl oxonium tetraflouroborate; c) tin(II) chloride; d) 2,4-dichloropyrimidine (38); e) MeI, rt.; f) sodium nitrite, acetic acid; g) trimethyl orthoformate, DMF, sulfuric acid, toluene, reflux; h) MeOH, Pd/C, H_2, paraformaldehyde, NaH, $NaBH_4$; i) 2,4-dichloropyrimidine (38); j) 5-amino-2-methylbenzenesulfonamide, reflux.*

produce the N2-methylated isomer. The nitro-derivative was subjected to hydrogenation (Pd/C catalyst) of (36) followed by methylation using Eschweiler-Clarke methylation reaction [62], providing N-2,3-dimethyl-6-nitro-2H-indazole intermediate (37) in 63.0% yield after recrystallization from the ethanol. Worth noting that in the patent disclosed Kumar and colleagues the reduction of the N2,3-dimethyl-6-nitro-2H-indazole (36) was converted to the amino-derivative using Raney nickel in 95% yield [63].

The secondary amine in N2,3-trimethyl-2H-indazol-6-amine (37a) was reacted with 2,4-dichloropyrimidine (38) in DMF under basic conditions and elevated temperature (100°C, 3 hr) afforded N-(2-chloropyrimidin-4-yl)-N,2,3-trimethyl-2H-indazol-6-amine (41) in 88.4% yield. The final product as hydrochloric salt was prepared as a result of reacting N-(2-chloropyrimidin-4-yl)-N,2,3-trimethyl-2*H*-indazol-6-amine (41) with 5-amino-2-methylbenzenesulfonamide (42) in isopropanol under acidic conditions in 76.0% yield. N-(2-chloropyrimidin-4-yl)-N,2,3-dimethyl-2*H*-indazol-6-amine (40) was prepared by allowing N2,3-trimethyl-2H-indazol-6-amine (37a) to react with 2,4-dichloropyrimidine (38) in a mixture of H2O/MeOH for 24 hr. at temperature ranges between 25 and 30°C in 86.7% yield.

The N-methylation was performed using methyl iodide in DMF under basic conditions (Cesium carbonate) with N-(2-chloropyrimidin-4-yl)-N,2,3-dimethyl-2*H*-indazol-6-amine (40) to afford N-(2-chloropyrimidin-4-yl)-N,2,3-trimethyl-2H-indazol-6-amine (41) in 90% yield.

Pazopanib hydrochloride salt was prepared by allowing N-(2-chloropyrimidin-4-yl)-N,2,3-trimethyl-2H-indazol-6-amine (41) to react with 5-amino-2-

methylbenzenesulfonamide (42) in refluxing acidified (4 M HCl) isopropanol for 10–12 hr. in 97% yield. In cases the last step was performed under acidic conditions (HCl in isopropanol) then Pazopanib hydrochloride salt (21) was collected.

3.1.1.2.2 Case study –2: synthesis of pyrimidine-based Bruton's Agammaglobulinemia tyrosine kinase (BTK) [Spebrutinib, CC-292, AVL2923, (5 N-[3-[[5-fluoro-2-[4-(2-methoxyethoxy)anilino]pyrimidin-4-yl]amino]phenyl]prop-2-enamide] (51)

Spebrutinib (51), an orally bioavailable, potent and selective covalent inhibitor of Bruton's agammaglobulinemia tyrosine kinase (BTK) [59]. BTK, a cytoplasmic tyrosine kinase and member of the Tec family of kinases, plays a central role in B lymphocyte development, activation, signaling, proliferation and survival [64]. Beside it potential in autoimmune related diseases, since BTK mediates the B-cell and Fc receptor signaling pathways [64], the drug is considered with great value for neoplastic disease and particularly in hematopoietic malignancies [65–68]. Upon administration, Spebrutinib targets and covalently binds to Cys 481 in BTK, blocking the ATP-binding pocket of the enzyme BTK, thereby preventing its downstream signaling. By irreversibly inhibiting BTK, administration of this agent may lead to an inhibition of B cell receptor (BCR) signaling and may inhibit cell proliferation of B-cell malignancies. Readers interested in a wider and deeper perceptions on BTK inhibitors and embodied role in malignant and non-malignant disease are referred to excellent reviews [65–68].

Synthetically, the molecule (49) is produced starting from 2,4-dichloro-5-fluoropyrimide (45). The coupling of the two different substituents to the pyrimidine core was performed under classical conditions. The first step was accomplished by reacting the 2,4-dichloro-5-fluoropyrimide with the mono-BOCylated *meta*-diaminobenzene (46). The selective displacement of chloride at C4 by *tert*-butyl N-(3-aminophenyl) carbamate (46) renders high ratio of regioselective synthon and yielded the intermediate (47), which was reacted with the second substrate at C2 by 4-(2-methoxyethoxy)aniline (48). Due to the presence of *tert*-BOC (acid cleavable) as protecting group the two steps have be conducted under basic conditions.

Following the cleavage of the protecting group *tert*-BOC under acidic conditions, the acryloyl chloride (prop-2-enoyl chloride) was coupled to the free amine under basic conditions (**Figure 6**).

3.1.1.2.3 Case study –3: synthesis of Dabrafenib [trade name Tafinlar, N-{3-[5-(2-Amino-4-pyrimidinyl)-2-(1,1-dimethylethyl)-1,3-thiazol-4-yl]-2-fluorophenyl}-2,6-difluorobenzenesulfonamide, GSK2118436, (60)]

Rheault and his colleagues reported the synthesis of Dabrafenib (GSK2118436, (60)) (**Figure 7**). The strategy depends on synthesizing the building blocks starting from the bromobenzoic acid derivative (52) and the sulfonamide (56). The coupling to pyrimidine moiety was performed via a nucleophilic substitution of 2-chloro-4-methylpyrimidine (61) facilitated by lithiation of the 4-methyl using LiHMDS. The yield of this step was recorded as high. The synthesis of the core substituted thiozole (59) was accomplished via bromination of the benzylic/alpha carbon of (58) using NBS and reacting the intermediate with 2,2,2- trimethylthioacetamide either in polar aprotic solvent like DMF or DMA [69].

The second substitution at N2-pyrimidine was accomplished by either concentrated ammonia (7 N NH_3) in methanol in sealed tube under heating of 100°C or under acidic facilitated substitution (using HCl in 2,2,2-trifluoroethanol as solvent) when the amine

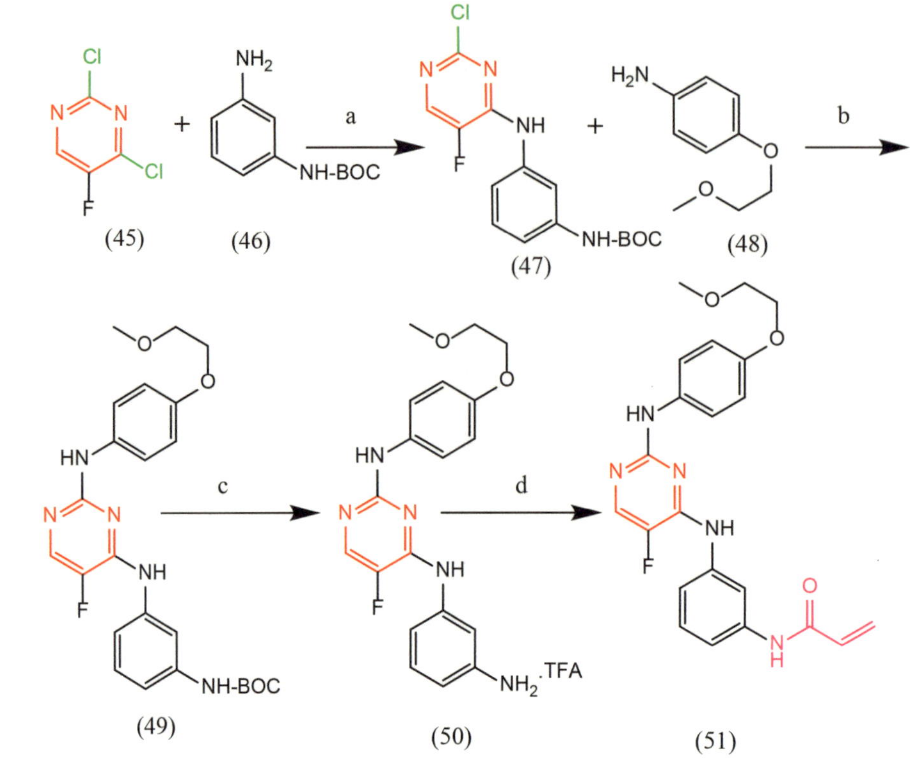

Figure 6.
Synthesis of Spebrutinib (51) [CC-292, AVL2923, (5 N-[3-[[5-fluoro-2-[4-(2-methoxyethoxy)anilino] pyrimidin-4-yl]amino]phenyl]prop-2-enamide]. Reagents and conditions: a) DMA, reflux; b) DMA, reflux; c) TFA, DCM, r.t, acyl halide, TEA, r.t.

(= R_2NH_2 amine) and microwave at elevated temperature of 180°C. This step of attaching 2-chloro-4-methylpyrimidine (61) to methyl 3-{[(2,6-difluorophenyl)sulfonyl]amino}-2-fluorobenzoate (57) was reported to proceed (71%) and afforded N-{3-[(2-chloro-4-pyrimidinyl)acetyl]-2-fluorophenyl}-2,6-difluorobenzenesulfonamide as title compound that is a mixture of keto-enol **Figures 8**, 58a ↔ 58b). The second substitution was performed following the formation of the thiazole central moiety and afforded N-{3-[5-(2-amino-4-pyrimidinyl)-2-(1,1-dimethylethyl)- 1,3-thiazol-4-yl]-2-fluorophenyl}-2,6-difluorobenzenesulfonamide (60) in 47% yield.

The 2,4-disubstitution regioisomer selectivity around the pyrimidine core was granted by using the 2-methyl-4-chloropyrimidine substrate [65–68, 70]. Other recent report confirmed the feasibility, high yield and regio-selective coupling using 2-methyl-4-chloropyrimidine [65–68].

3.1.1.2.4 Case study –4: synthesis of the antiviral drug Rilpivirine hydrochloride [(Edurant®), 4-[[4-[4-[(E)-2-cyanoethenyl]-2,6-dimethylanilino]pyrimidin-2-yl]amino]benzonitrile; hydrochloride (62)]

The non-nucleoside reverse transcriptase inhibitor (NNRTI), Rilpivirine Hydrochloride (RPV, Edurant®) (62) got the approval both from the U.S. FDA and E.U. EMA in 2011 for the treatment of HIV-1 infection in treatment-naïve adult patients (**Figure 9**) [71, 72].

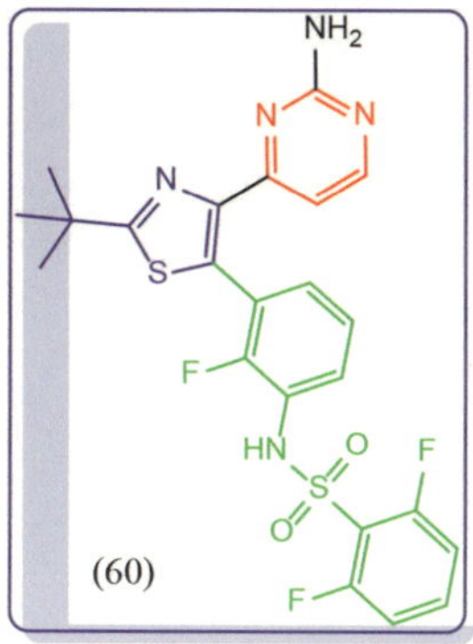

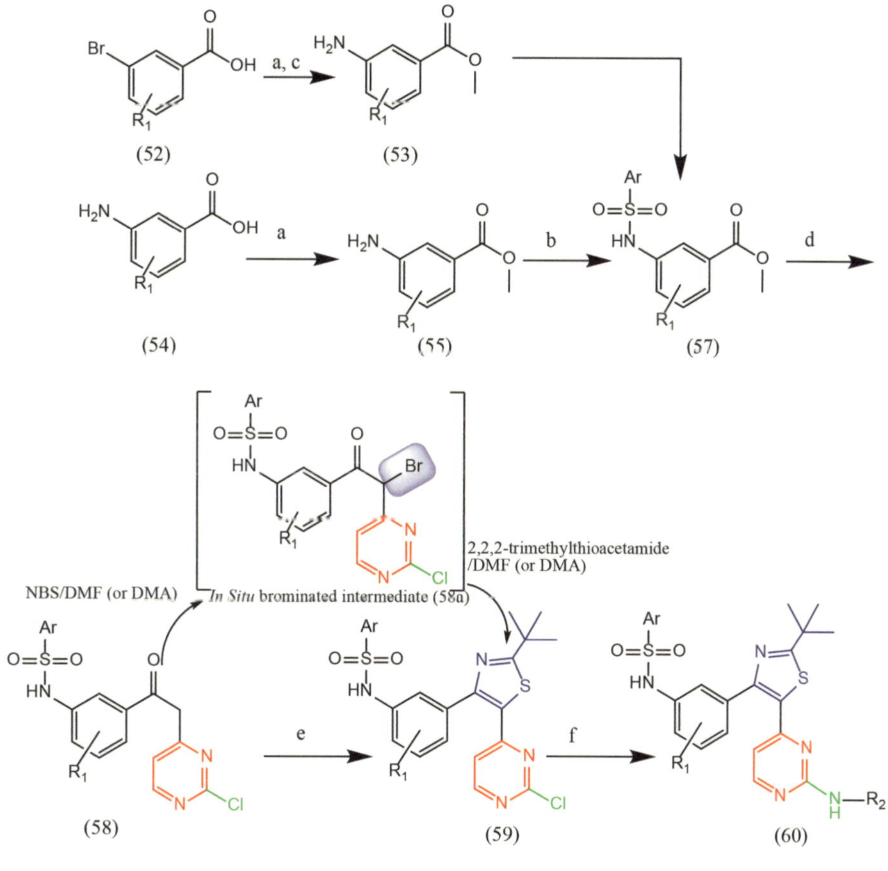

Figure 7.
Synthesis of Dabrafenib (GSK2118436, (60)). Reagents and conditions: a) H_2SO_4, MeOH; b) $ArSO_2Cl$ (43), pyridine, DCM; c) 10% Pd/C, H_2, rt., 100%, d) 2-chloro-4-methylpyrimidine (61), LiHMDS, 0°C to rt., 1 h, 92%;; e) NBS, DCM then 2-aminopyridine (61), dioxane; f) NBS, DMF (or DMA) then 2,2,2-trimethylthioacetamide, rt. → 60°C, 1 h, 30–44%; g) 7 N ammonia in methanol, sealed tube 100°C or R_2NH_2 amine, HCl, 2,2,2-trifluoroethanol, microwave, 180°C.

Structurally, the drug Rilpivirine hydrochloride (62) belongs to diarylpyrimidine (DAPY) family of compounds that re defined as the second-generation non-nucleoside reverse transcriptase inhibitors (NNRTIs) targeting reverse transcriptase,

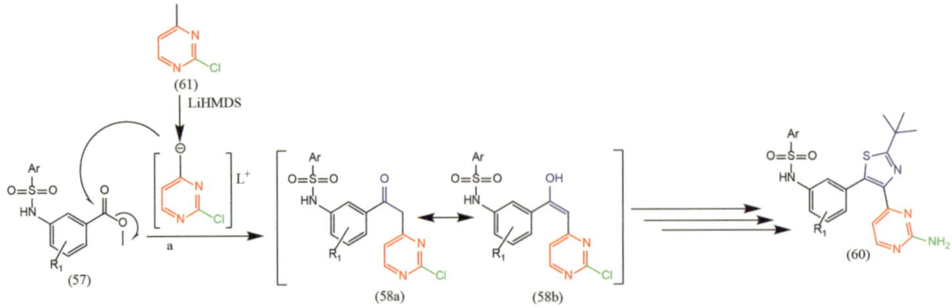

Figure 8.
Coupling of the 2,4-disubstituted pyrimidine (60) to the sulfonamide benzoic acid ester (57). Reagents and conditions: a) 2-chloro-4-methylpyrimidine (61), LiHMDS, 0°C → r.t, 1 h, 90%.

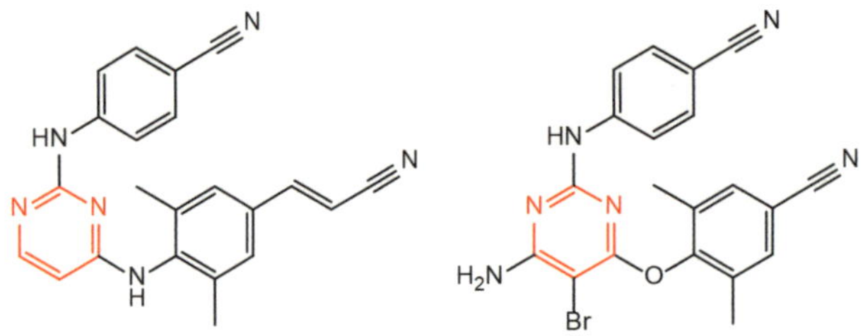

Figure 9.
Structures of FDA approved diarylpyrimidines (DAPYS) of pyrimidine-based non-nucleoside reverse transcriptase inhibitors (NNRTI) Rilpivirine (62) and Etravirine (63).

playing a great irreplaceable role in HIV transcriptional therapy [73–75]. Other antiviral agents like Etravirine (63) is also defined as DAPYs and got approved by US-FDA. Some of the DAPYs also exert anticancer action [76].

A large-scale synthetic process starting from the commercially available 2-thiouracil (65) that could be converted to 2-methylthio-4-pyrimidinone (64) following methylation using methyl iodide under basic conditions (r.t., overnight, 88%) was developed. Otherwise, 2-methylthio-4-pyrimidinone (64) could be used as a starting material. The reported synthetic process comprises from 6 steps [77].

The condensation of thioether (66) with neat 4-cyanoaniline (67) at elevated temperature to afforded the substituted pyrimidone (71) in 77% yield, which upon subsequent refluxing in $POCl_3$ provided the corresponding 4-chloropyrimidine (72) in 77% yield. 4-chloropyrimidine derivative (72) was treated with the (*E*)-cinnamonitrile aniline (72) under basic conditions (K_2CO_3) to give rilpivirine hydrochloride (62) in good yield. The final step of amination was particularly challenging and required longer time and elevated temperatures (**Figure 10**).

(*E*)-3-(4-Amino-3,5-dimethylphenyl) acrylonitrile (73) was prepared via a Heck reaction starting from the of commercially available reagents either 4-iodo- or 4-bromo-2,6-dimethyl-benzeneamine (74) or (75) and acrylamide (76) affording compound (78) as a 4:1 mixture of *E/Z* isomers. The distribution of *E/Z* olefins was increased to 98:2 by salt formation and recrystallization to ultimately provide pure (*E*)-(62) in an overall 64% yield for the two steps. The final yield was

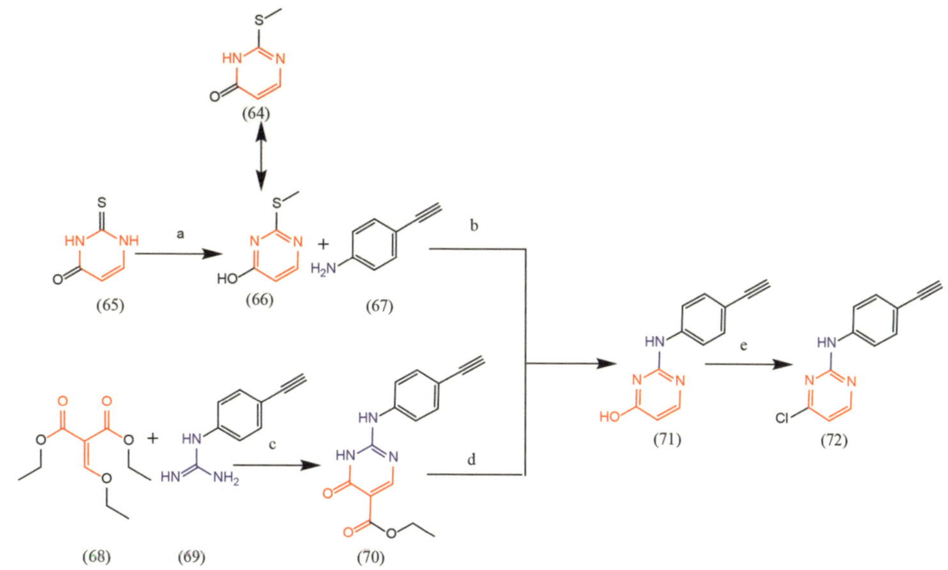

Figure 10.
Synthesis of 4-[(4-chloropyrimidin-2-yl)amino] benzonitrile intermediate (72) from 2-thioxo-2,3-dihydropyrimidin-4(1H)-one (65). Reagents and conditions: a) CH_3I, NaOH, r.t., overnight, 88%; b) DME, reflux, 18 h, 68%; c) 180–190°C, 10 h, 70–74%; d) 180°C, 8 h, 73.6%; e) $POCl_3$, reflux, 20 min, 77%.

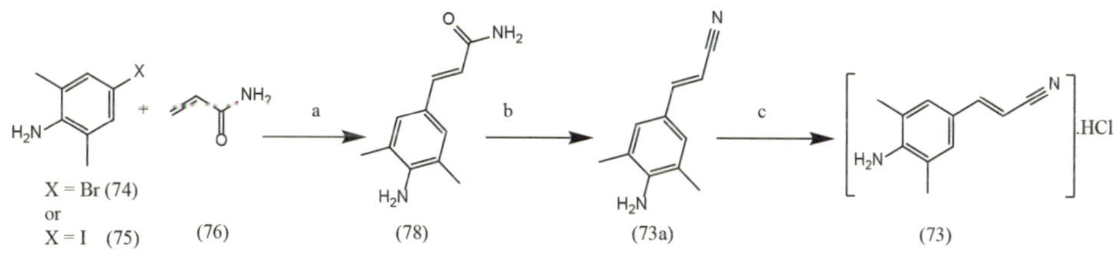

Figure 11.
Synthesis of intermediate (73) from 4-bromo-2,6-dimethylaniline (74) and acrylamide (76) as starting materials [78]. Reagents and conditions: a) $Pd(OAc)_2$, $P(C_6H_5CH_3)_3$, Et_3N, CH_3CN, N_2, 79°C, overnight, 79.5%; b) $POCl_3$, 0°C, 30 min; 20°C, overnight, 84%; c) EtOH, $((CH_3)_2CH_2)_2O$, N_2, 60°C, 30 min; HCl, 2-propanol, 60°C, 30 min, 77%. When starting from 4-iodo-2,6-dimethylaniline (75) and acrylonitrile, reagents and conditions: a) CH_3COONa, Pd/C, DMAC, N_2, 140°C, 21 h, 81%; b) EtOH, HCl, 2-propanol, 60°C, 1 h, 64.5%.

slightly improved when 4-bromo-2,6-dimethyl-benzeneamine (74) was used compared to or 4-iodo-2,6-dimethyl-benzeneamine (75) (**Figure 11**).

The conventional way to prepare the drug Rilpivirine (62) was accomplished by nucleophically displacing the 4-chloro in 4-[(4-chloropyrimidin-2-yl)amino] benzonitrile (72) with (2*E*)-3-(4-amino-3,5-dimethylphenyl)prop-2-enenitrile hydrochloride (73). The reaction was performed in acetonitrile under reflux condition for 69 h (**Figure 12**, yield: 68.6%). Connecting the two previously prepared building blocks (72) and (73) resulted in the desired products. However, the elongated refluxing of acetonitrile resulted in extended industrial process, high demand of energy, and reduced quality and purity of the final product. Using NMP at 95°C shorten the reaction time but ended in increased ration of the *cis*-(*E*)- undesired isomer byproduct. Additionally, the high boiling point of NPM renders reclaiming the solvent in industrial process unfavorable.

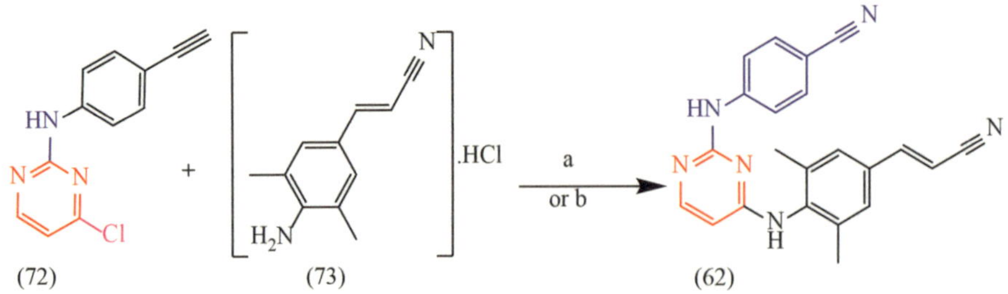

Figure 12.
Synthesis of Rilpivirine from intermediates (72) and (73). Reagents and conditions: a) CH_3CN, reflux or NMP, 95° C, or b) microwave-irradiation, CH_3CN, 140°C, 90 min, 71% [79].

Zhang *et al* Noted four drawbacks in the traditional synthetic methods: (a) the preparation of intermediate (73) via Heck reaction turned to be expensive due to the required catalyst (palladium acetate) and its ligands; (b) the preparation of intermediate (72) is also expensive and the reaction temperature is high; (c) when using uracil as a starting material instead, the reaction process and workup was rather tedious with reduced yield and (d) the final step in the synthesis of Rilpivirine, is too long (69 h) and causing energy consumption. Due to reported shortcomings of the previously reported synthesis Zhang et al. reported an optimized conditions for the synthesis of Rilpivirine and required building blocks, 4-[(4-hydropyrimidin-2-yl) amino] benzonitrile (71) and (2*E*)-3-(4-amino-3,5-dimethylphenyl)prop-2-enenitrile hydrochloride (73) employing microwave-irradiation reaction (see below more details).

Hence there is an urgent need to find more efficient and practical methods for synthesizing Rilpivirine in the pharmaceutical industry. Herein, we represent our efforts to develop an efficient synthetic route with increased overall yield and reasonable reaction time. An alternative six-step process was proposed. The improvement was primarily in preparing the intermediate (64) and in the conditions and yield on the final step. Zhang *et al* reported the solvent free fusion reaction between fusion 2-(methylthio)-4(3H)-pyrimidinone (64) and *p*-aminobenzonitrile (67) under an inert atmosphere that afforded the intermediate (72) in 70%. The intermediate 2-(methylthio)-4(3H)-pyrimidinone (64) was converted to 4-chloropyrimidine (72) form by reflux in $POCl_3$. Though the synthesis on the second building block (intermediate (73)) was still dependent of Heck reaction conditions, the final nucleophilic step was performed under microwave-assisted conditions. After trying different solvents (dioxane, acetonitrile, and NMP) and temperatures they reported a slight improvement in the yield of the final amination product (71%) for microwave-assisted in microwave-irradiation reaction with acetonitrile solvent at 140°C for 90 min compared to (69%) via traditional method [79].

Recently, it was reported by that amination of 2-chloro-4-aryloxypyrimidines using palladium catalyzed transformation (Xantphos, $Pd(AcO)_2$, Cs_2CO_3, 1.4-dioxane, reflux in N_2 atmosphere, 80°C) affording the 2,4-disubtituted product (62) in 50% yield [80]. A group of aryl-2-[(4-cyanophenyl)amino]-4-pyrimidinone hydrazones reported as potent mon-nucleoside reverse transcriptase inhibitors were prepared by Ma et al. [81]. The 2-amino-hydrazone derivatives were synthesized in a yield that ranges between 40 and 50%.

3.1.2 Synthesis of 4,5,6-Trisubstituted pyrimidines

3.1.2.1 Case study 1: synthesis of Remibrutinib [LOU064, N-[3-[6-amino-5-[2-[methyl (prop-2-enoyl)amino]ethoxy]pyrimidin-4-yl]-5-fluoro-2-methylphenyl]-4-cyclopropyl-2-fluorobenzamide, (31)]

Remibrutinib (LOU064, (31)), a highly selective and potent oral BTK inhibitor, with best-in-class profile, under investigation for a number of immune-mediated conditions [66, 68, 82]. Novartis reported rapid and effective disease activity control of Remibrutinib (31) that resulted in significant improvement in quality of life in patients with chronic spontaneous urticaria that were treated with the drug [83].

Structurally, Remibrutinib (31) belongs to a group of 4-N-5-O-6-C pyrimidine derivatives. Synthetically, the process was divided into two main parts".

i. Synthesizing the substrates (substituents) to be introduced.

ii. Step-wise substitution of the pyrimidine heterocycle core.

Three substituents are required for the construction of Remibrutinib (31):

Ammonia (NH_3, **Figure 13**, violet part), N-methyl-N-(2-hydroxyethyl) acrylamide (**Figure 13**, black and brow parts), and 5-fluoro-2-methylphenyl]-4-cyclopropyl-2-fluorobenzamide (**Figure 13**, blue and cyan parts). The three should be prepared with the proper functionalization site and reactivity for the successive substitution to work.

Starting from, 4,6-dichloro-5-methoxypyrimidine (79), the synthesis reported used the commercially available ammonium hydroxide (NH_4OH) for performing the first amination in heated 2-propanol, 70°C for 48 h. This produced the 4-amino-5-methoxy-6-chloropyrimidine (80) in high yield of 94%. Following the cleavage of 5-methoxy group the 4-amino-5-hydroxy-6-chloropyrimidine (81) was produced in 59% using conventional conditions of (BBr_3, DCM, 40°C, 3 h). The attachment of the derivatizable N-Boc-N-methyl-2- hydroxyethylamine (82) was attached to the pyrimidine core at 5 hydroxy using Mitsunobu reaction conditions (DIAD, Smopex-301, THF, 60°C, 2 hr) which afforded the intermediate (83) in 53% yield. Prior to

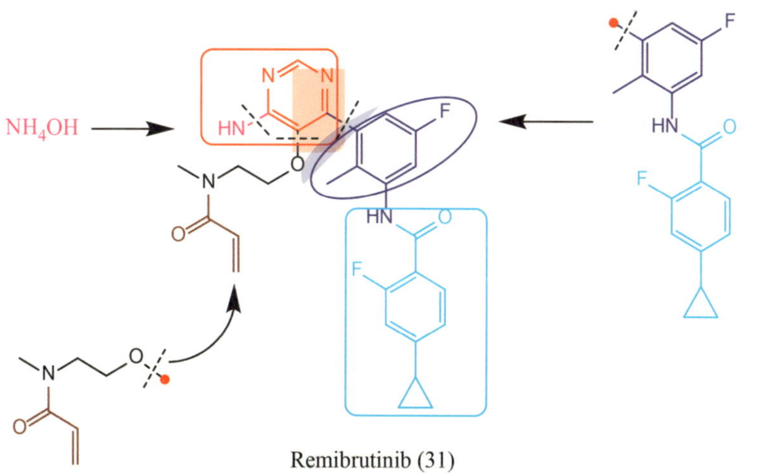

Figure 13.
Structure of Remibrutinib (LOU064, (31)) and the possible disconnections: Pyrimidine core and three different substituents.

coupling to the central pyrimidine at C6, the third substituent boronic ester intermediate (89) ought to be synthesized following the procedure depicted in the **Figure 14**.

2-Bromo-4-fluoro-6-nitrotoluene (84) was activated under Miyaura borylation reaction conditions (cross-coupling of bis (pinacolato) diboron (B_2pin_2) with aryl halides and vinyl halides using BISPIN, Pd(dppf)Cl_2·DCM, KOAc, dioxane, 100°C, 3.5 h) to afford 2-(5-fluoro-2-methyl-3-nitrophenyl)-4,4,5,5-tetramethyl-1,3,2-dioxaborolane (85) in 92%, which was subjected to reduction using hydrogen gas over Pd-C catalyst to produce the amino derivative (86) in 93%. In parallel, 3-fluoro-4-bromobenzoate ester (87) was coupled to cyclopropyl moiety through Suzuki reaction. The sodium bis(trimethylsilyl) amide mediated coupling with 2-(5-fluoro-2-methyl-3-minophenyl)-4,4,5,5-tetramethyl-1,3,2-dioxaborolane (86) afforded the desired boronate intermediate (89). Having the 4-cyclopropyl-2-fluoro-N-(5-fluoro-2-methyl-3-(4,4,5,5-tetramethyl-1,3,2-dioxaborolan-2-yl)phenyl) benzamide intermediate (89) in hand paved the road to the third substitution at the pyrimidine core in sequence.

To synthesize the desired intermediate *tert*-butyl N-[2-[4-amino-6-[3-[(4-cyclopropyl-2-fluorobenzoyl)amino]-5-fluoro-2-methylphenyl]pyrimidin-5-yl] oxyethyl]-N-methylcarbamate (93) (**Figure 15**), the previously synthesized 4-cyclopropyl-2-fluoro-N-(5-fluoro-2-methyl-3-(4,4,5,5-tetramethyl-1,3,2-dioxaborolan-2-yl)phenyl) benzamide (89, **Figure 14**) was coupled under Suzuki conditions to the BOC-protected *tert*-butyl-N-[2-(4-amino-6-chloropyrimidin-5-yl)-oxyethyl]-N-methylcarbamate (83) that was prepared utilizing Mitsunobu reaction (step c in **Figure 15**). The reaction proceeded under microwave and catalyst assisted ($PdCl_2(PPh_3)_2$, aq Na_2CO_3, DME, water, microwave, 110°C, 25 min) in 74% yield. The last two steps of de-BOCylation and coupling to acrylic acid undergone in feasible conditions. Worth noting the exploitation of Mitsunobu reaction (DIPEA, T_3P (50% in DMF), DMF, RT, 2 h, 45% over 2 steps).

3.1.2.2 Case study 2: synthesis of 2,4-Diamino-6-alkyl- (or 6-aryl-) pyrimidine derivatives

Wang and colleagues reported the synthesis of 2,4-diamino-6-alkyl- or 6-aryl-Pyrimidine Derivatives [84]. In attempt to develop a general method, two approaches

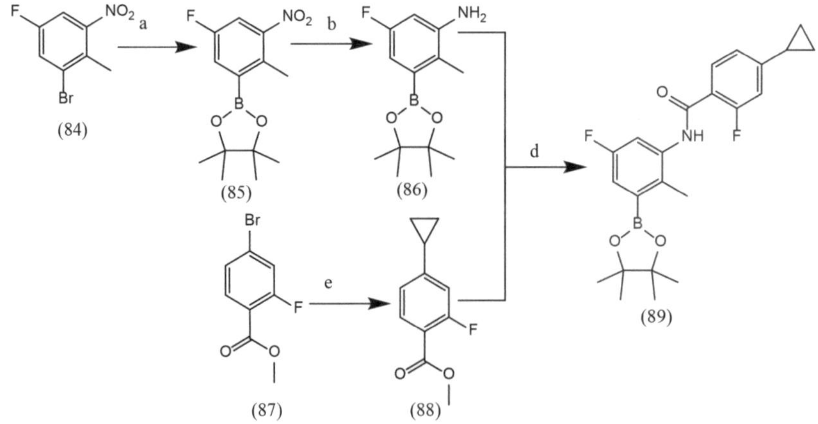

Figure 14.
Synthesis of Boronic Ester building block (89). Reagents and conditions: a) BISPIN, Pd(dppf)Cl2·DCM, KOAc, dioxane, 100°C, 3.5 h, 92%; b) H_2, Pd/C, MeOH, RT, 7 h, 93%; c) cyclopropylboronic acid, $Pd(OAc)_2$, tricyclohexylphosphine, K_3PO_4, water, toluene, 100°C, overnight, 99%; d) (86), NaHMDS (1 M in THF), THF, RT, 4 h, 76%; e) cyclopropylboronic acid, $Pd(PPh_3)4$, K_3PO_4, water, toluene, 110°C, 30 h, 96%.

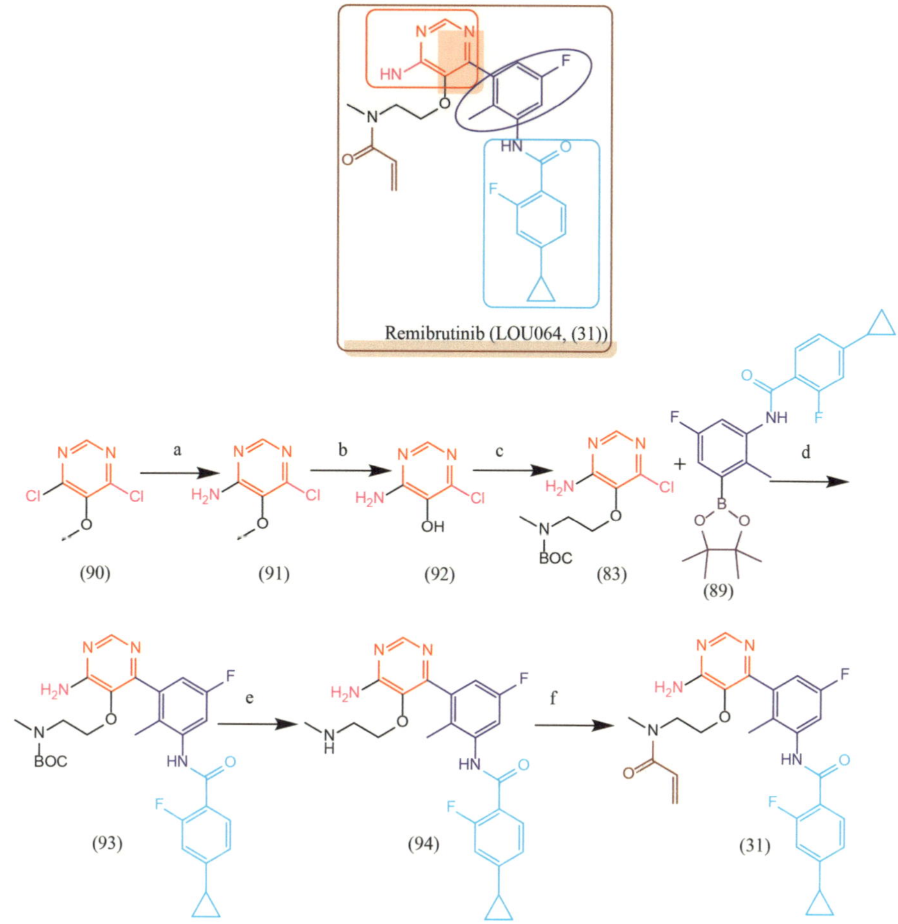

Figure 15.
Synthesis of Remibrutinib (LOU064,)31(). Reagents and conditions: a) NH_4OH, 2-propanol, 70°C, 48 h, 94%; b) BBr_3, DCM, 40°C, 3 h, 59%; c) N-Boc-N-methyl-2-hydroxyethylamine, DIAD, Smopex-301, THF, 60°C, 2 h, 53%; d) (89), $PdCl_2(PPh_3)_2$, aq Na_2CO_3, DME, water, microwave, 110°C, 25 min, 74%; e) TFA, DCM, RT, 12 h; f) acrylic acid, DIPEA, T_3P (50% in DMF), DMF, RT, 2 h, 45% over 2 steps.

were tried. First approach, the predetermined substituents were incorporated while constructing the heterocyclic pyrimidine core. This was achieved by condensing proper derivatives of 1,3-dicarbonyl with amidine or guanidine derivatives (see **Figure 16**).

The method depicted in **Figure 16** above incorporated 2-amino group (R1-NH-C2) and the 6-phenyl (R2-C6) during the assembling of the heterocyclic core. The 2-amino-6-phenylpyrimidine (101) was readily amminated at site-4 of the pyrimidine basic conditions. Relatively, mild conditions where needed for the step-wise introduction of the three substituents (**Figure 17**).

The researchers also reported the synthesis of 2,4,6-trisbustituted pyrimidines (108) using the commercially available 4,6-dichloro-2-methylthiopyrimidine (104) as a starting material for the sequential substitution under Suzuki conditions (phenylboronic acid in the presence of triphenylphosphine and palladium acetate) afforded the desired 4-chloro-2-methylthio-6-phenylpyrimidine (106a or 106b) in high yield [84] compared to the case when 2,4,6-trichloropyrimidine (109) was used

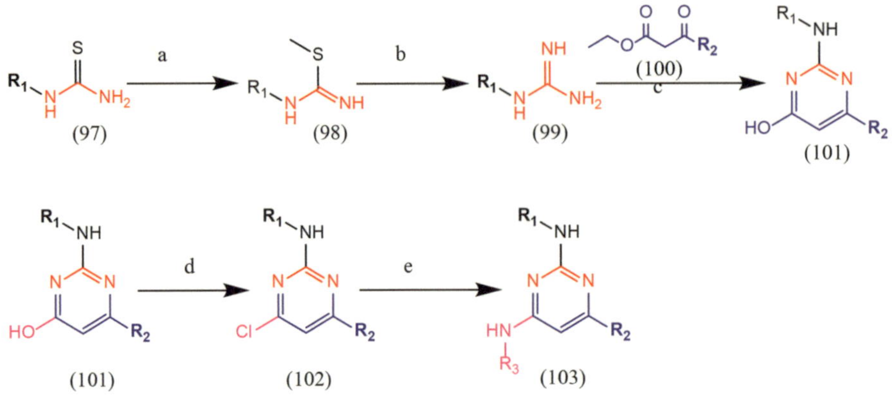

Figure 16.
Synthesis of trisubstituted pyrimidine (103) via construction of the heterocyclic core. Reagents and conditions: a) CH_3I, acetone, reflux; b) NH_3, EtOH, 100°C; c) R_2COCH_2COOEt (100), DMF, 100°C, 48 hr.; d) $POCl_3$; e) R_3NH_2, 110°C.

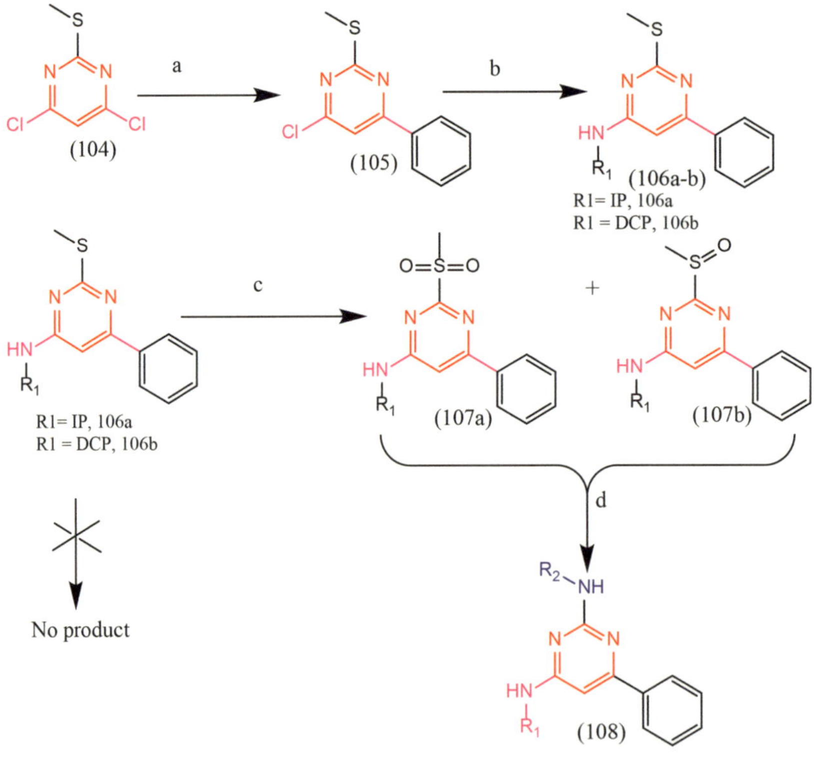

Figure 17.
Synthesis of 2,4,6-trisubstituted pyrimidines using the commercially available 4,6-dichloro-2-methylthiopyrimidine (104). Reagents and conditions: a) $C_6H_5B(OH)_2$, Pd(Oac)2II, TPP, Na_2CO_3, Glyme, reflux, 18 hr.; b) $R1NH_2$, 1-butanol, reflux, 6 hr.; c) 30%H_2O_2, NaWO4, EtOAc/toluene (1,1 v/v) 0°C for 30 min then RT for 2 hr.; d) R_2NH_2, neat, 140°C. DCP = 3,4-dichlorophenyl, IP = isopropyl, R1 = IP, R2 = DCP, R1 = DCP, R2 = IP.

as a starting material. Though the first substitution under Suzuki conditions afforded the 2,4-dichloro-6-phenylpyrimidine (106a or 106b) in high yield. Underlying the difference in the reactivity between 2-, 4- or 6-chloro groups toward amination reactions the following amination proved to be challenging with almost equal amounts of 2-amino- and 4-aminopyrimidine derivatives due to reduced selectivity. Using 4,6-dichloro-2-methylthiopyrimidine (104) as a starting material proved to be efficient in facilitating the first amination while the second amination was not accomplished in high yield.

3.1.2.3 Case study 3: synthesis of Buparlisib [NVP-BKM120; BKM120; 1,202,777: 78-3, 5-(2,6-Dimorpholinopyrimidin-4-Yl)-4-(trifluoromethyl)pyridin-2-amine (115)]

Buparlisib (BKM120, NVP-BKM120), belongs to a family of 2-morpholino, 4-substituted, 6-heterocyclic pyrimidine derivatives, that was developed by Novartis as a pan-PI3K inhibitor [85]. Recent studies indicated that it also targets tubulin [86] and it is a brain penetrable [87].

Burger *et al*. reported the synthesis of Buparlisib (BKM120; NVP-BKM120, 115) as a pan-class I PI3K inhibitor [88]. The synthesis was accomplished in a four-step process starting from 2,4,6-trichloropyrimidine (109). The first step encompasses a nucleophilic substitution reactions to form two C-N bonds of morpholine substituents with the pyrimidine core [4,6-dimorpholino-2-chloropyrimidine (110) and 2,4-dimorpholino-6-chloropyrimidine or 4,4′-(6-chloropyrimidine-2,4-diyl) dimorpholine (111), **Figure 18**]. Reacting (109) with 2.5 equivalent of morpholine ended in producing the 2,4-disubtituted intermediate 4,4′-(6-chloropyrimidine-2,4-diyl) dimorpholine (111) in a 80% yield. The reaction was highly regioselective, however, a minor amount (8%) of the 4,6-regioiomer (110) was also detected. In parallel, 2-amino-4-trifluoromethyl pyrimidine (112) was converted to 5-bromo-4-(trifluoromethyl)-2-pyrimidylamine (113) using N-bromosuccinimide (NBS) via an oxidative radical electrophilic brominating reaction [89]. The functionalized 5-(4,4,5,5-tetramethyl-1,3,2-dioxaborolan-2-yl)-4-(trifluoromethyl)pyridin-2-amine (114) was prepared by reacting 5-bromo-4-(trifluoromethyl)-2-pyridylamine (113) with bis (pinacolato) diboron under conventional conditions (potassium acetate, bis (pinacolato) diboron and bis (diphenylphosphino) ferrocene palladium (II) chloride) in a 5:1 mixture with the starting material. The dioxaborolanated intermediate (114) was introduced to the pyrimidine core via palladium-mediated C-C cross-coupling

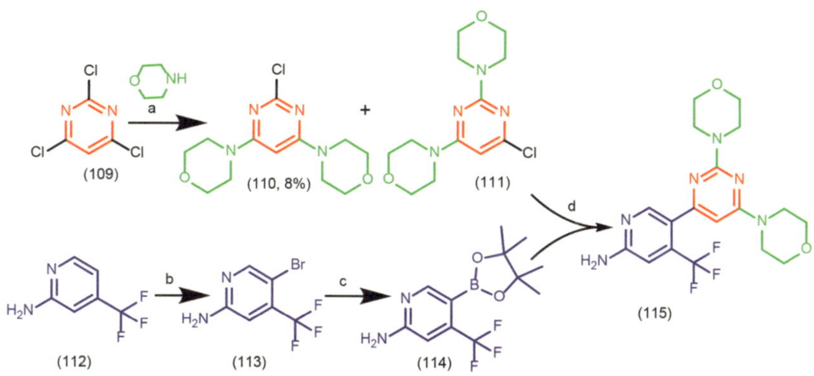

Figure 18.
Synthesis of Buparlisib (BKM120; NVP-BKM120) (115). Reagents and conditions: a) morpholine, DIEA, EtOH, 80%; b) NBS, CH_2Cl_2, 80%; c) bispiacolatodoron (1,1'-Bis(diphenylphosphino) ferrocene palladium (II) chloride, Pd(dppd)$_2$CleDCM), DME, 2 N Na_2CO_3, 95°C, 15 h, 48%; d) bispiacolatodoron, Pd (dppd)$_2$CleDCM, DME, KOAc, (potassium acetate, bis(pinacolato) diboron (Bis(diphenylphosphino) ferrocene palladium (II) chloride), 95°C, 15 h, 95%.

(step d in **Figure 18**) applying Suzuki reaction conditions affording the final product (115) in 95% yield [88].

The synthesis of Buparlisib (115) was also disclosed by Xu et al. employing similar conditions. A high yield and purity cross-coupling was reported (yield >94% and HPLC purity: >99%) [90].

3.1.3 Synthetic methods in preparing 2,4,6-Trisubstituted pyrimidines

Step-wise replacement of chloride groups employing nucleophilic substitution conditions, Suzuki coupling, Ulman coupling or Grignard reaction was reported to afford di-, tri- and tetra-substituted derivatives of pyrimidine (114) (**Figure 19**). The conditions used depend in many cases on the order and type of the linkage emerging between the substituents and the heterocyclic pyrimidine. It has also to consider the physicochemical properties of the substituent.

The final 2,4,6-trisubstituted product (115) can vary according to the type of linkage with the core heterocyclic pyrimidine. There are various versions of linkages (see **Figure 20**, for possible combination of linking moieties with the pyrimidine heterocyclic core) (**Figure 21**).

3.1.3.1 Methods to synthesized N-2,4,6*-trisubstituted (at C2, C4 and C6 respectively) pyrimidine derivatives*

One of the early identified S-2-*N*-4,6-trisubstituted pyrimidine aurora kinase inhibitor is the compound VX-680 (**Figure 22**). Luo et al. reported the synthesis of a group of N-2,4,6-trisubstituted pyrimidine derivatives and evaluated their action as

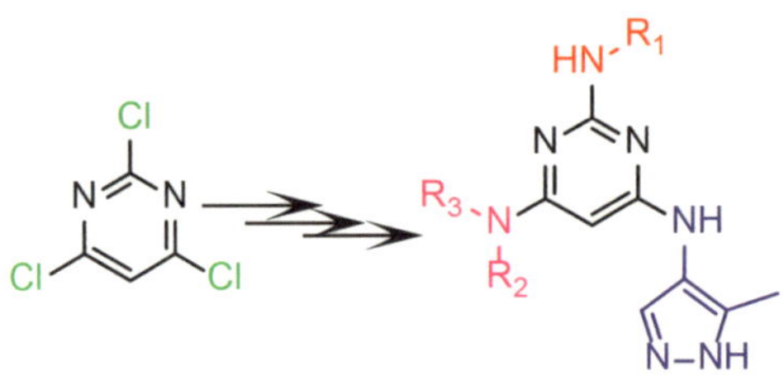

Figure 19.
Synthesis of unsymmetrical trisubstituted pyrimidines.

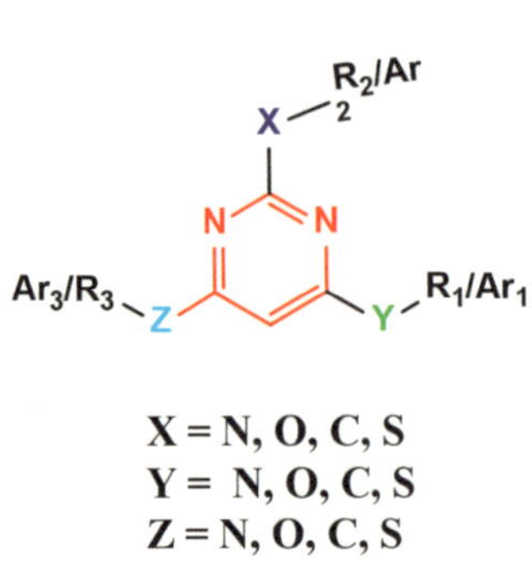

Figure 20.
Possible linkages between the substituents and pyrimidine heterocyclic core.

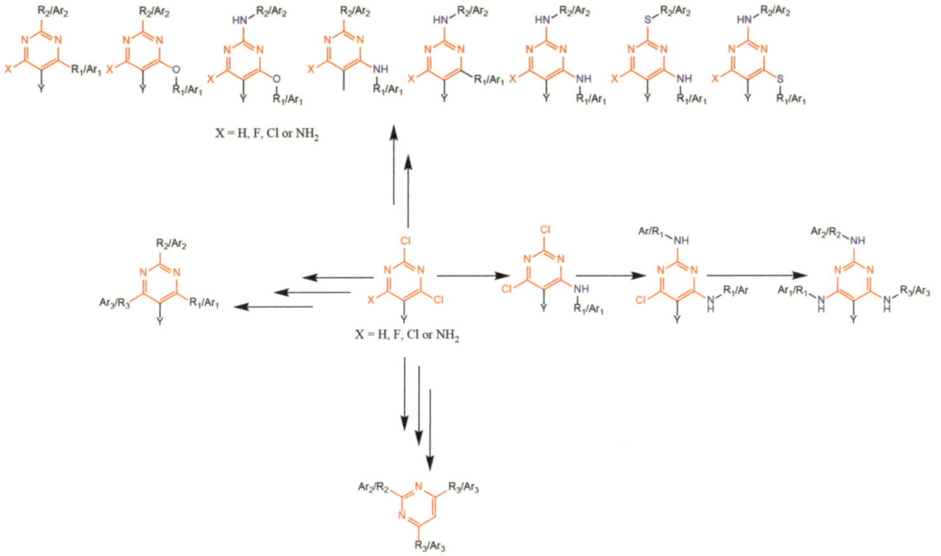

Figure 21.
Possible derivatives of trisubstituted pyrimidine heterocyclic core.

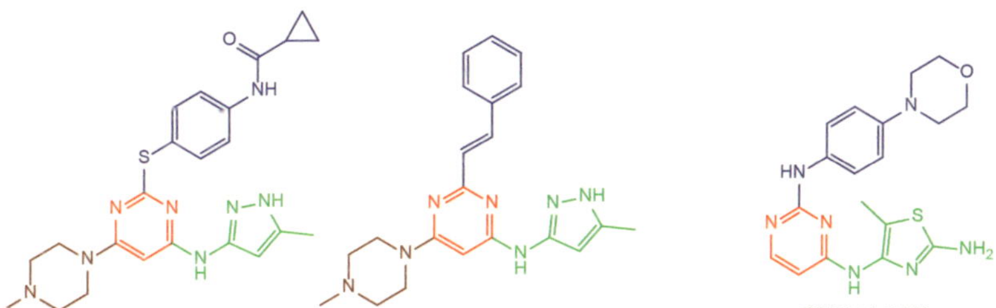

VX-680 (116) ENMD-2076 (117) CYC116 (118)

Figure 22.
Examples of bioactive trisubstituted pyrimidine.

selective aurora A kinase inhibitors [91]. Luo et al. reported synthesis of N-trisubstituted pyrmidines starting from 2,4,6-trichloropyrimidine (109).

Starting from 2,4,6-trichloropyrimidine (109), and in a sequential nucleophilic substitution the three substrates were introduced at sites 4-, 2- and 6- of the pyrimidine core. Aminopirazole 5-methyl-1H-pyrazol-3-amine (119) was allowed to react with 2,4,6-trichloropyrimidine (109) under basic conditions at cold temperature and afforded the 4-substitited pyrimidine (120) in high yield. The acid facilitated second amination was performed under elevated temperature afforded the 2,4-diaminated intermediate (121). The third amination afforded the final 2,4,6-trisubstituted product (122) took place under microwave assisted and elevated temperatures (1,4-dioxane, under microwave, 150–180°C) (**Figure 23**) [91].

3.1.4 Synthetic methods in preparing 2,4,5,6-Tera-substituted pyrimidines

One main strategic approach employs the step-wise replacement of leaving groups that exist on the already constructed pyrimidine (like 2,4,6-trichloropyrimidine (109)). The selection of the substituted pyrimidine as starting material depends on the

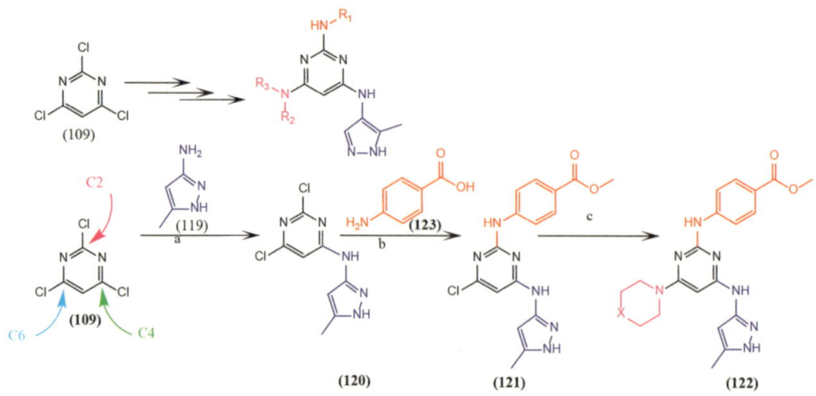

Figure 23.
Synthesis of N*-2,4,6-trisubstituted pyrimidine derivatives as potent aurora a kinase inhibitor. Reagents and conditions: a)* Et_3N*, EtOH, 0°C; b) TsOH.*H_2O*, n-BuOH or 1,4-dioxane, 100–140°C; c) 1,4-dioxane, under microwave, 150–180°C.*

desired product i.e. in case of 2,4-disbustitited pyrimidine the synthesis starts with 2,4-dihalopyrimidine and in case of 2,4,6-trisubstitutedpyrimidine the 2,4,6-trichloropyrimidine (109) is picked. Additionally, the reaction conditions and reagents used in the chemical process is by virtue reliant on the type of chemical bond formed or the type reaction employed. The order of applying the sequential displacement of the leaving groups is contingent on the type and properties of the functional group and conditions that ensure better purity and higher yield of the intermediates, building blocks or the final products. A special attention will be devoted to recent reports related to synthesis of FDA approved drugs or in few cases such derivatives with high potential bio-activities.

Regioselectivity is mostly guaranteed by picking the proper starting material and the fitting sequence of substitution (**Figure 24**).

In a recent study Zhang and colleagues reported the synthesis of anisole containing 2,4,5,6-tetrasubstituted pyrimidines [92]. The team exploited the commercially available reagents like properly substituted malonic acid diesters (124) and guanidine hydrochloride (125) that were condensated in anhydrous methanol under slightly

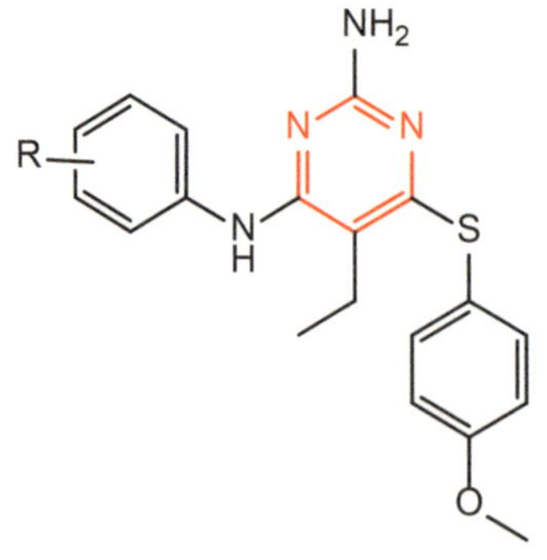

R = H, Methyl-, Ethyl-, Propyl-, Isopropyl-, Propargyl-
R = Allyl-, Butyl-, *Sec*-Butyl-, Phenyl-, Benzyl-, Fluoro-

Figure 24.
Examples of 2,4,5,6-tetrasubstituted pyrimidines. R = H, methyl-, ethyl-, propyl-, isopropyl-, Propargyl-., R = allyl-, butyl-, sec*-butyl-, phenyl-, benzyl-, Fluoro-.*

basic conditions (sodium methoxide) to produce the 5-substituted 4,6-dihydroxy-2-aminopyrimidine intermediate (126). (126) was converted to the 5-substituted 4,6-dichloro-2-aminopyrimidine derivative (127) upon treatment with Vilsmeier–Haack–Arnold (VHA) reagent [93]. Cations should be taken prior to treating 5-substituted 4,6-dichloro-2-aminopyrimidine derivatives with (chloromethylene) dimethyliminium chloride (VHA, 130) (**Figure 26**).

Using VHA reagent, drying of the starting materials and conducting the reaction under inert conditions helped in affording the final products in higher yields and purity compared to the previously reported reaction conditions (such as chloride donating mineral acids such as $POCl_3$, PCl_5, $SOCl_2$, or $COCl_2$ with diverse additives like DMF, pyridine, 2-methylpyridine, diphenylamine, or triethylamine) reported to end in less than 30% yields and complicated purification procedures [94]. Following the deprotection of the 2-(dimethylamino) methylene protecting group using hydrochloric acid the desired 4,6-dichloro-5-substituted-2-aminopyrimidines were isolated and purified. The 4,6-dichlorides were sequentially displaced under nucleophilic substitution conditions. The first 4-chloro was substituted by aniline derivatives in refluxing ethanol (see step c in **Figure 25**) while the second chloride was exchanged for anisole (4-methoxythiophenol) employing sodium tert-butoxide (1.33 mmol), under heating at 82°C (see step d in **Figure 25**, [92].

3.1.4.1 Synthesis of bay 41–4109 racemate [methyl 4-(2-chloro-4-fluorophenyl)-2-(3,5-difluoropyridin-2-yl)-6-methyl-1,4-dihydropyrimidine-5-carboxylate (135)]

The synthesis of Bay 41–4109 (135) (**Figures 27**) was initiated by allowing β-ketoacetate (133) to react with pyridine-2-carboximidamide salt and phenylacetaldehydes (134) via Biginelli cyclocondensation [95]. The reaction proceeded in isopropanol under microwave irradiation to produce the corresponding products (135) (**Figures 28** and **29**). On note, Bay 41–4109 (135) can be used for further derivatization. For example, the bromination of (140) using *N*-bromosuccinimide (NBS) can lead to the brominated intermediate (141), which can be easily substituted with nucleophiles like morpholine, *N*-methylpiperazine, methoxyethanol or

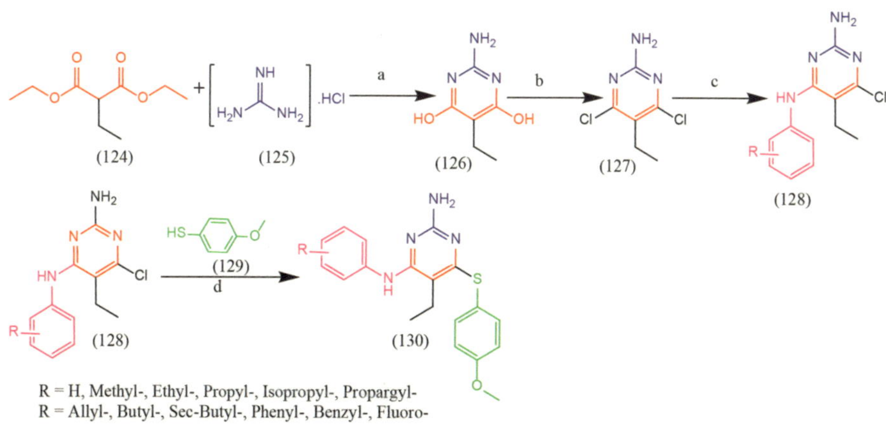

Figure 25.
Conversion of 5-substituted-2-amino-4,6-dihydroxypyrimidines (126) to 5-substituted 2-amino-4,6-dichloropyrimidines (127) using (chloromethylene) dimethyliminium chloride (130).

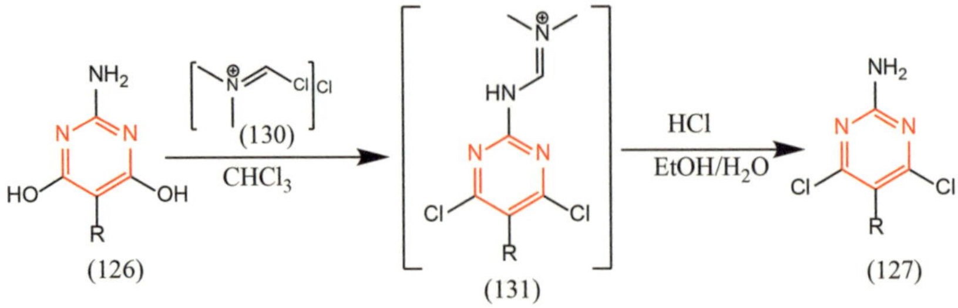

Figure 26.
Synthesis of 2,4,5,6-tetrasubstituted pyrimidine derivatives. Reagents and conditions: a) EtONa/EtOH; b) (chloromethylene) dimethyliminium chloride/CHCl$_3$*; c) different anilines, ethanol, 100°C, 4 h; d) isopropanol, sodium* tert*-butoxide, 82°C, 6 h.*

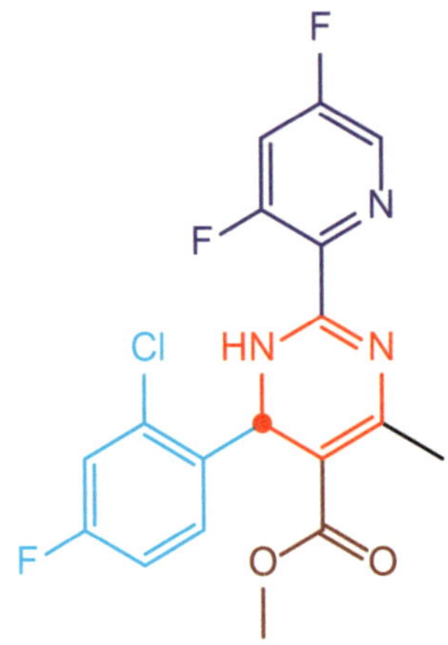

Figure 27.
Bay 41–4109 racemate (135).

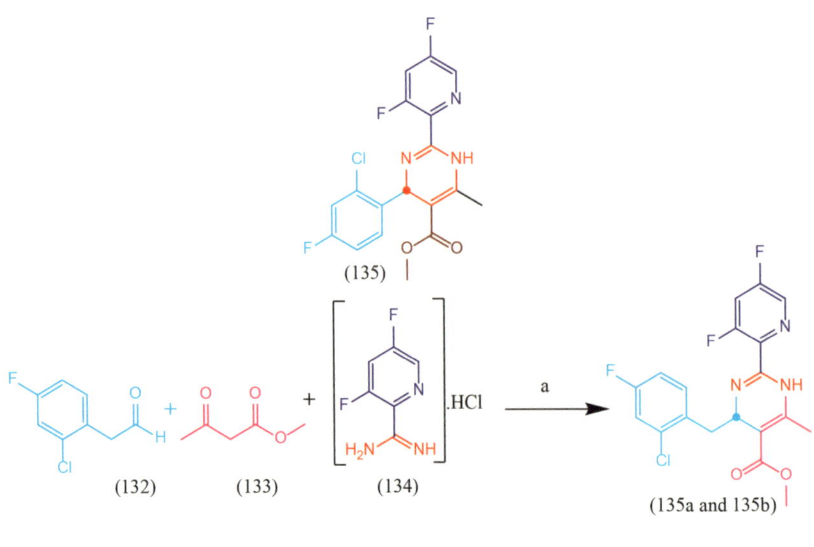

Figure 28.
Synthesis of bay 41–4109 (135). Method-1: One-pot three-component Biginelli condensation using aldehyde, β-ketoester and amidine. Reagents and conditions: a) piperidine, AcOH, iPrOH, 12 h, 11–36%.

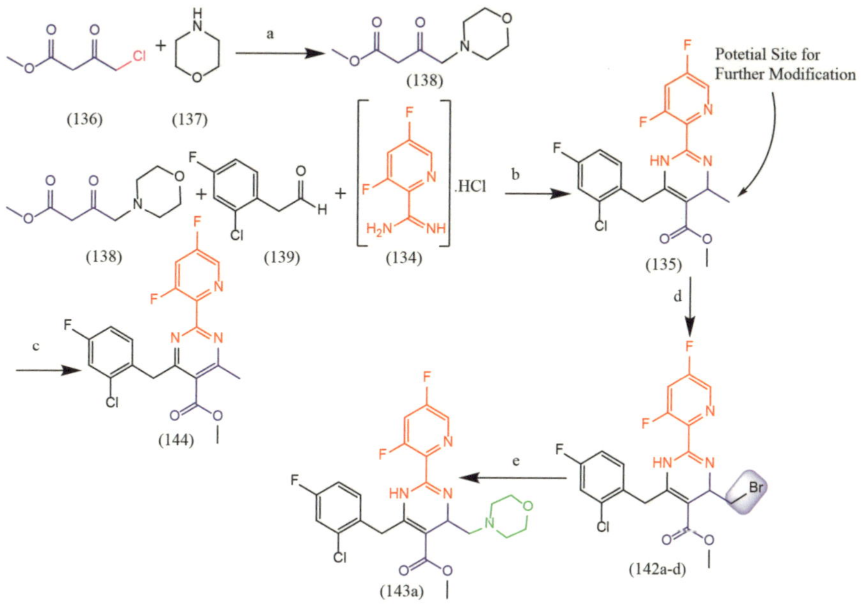

Figure 29.
Synthesis of bay 41–4109 (135). Method-2: Reagents and conditions: a) NaH, DMF, 0°C → rt., 1 h, 41–60%, b) pyridine-2-carboximidamide, Et_3N, μW, 10 min, 14%; c) DDQ, toluene, rt., 1 h, 42–68%; d) NBS, 1,2-DCE, 50°C, 30 min, 80%; e) morpholine, Et_3N, DMF, 0°C, 1 h, 29–72%.

thiobenzene producing the corresponding derivatives (142a-d). Oxidation of these dihydropyrimidines using 2,3-dichloro-5,6-dicyano-1,4-benzoquinone (DDQ) easily led to the desired pyrimidines (144).

3.1.5 Synthesis of 2-amino-4,6-disubstituted pyrimidine

Altenbach et al. initiated the synthesis of a group of 4-*tert*-butyl-6-substituted pyrimidin-2-ylamine derivatives (**Figure 30**) starting 2,4-dichloropyrimidine (38) that was alkylated with pivalic acid (151) via nucleophilic substitution of 4-chloro- using silver nitrate $AgNO_3$ and ammonium persulfate. The 4-*tert*-butyl-6-chloropyrimidin-2-ylamine intermediate (151) was subjected to sequential nucleophilic amination to end in the desired group of compounds (154a-e). At the first step the displacement of the 2-chloro with N-Me-piperazine (155) afforded 4-*tert*-butyl-2-chloro-6-(4-methylpiperazin-1-yl) pyrimidine (152) as a mixture with second regioisomer 4-*tert*-butyl-6-chloro-2-(4-methylpiperazin-1-yl) pyrimidine (153). Following chromatography, (152) was treated with the second nucleophile to result in the desired group of compounds (154a-e) [94].

The same group also reported the synthesis of a group of 2-amino-4,5,6-trisubstituted pyrimidines (160a-m and 161a-o) starting from 2-amino-4,6-dichloro-5-substituted intermediate (156) (**Figure 31**). Evidently, 2-amino-4,6-dichoro-5-substituted intermediate (156) was treated with 1-methylpiperazine (155) under basic conditions (TEA or DIEA) while refluxed for 16 hours [94]. The second substitution was dependent on the type of connecting bond. In case of forming a C-C bond at C6 of the pyrimidine core, Suzuki conditions boronic acid derivatives (4-cyanophenylboronic acid or 4-methylphenylboronic acid), tetrakis (triphenylphosphine)-palladium (0) ($Pd(PPh_3)_4$) in 2-dimethoxyethane under basic (2 M Na_2CO_3) and inert

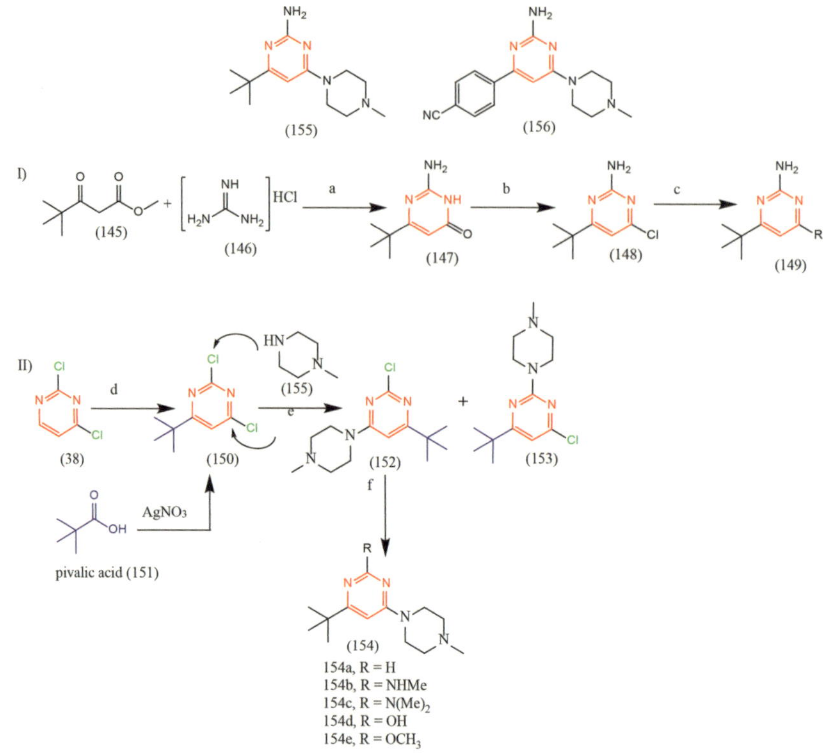

Figure 30.
Synthesis of 2-amino-4,6-disubstituted pyrimidine derivative. Reagents and conditions: I): a) guanidine hydrochloride, EtONa, EtOH, Δ; b) $POCl_3$, Δ; water; and c) amine, Δ, EtOH, Et_3N; II) reagents and conditions: d) pivalic acid, $AgNO_3$, ammonium persulfate, CH_3CN/H_2O; e) N-me-piperazine, Et_3N, EtOH, Δ, chromatography; (154a): R = H): H_2, Pd/C, MeOH; (154b): R = NHMe) 40% aqueous $MeNH_2$, 2-MeOEtOH, Δ; (154c): R = NMe_2): 40% aqueous Me_2NH, 2-MeOEtOH, Δ; (154d): R = OH): 1 M HCl, 16 h, Δ; (154e): R = OMe): Excess NaOMe, MeOH, Δ.

conditions. In the case of forming C-N bond at C6, Ullmann nucleophilic substitution conditions were applied. For example for synthesizing 4-(4-methylpiperazin-1-yl)-6-(4-phenylimidazol-1-yl)pyrimidin-2-ylamine (160). The corresponding substituent 4-phenylimidazole (87 mg, 0.6 mmol) (159) was added to 4-(4-methylpiperazin-1-yl)- 6-chloropyrimidin-2-ylamine (157) in presence of catalytic copper iodide (CuI, 0.13 mmol ratio) under basic condition (K_2CO_3) in DMF that was heated to 135°C overnight. Generally, the yields reported for Suzuki conditions (C-C bond) were higher than those reported under Ullmann conditions.

2-Amino-4,6-dichloropyrimidines (156) were also used as starting materials for preparing 2-amino-4,5,6-trisubstituted derivatives (160a-m and 161a-o) (see **Figure 31**).

3.1.6 Synthesis of etravirine; (TMC 125); 2,4-[[6-amino-5-bromo-2-[(4-cyanophenyl) amino]-4-pyrimidinyl] oxy]-3,5-dimethylbenzonitrile (63)]

Etravirine is a novel diarylpyrimidine, second-generation non-nucleoside reverse transcriptase inhibitor (NNRTI) for the treatment of human immunodeficiency virus type 1 infection, which has been approved by the U.S. Federal Drug Administration for the treatment of AIDS in the year 2008 [80, 96–98].

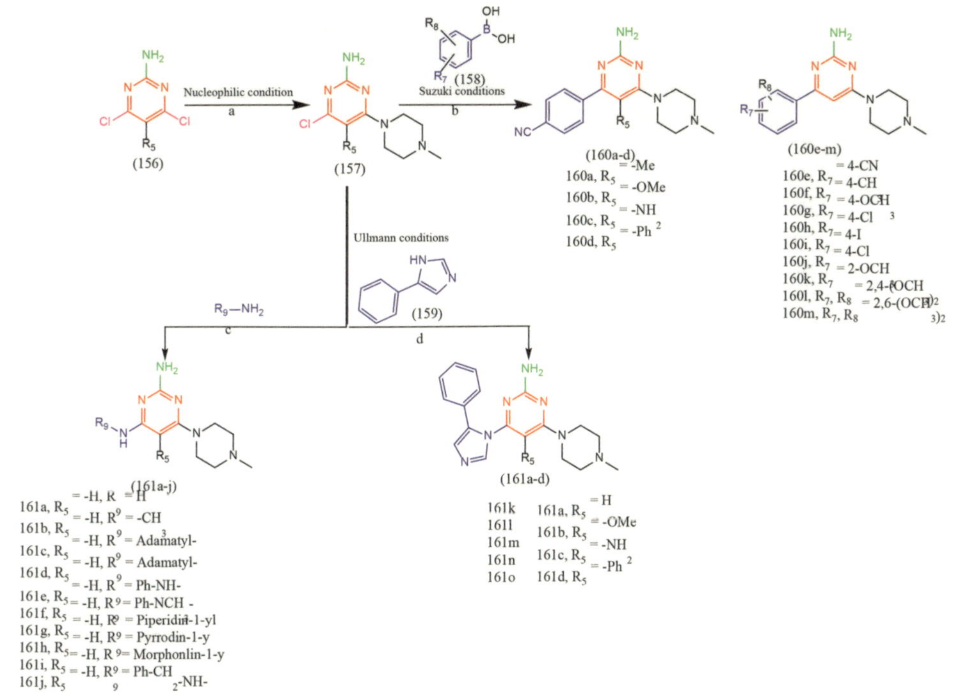

Figure 31.
Synthesis of 2-Amino-4,5,6-Trisubstituted pyrimidine derivatives. Reagents and conditions: a) N-me-piperazine, Et_3N, EtOH, Δ; b) compounds (160a-160 m): Intermediate (157), R7/R8-Ph-B(OH)2, $Pd(PPh_3)_4$, Na_2CO_3, 1,2-dimethoxyethane, Δ; c) compounds (161a-160o): Intermediate (157), the corresponding amine, 2-ethoxyethanol, Hunig's base, 110–140°C; d) Ullmann conditions: Intermediate (157), corresponding heterocycle, CuI, K_2CO_3, DMF, 130°C; and (iii) NaI, 47% aqueous HI, 80°C, 98% [94].

It is a potent inhibitor of HIV reverse transcriptase [98], and active against NNRTI-Resistant Strains of HIV, by its ability to adapt its binding orientation and overcome common NNRTI resistance associated mutations (RAMs) such as K103N and achieving viral suppression and improving the immune function in treatment-experienced HIV-infected patients [99].

The synthesis of Etravirine (63) is fraught with many difficulties, the foremost being the poor yield and long reaction time required at the aminolysis stage [100].

The most efficient approach to overcome the mentioned disadvantages is four linear steps where the microwave promoted amination is the most critical one of these routes, as it shortens the amination reaction time from 12 h to 15 min, and improve the overall yield of the synthetic route from 30.4 to 38.5% (**Figure 32**) [100].

2,4,6-trichloropyrimidine (109) was first reacted with 4-hydroxy-3,5-dimethyl-benzonitrile (162) in the presence of diisopropylethylamine in refluxing dioxane and heated at 70°C for 2 hours, to give the biaryl ether derivative: 4-[(2,6-dichloro)-4-pyrimidinyloxy]-3,5-dimethylbenzonitrile (163) [96].

The second substitution reaction with aniline derivative 4- aminobenzonitrile (164) using potassium tert-butoxide as a base and N-methylpyrrolidone as a solvent over a period of 30 min and stirred for another 2 h at 0–5°C, to give compound 4-[[6-Chloro-2-[(4-cyanophenyl) amino]-4-pyrimidinyl] oxy]3,5-dimethylbenzonitrile (165) with a yield of 60.6%. The aminolysis of (165) went smoothly using 25% aq ammonia (15 mL), and N-methylpyrrolidone (20 mL) at temp 130°C, for 15 min in a microwave reactor, then the reaction mixture was brought to 5–10°C, 100 mL water

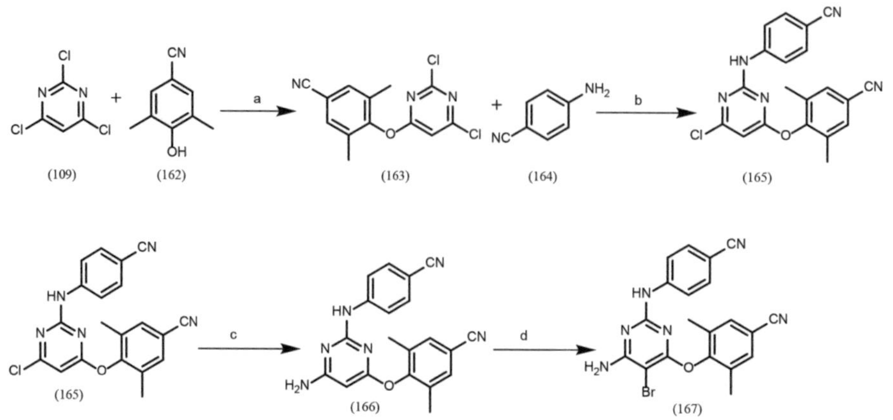

Figure 32.
Synthesis of Etravirine (63). Reagents and conditions: a) DIEA in refluxing dioxane, 70°C, 2 hours, 92.5%; b) potassium tert-butoxide, N-methylpyrrolidone, 30 min and stirred for another 2 h at 0–5°C, 60.6%; c) 25% aq NH_3, N-methylpyrrolidone, 130°C, 15 min in a microwave reactor, 5–10°C, 100 mL water, stir 30 min. Filter, wash with 100 mL H_2O, dry at 45–50°C, 85.6%; d) Br_2, DCM, 0–5°C, stirring 5 h, 80.2%.

was added to this solution followed stirring another 30 min. The generated solid was filtered, washed with 100 mL of water and dried at 45–50°C to give the amine: 4-[[6-amino-2-[(4-cyanophenyl) amino]-4-pyrimidinyl] oxy]3,5-dimethylbenzonitrile (166) in good yield (85.6%).

Finally, bromination of (166) was conducted in the presence of liquid bromine in DCM at 0–5°C, the reaction was stirred at this temperature for 5 h to give etravirine (63) in 80.2% yield [96].

3.1.7 Synthesis of Minoxidil; (MNX); [(2,4-diamino-6-piperidinopyrimidine 3-oxide) (171)]

Minoxidil (MNX) is a direct vasodilator introduced in the early 1970s for the treatment of hypertension [101]. Its first literature appearance was in 1968 and preliminary trials were first described in man in 1969 [102].

Coincidentally, physicians observed hair regrowth and generalized hypertrichosis in balding patients, which led to the development of a topical minoxidil formulation for treating androgenetic alopecia (AGA) first in male and then in female individuals [26, 103, 104].

An easy protocol for the synthesis of minoxidil drug in a two-step procedure is done, using magnetic nanoparticles of ferrites, which have been widely used as green and efficient heterogeneous catalysts in the synthesis of organic compounds, where these nano catalysts provide prominent advantages such as simple synthetic procedure, high catalytic activity, chemical reactivity and perfect reusability.

Magnetic nanoparticles (MNPs) of $CoFe_2O_4$ were prepared through a solid-state grinding procedure, in an agate mortar, where a mixture of $CoCl_2$, Fe $(NO_3)_3{\cdot}9H_2O$, NaOH, and NaCl in a molar ratio of 1:2:8:2 was mixed, ground for 45 min at room temperature, then the excess amount of salt was removed from the reaction mixture, the obtained residue was dried in an oven at 80°C for 60 min, calcinated at 300, 500, 700, and 900°C within 80 min (20 min at each temperature) (**Figure 33**) [105].

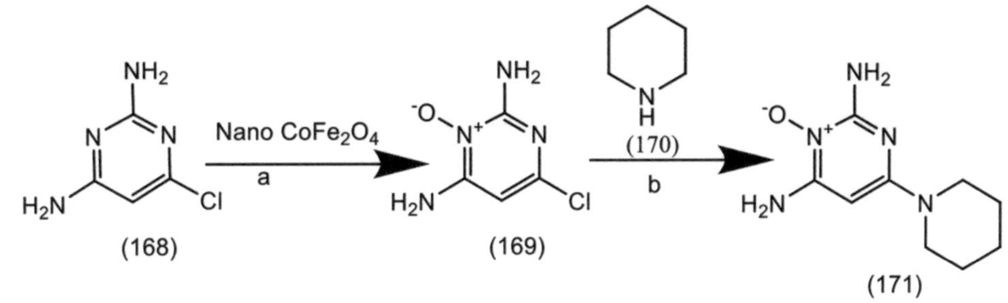

Figure 33.
Synthesis of Minoxidil (171). Reagents and conditions: a) Nano $CoFe_2O_4$ (5 Mol%), H_2O_2 (0.5 ml), ETOH, reflux, 60 min, 95%; b) Piperidine, reflux, 120 min, 80%.

First step was carried out by N-oxidation of 6-chloro-2,4-diaminopyrimidine (168) using (30%, 0.5 mL) H_2O_2 in the presence of $CoFe_2O_4$ magnetic nanocatalyst (5 mol %) in 5 ml ethanol under reflux conditions for 60 min to get 2,6-diamino-4-chloro-pyrimidine N-oxide (169) in yield (95%).

A nucleophilic substitution reaction using boiling piperidine (106°C) under neat conditions with 2,6-diamino-4-chloro-pyrimidine N -oxide (169), for 120 min was then carried as step two, affording 2,4-diamino-6-piperidinopyrimidine 3-oxide (minoxidil, (171)) product in high yield (80%) and purity without requiring any further purification.

4. Conclusion

This manuscript brings a focused perspective of synthetic methods employed in producing bioactive pyrimidine-based derivatives. A special consideration is given to the FDA approved pyrimidine-based drugs, however, approaches to synthesize bio-active synthons endowed with in interesting bioactivities are also included.

Synthetic approaches used for preparing pyrimidine-cored structures varied from nucleophilic substitution to C-C, C-N cross-coupling or heterocyclization of complementary dielectrophilic (+)C–C–C(+) and dinucleophilic (−)N-C-N(−) fragments. Converting the carbonyl/or hydroxyl group to chloride using $POCl_3$ was often employed in facilitating the substitution reaction at the desired site. In other cases, oxidizing methyl thioether to sulfone offers a convenient option for substitution reactions.

It was noted that reactivity of the four possible sites (C2, C4, C4 and C6) is affected by a prior existence of substitutes or the type of the linkages (C-C, C-N, C-S or C-O).

4.1 Decorated pyrimidines: privileged scaffolds meeting the Mission

The 2,3-diazine (pyrimidine) is found in the core of wide range of bioactive drugs and drug candidates. That includes natural products (from bacteria all along until mammalians) sources and synthetic pharmaceuticals. A wide-range of activities were associated with pyrimidine and derivatives (antibacterial, antitumor, antiviral, analgesic, antiarrhythmic, antifungal [106], antimalarial, anticonvulsant, sought be. Thus, the structure is considered by medicinal chemists, drug discovery researchers and pharmacologist as medicinally privileged scaffold.

4.2 Two strategies to afford diverse derivatives

4.2.1 Post Heterocyclization

One main strategy in derivatizing pyrimidine get advantage of the availability of halogenated core synthons [mono-, di- or trichloro-pyrimidines]. An approach defines as "post- heterocyclization modification. In such cases, halogenated pyrimidine (frequently 2,4-dichloropyrimidine (38) or 2,4,6-trichloropyrimidine (109)) proved to be highly treasured when subjected to modifying reaction conditions and reagents and afforded the desired products. 2,4-dichloropyrimidine (38), 2,4,6-trichloropyrimidine (109) or similar analogues are made use of in preparing long list of modified pyrimidines. In such case, synthetic methods used in decorating pyrimidine-cored analogues were diverse and include nucleophilic substitution, C-C, C-N, C-O and C-S cross-coupling employing Suzuki or Ullmann conditions or amide coupling.

4.2.2 Constructing while Heterocyclization

In the cases where the desired product is not feasible via coupling or substitution, cross-coupling to reactive form of the heterocyclic core, alternative approaches were implemented. Constructing of the substituted pyrimidines via heterocyclization of predetermined "designed" components i.e. starting from derivatized parts that upon applying matching reaction conditions a "merged" modified core is amalgamated. For example, the syntheses of 2,4,5,6-terasubstituted pyrimidines were made possible by "fusion" of ß-ketoester derivatives with the corresponding amidine or guanidine elements. All should be conducted under carefully designed and appropriate implemented reaction conditions.

4.3 Cases elaborated

Four examples of FDA approved 2,4-disubstitited pyrimidine drugs Pazopanib (21), Remibrutinib (31), Dabrafenib (60) and Rilpivirine (62) were discussed in particular. Examples 4,5,6-trisubstituted pyrimidines like Remibrutinib (31), 2,4-diamino-6-alkyl- or 6-aryl-pyrimidine derivatives were also presented. The approach starting from 2,4,6-trichloropyrimidine (109), giving rise to N-trisubstituted pyrimidine derivatives like Buparlisib [NVP-BKM120] (115) were discussed. A light was shed on 2,4,5,6-tetrasubstituted pyrimidines like Bay 41–4109 (135) and 2-amino-4,6-disubstituted pyrimidines, Etravirine (TMC 125) (63), 2,4-diamino-6-piperidino-pyrimidine 3-oxide Minoxidil (171).

4.4 No means to delineate all

This manuscript aimed at briefing the reader, in an elaborative manner, with some instances and show-case of chemical process affording selected examples of FDA-approved therapeutics. The focus is on approaches employed the "post-heterocyclization" modification methods.

Acknowledgements

Yousef Najajreh wish to acknowledge my home institution Al-Quds University, Jerusalem-Palestine, for the support and encouragement. In addition, I wish to thank the editor and high value the editor's work, reading, and comments. The accomplishment of this chapter was made possible thanks to such support and help.

Conflict of interest

The authors declare that no conflict of interest exists.

Notes/thanks/other declarations

Yousef Najajreh would like to express his special thanks, gratitude and truthful appreciation to my dearest family: my wife Muna, my two daughters Aseel and Layan and my two sons Mulham and Majd for all the love, compassion, and support received from them all along.

This chapter is dedicated to the respectable memories of my mother Jaleelah and my father Salem who passed a way of old age and to the reminiscence of my dearest brothers Mohammed, Iqab and my sister Zahra who left us to a better world. Peace Be Upon Them All.

Maha Awwad-Khoury would like to express special thanks my family as a whole for their support and love. I would like also to extend my sincere gratitude to my supervisor Dr. Yousef Najajreh for giving me the opportunity to participate in the research. It was a great privilege and honor for me.

Abbreviations

PI3K	phosphoinositide-3-kinase
DNAPK	DNA dependent protein kinase
PI4K	1-phosphatidylinositol-4-kinase
mTOR	mammalian target of rapamycin
PTEN	phosphatase and tensin homolog
NBS	N-bromosuccinimide
DIEA	diisopropylethylamine
Pd(dppf)Cl2-DCM	dichloro [1,1′-bis(diphenylphosphino)ferrocene] palladium (II) dichloromethane adduct
DIEA	diisopropylethylamine
DME	dimethoxyethane
CYP	cytochrome P450
CL	clearance
PK	pharmacokinetics
ABCB1	ATP-binding cassette sub-family B member 1
ABCG2	ATP-binding cassette sub-family G member 2
Ac	Acetyl
ALK	anaplastic lymphoma kinase
BMS	Bristol–Myers Squibb

Bn	benzyl
Boc	*N-tert*-butoxycarbonyl
CbzCl	benzyl chloroformate
CDI	*N,N′*-carbonyldiimidazole
CFDA	Chinese Food and Drug Administration
CNS	central nervous system
COD	1,5-cyclooctadiene
Cy	cyclohexyl
Dba	dibenzylideneacetone
DBU	1,8-diazabicycolo [5.4.0] undec-7-ene
DCC	1,3-dicyclohexylcarbodiimide
DCHA	dicyclohexylamine
DCE	1,2-dichloroethane
DCM	dichloromethane
DDQ	2,3-dichloro-5,6-dicyano-1,4-benzoquinone
3,4-DHP	3,4-dihydropyran
$(DHQ)_2PHAL$	hydroquinine 1,4-phthalazinediyl diether
DIAD	diisopropyl azodicarboxylate
DIBAL	diisobutylaluminium hydride
DIPEA	*N,N*-diisopropylethylamine
DMA	dimethylacetamide
DMAC	dimethylacetamide
DMAP	4-dimethylaminopyridine
DMD	Duchenne's muscular dystrophy
DME	dimethoxyethane
DMF	*N,N*-dimethylformamide
DMM	maleic acid dimethyl ester
DMPU	1,3-dimethyl tetrahydropyrimidin-2(1*H*)-one
DMSO	dimethyl sulfoxide
DPPA	diphenylphosphoryl azide
EDAC	1-ethyl-3-(3-dimethylaminopropyl)carbodiimide
EDC	*N*-(3-dimethylaminopropyl)-*N′*-ethylcarbodiimide
ee	enantiomeric excess
EMA	European Medicine Agency
Et	ethyl
EU	European Union
EtOAc	ethyl acetate
GD1	Gaucher disease 1
GT1	genotype 1
HATU	*o*-(7-azabenzotriazol-1-yl)-*N,N,N′,N′*-tetramethyluronium hexafluorophosphate
HBTU	*N,N,N′,N′*-tetramethyl-*O*-(1*H*-benzotriazol-1-yl)uronium hexafluorophosphate
HCV	hepatitis C virus
HDAC	histone deacetylase
HOAc	acetic acid
HOBt	1-hydroxybenzotriazole hydrate
HONB	*N*-hydroxy-5-norbornene-2,3-dicarboximide
HPLC	High performance liquid chromatography
i-Pac	isopropyl acetate

IPF	idiophathic pulmonary fibrosis
i-Pr	isopropyl
LiHMDS	lithium hexamethyldisilazide
LDA	lithium diisopropylamide
*m*CPBA	3-chloroperoxybenzoic acid
Me	methyl
MeCN	acetonitrile
MEK	methyl ethyl ketone
MEMCl	2-methoxyethoxymethyl chloride
2-MeTHF	2-methyltetrahydrofuran
MIBK	methyl isobutyl ketone
Moc	methoxycarbonyl
MsCl	methanesulfonyl chloride
MsOH	methanesulfonic acid
MSRA	methicillin-resistant *Staphylococcus aureus*
MTBE	methyl *tert*-butyl ether
MVK	methyl vinyl ketone
MW	microwave
N-Ac-Leu	*N*-acetyl leucine
n-BuLi	*n*-butyllithium
NBS	*N*-bromosuccinimide
NFSI	*N*-fluorobenzenesulfonimide
NHS	*N*-hydroxysuccinimide
NK_1	Neurokinin-1
NMP	*N*-methyl-2-pyrrolidone
NMM	*N*-methyl morpholine
NMMO	4-methylmorpholine *N*-oxide
NS5A/B	nonstructural 5A/B
NsCl	2-nitrobenezenesulfonyl chloride
NSCLC	non-small cell lung cancer
$(o\text{-tol})_3P$	tris(2-methylphenyl)phosphine
$Pd_2(dba)_3$	tris(dibenzylideneacetone)dipalladium
$Pd(dppf)_2Cl_2{\cdot}CH_2Cl_2$	[1,1′-bis(diphenylphosphino)ferrocene]dichloropalladium (II), complex with dichloromethane
PDE-4	phosphodiesterase IV
$Pd(OAc)_2$	palladium acetate
Ph	phenyl
PhMe	toluene
PI3K	phosphatidylinositol 3-kinase
PMDA	Pharmaceuticals and Medical Devices Agency
PPAR	peroxisome proliferator-activated receptor
PPTS	pyridinium *p*-toluenesulfonate
PSA	psoriatic arthritis
PTCL	peripheral T-cell lymphoma
p-TsOH	*p*-toluenesulfonic acid
PTSA	*p*-toluenesulfonamide
Py	pyridine
Red-Al	sodium bis(2-methoxyethoxy)aluminum dihydride
$[Rh(COD)_2]$ OTf	bis (1,5-cyclooctadiene)rhodium(I)trifluoromethane-sulfonate

Rt	room temperature
SGLT2	sodium-glucose co-transporter 2
TB	tuberculosis
TBAF	tetrabutylammonium fluoride
TBAHS	tetrabutylammonium hydrogen sulfide
TBME	*tert*-butylmethyl ether
t-Bu	*tert*-butyl
TEA	triethylamine
TEPA	triethylphosphonoacetate
TIPS	triisopropylsilyl
TFA	trifluoroacetic acid
TFE	trifluoroethanol
TfOH	trifluoromethanesulfonic acid
THF	tetrahydrofuran
TMDS	1,1,3,3-tetramethyldisiloxane
TMS	trimethylsilyl
TNF	tumor necrosis factors
THP	tetrahydropyranyl
THF	tetrahydrofuran
TMSCl	trimethylsilyl chloride
TNF	tumor necrosis factor alpha
T_3P	2,4,6-tripropyl-1,3,5,2,4,6-trioxatriphosphorinane-2,4,6-trioxide
TPAP	tetrapropylammonium perruthenate
Ts	4-toluenesulfonyl
USA	United States of America
US FDA	United States Food and Drug Administration
VEGFR2	vascular endothelial growth factor 2
Xantphos	4,5-bis(diphenylphosphino)-9,9-dimethylxanthene
XPhos	2-dicyclohexylphosphino-2′,4′,6′-triisopropylbiphenyl

Author details

Yousef Najajreh* and Maha Awwad Khoury
Anticancer Drugs Research Lab, Faculty of Pharmacy, Al-Quds University, Jerusalem, Palestine

*Address all correspondence to: ynajajreh@staff.alquds.edu; y.s.najajreh@gmail.com

References

[1] Hunter T. Treatment for chronic myelogenous leukemia: The long road to imatinib. The Journal of Clinical Investigation. 2007;**117**(8):2036-2043

[2] Lindauer M, Hochhaus A. Dasatinib. In: Recent Results Cancer Research. Vol. 212. Cham: Springer; 2018. pp. 29-68. DOI: 10.1007/978-3-319-91439-8_2

[3] Chiacchio MA, Iannazzo D, Romeo R, Giofrè SV, Legnani L. Pyridine and pyrimidine derivatives as privileged scaffolds in biologically active agents. Current Medicinal Chemistry. 2018;**26** (40):7166-7195

[4] Sharma V, Chitranshi N, Agarwal AK. Significance and biological importance of pyrimidine in the microbial world. International Journal of Medical Chemistry. 2014;**2014**:1-31. DOI: 10.1155/2014/202784

[5] Sun L, Wu J, Zhang L, Luo M, Sun D. Synthesis and antifungal activities of some novel pyrimidine derivatives. Molecules. 2011;**16**(7):5618-5628

[6] Poletto J, da Silva MJV, Jacomini AP, Bidóia DL, Volpato H, Nakamura CV, et al. Antiparasitic activities of novel pyrimidine N-acylhydrazone hybrids. Drug Development Research. 2021;**82** (2):230-240

[7] Majeed J, Shaharyar M. Synthesis and in vivo diuretic activity of some novel pyrimidine derivatives. Journal of Enzyme Inhibition and Medicinal Chemistry. 2011;**26**(6):819-826

[8] Abbas N, Swamy PMG, Dhiwar P, Patel S, Giles D. Development of fused and substituted pyrimidine derivatives as potent anticancer agents (a review). Pharmaceutical Chemistry Journal. 2021; **54**(12):1215-1226. DOI: 10.1007/s11094-021-02346-8

[9] Ayati A, Moghimi S, Toolabi M, Foroumadi A. Pyrimidine-based EGFR TK inhibitors in targeted cancer therapy. European Journal of Medicinal Chemistry. 2021;**221**:113523

[10] Matada GSP, Abbas N, Dhiwar PS, Basu R, Devasahayam G. Design, synthesis, In Silico and In vitro evaluation of novel pyrimidine derivatives as EGFR inhibitors. Anti-Cancer Agents in Medicinal Chemistry. 2020;**21**(4):451-461

[11] Jadhav M, Sankhe K, Bhandare RR, Edis Z, Bloukh SH, Khan TA. Synthetic strategies of pyrimidine-based scaffolds as aurora kinase and polo-like kinase inhibitors. Molecules. 2021;**26**(17):5170

[12] Sun W, Hu S, Fang S, Yan H. Design, synthesis and biological evaluation of pyrimidine-based derivatives as VEGFR-2 tyrosine kinase inhibitors. Bioorganic Chemistry. 2018;**78**:393-405

[13] Wang S, Yuan XH, Wang SQ, Zhao W, Chen XB, Yu B. FDA-approved pyrimidine-fused bicyclic heterocycles for cancer therapy: Synthesis and clinical application. European Journal of Medicinal Chemistry. 2021;**214**:113218

[14] Kumar A, Saxena J, Chauhan P. Synthesis of 4-amino-5-cyano-2, 6-Disubstituted pyrimidines as a potential Antifilarial DNA topoisomerase II inhibitors. Journal of Medicinal Chemistry (Los Angeles). 2008;**4**(6): 577-585

[15] Uria-Nickelsen M, Neckermann G, Sriram S, Andrews B, Manchester JI, Carcanague D, et al. Novel

topoisomerase inhibitors: Microbiological characterisation and in vivo efficacy of pyrimidines. International Journal of Antimicrobial Agents. 2013;**41**(4):363-371

[16] Liu P, Yang Y, Tang Y, Yang T, Sang Z, Liu Z, et al. Design and synthesis of novel pyrimidine derivatives as potent antitubercular agents. European Journal of Medicinal Chemistry. 2019;**163**:169-182

[17] Yuthavong Y, Tarnchompoo B, Kamchonwongpaisan S. Antimalarial pyrimidine derivatives and methods of making and using them. U.S. Patent 7,371,758 B2. 2008;**2**:12

[18] Singh K, Kaur T. Pyrimidine-based antimalarials: Design strategies and antiplasmodial effects. Medchemcomm. 2016;**7**(5):749-768. DOI: 10.1039/C6MD00084C

[19] Sondhi SM, Singh N, Johar M, Kumar A. Synthesis, anti-inflammatory and analgesic activities evaluation of some mono, bi and tricyclic pyrimidine derivatives. Bioorganic Medicinal Chemistry. 2005;**13**(22):6158-6166

[20] Jansa P, Holý A, Dračínský M, Kolman V, Janeba Z, Kostecká P, et al. 5-substituted 2-amino-4, 6-dihydroxypyrimidines and 2-amino-4, 6-dichloropyrimidines: Synthesis and inhibitory effects on immune-activated nitric oxide production. Medicinal Chemistry Research. 2014;**23**(10):4482-4490

[21] Yejella RP, Atla SR. A study of anti-inflammatory and analgesic activity of new 2,4,6-trisubstituted pyrimidines. Chemical and Pharmaceutical Bulletin. 2011;**59**:1079-1082

[22] Mordant C, Schmitt B, Pasquier E, Demestre C, Queguiner L, Masungi C, et al. Synthesis of novel diarylpyrimidine analogues of TMC278 and their antiviral activity against HIV-1 wild-type and mutant strains. European Journal of Medicinal Chemistry. 2007;**42**(5):567-579

[23] Liang YH, Feng XQ, Sen ZZ, Chen FE, Balzarini J, Pannecouque C, et al. Design, synthesis, and SAR of naphthyl-substituted diarylpyrimidines as non-nucleoside inhibitors of HIV-1 reverse transcriptase. Chem. Medicinal Chemistry. 2009;**4**(9):1537-1545

[24] Irshad N, Khan A, ullah, Alamgeer, Khan SUD, Iqbal MS. Antihypertensive potential of selected pyrimidine derivatives: Explanation of underlying mechanistic pathways. Biomedicine & Pharmacotherapy. 2021;**139**:111567. DOI: 10.1016/j.biopha.2021.111567

[25] Neef S, Steffens A, Pellicena P, Mustroph J, Lebek S, Ort KR, et al. Improvement of cardiomyocyte function by a novel pyrimidine-based CaMKII-inhibitor. Journal of Molecular and Cellular Cardiology. 2018;**115**(April 2017):73-81

[26] Suchonwanit P, Thammarucha S, Leerunyakul K. Minoxidil and its use in hair disorders: A review. Drug Design, Development and Therapy. 2019;**13**:2777-2786. DOI: 10.2147/DDDT.S214907

[27] Peters JU, Weber S, Kritter S, Weiss P, Wallier A, Boehringer M, et al. Aminomethylpyrimidines as novel DPP-IV inhibitors: A 105-fold activity increase by optimization of aromatic substituents. Bioorganic Medicinal Chemistry Letters. 2004;**14**(6):1491-1493

[28] Peters JU, Hunziker D, Fischer H, Kansy M, Weber S, Kritter S, et al. An aminomethylpyrimidine DPP-IV inhibitor with improved properties.

Bioorganic Medicinal Chemistry Letters. 2004;**14**(13):3575-3578

[29] Oh H, Nguyen HD, Yoon IM, Ahn BR, Kim MS. Antidiabetic effect of gemigliptin: A systematic review and meta-analysis of randomized controlled trials with Bayesian inference through a quality management system. Scientific Reports. 2021;**11**(1):20938. DOI: 10.1038/s41598-021-00418-z

[30] Yang Z, Li L, Zheng J, Ma H, Tian S, Li J, et al. Identification of a new series of potent adenosine A2A receptor antagonists based on 4-Amino-5-carbonitrile pyrimidine template for the treatment of Parkinson's disease. ACS Chemical Neuroscience. 2016;7(11):1575-1584

[31] Trifonov L, Yurchenko M, Skjesol A, Cohen G, Espevik T, Korshin EE, et al. Benzyl-Para-di-[5-methyl-4-(n-octylamino) pyrimidin-2(1H)one] as an interferon beta (IFN-β) modulator. Molecular Diversity. 2022;**26**(4):2175-2188

[32] Sato N, Yuki Y, Shinohara H, Takeji Y, Ito K, Michikami D, et al. Cyanopyrimidine Derivative 1. Vol. 19. United States: Otsuka Pharmaceutical Co., LTD. Tokyo, JP; 2017. pp. 1-34, 9708292. Available from: https://www.freepatentsonline.com/9708292.html

[33] Vempala N, Matta B, Rao SV, Maddirala SJ, Shree AJ. An efficient, cyanide free total synthesis of rosuvastatin calcium. Tetrahedron. 2022; **111**:132717. DOI: 10.1016/j.tet.2022.132717

[34] Shao H, Shi S, Huang S, Hole AJ, Abbas AY, Baumli S, et al. Substituted 4-(thiazol-5-yl)-2-(phenylamino) pyrimidines are highly active CDK9 inhibitors: Synthesis, X-ray crystal structures, structure-activity relationship, and anticancer activities. Journal of Medicinal Chemistry. 2013;**56**(3):640-659

[35] Mahapatra A, Prasad T, Sharma T. Pyrimidine: A review on anticancer activity with key emphasis on SAR. Future Journal of Pharmaceutical Sciences. 2021;7:123. DOI: 10.1186/s43094-021-00274-8

[36] Abbas N, Matada GSP, Dhiwar PS, Patel S, Devasahayam G. Fused and substituted pyrimidine derivatives as profound anti-cancer agents. Anti-Cancer Agents in Medicinal Chemistry. 2020;**21**(7):861-893

[37] Zhao L, Ma XD, Chen FE. Development of two scalable syntheses of 4-amino-5-aminomethyl-2-methylpyrimidine: Key intermediate for vitamin b 1. Organic Process Research and Development. 2012;**16**(1):57-60

[38] Zhang SL, Zhang W, Xiao Q, Yang Z, Hu X, Wei Z, et al. Development of dichloroacetamide pyrimidines as pyruvate dehydrogenase kinase inhibitors to reduce cancer cell growth: Synthesis and biological evaluation. RSC Advances. 2016;**6**(82):78762-78767. DOI: 10.1039/C6RA14060B

[39] Rodrigues ALS, Rosa JM, Gadotti VM, Goulart EC, Santos MM, Silva AV, et al. Antidepressant-like and antinociceptive-like actions of 4-(4′-chlorophenyl)-6-(4″-methylphenyl)-2-hydrazinepyrimidine Mannich base in mice. Pharmacology, Biochemistry, and Behavior. 2005;**82**(1):156-162

[40] Desai NC, Kotadiya GM, Trivedi AR. Studies on molecular properties prediction, antitubercular and antimicrobial activities of novel quinoline based pyrimidine motifs. Bioorganic Med Chem Lett. 2014;**24**(14): 3126-3130

[41] Das S, Akbar S, Ahmed B, Dewangan RP, Iqubal A, Pottoo FH, et al. Structural activity relationship based medicinal perspectives of pyrimidine derivatives as anti-Alzheimer's agents: A comprehensive review. CNS Neurol Disord - Drug Targets. 2021;**21**(10):926-939

[42] Singh S, Dhanawat M, Gupta S, Kumar D, Kakkar S, Nair A, et al. Naturally inspired pyrimidines analogues for Alzheimer's disease. Current Neuropharmacology. 2020;**19**(2):136-151

[43] Ben WS, Piao GC, Zhang HJ, Quan ZS. Synthesis of 5-alkoxythieno [2,3-e] [1,2,4] triazolo [4,3-c] pyrimidine derivatives and evaluation of their anticonvulsant activities. Molecules. 2015;**20**(4):6827-6843

[44] Ban M, Taguchi H, Katsushima T, Aoki S, Watanabe A. Novel antiallergic agents. Part I: Synthesis and pharmacology of pyrimidine amide derivatives. Bioorganic. Medicinal Chemistry. 1998;**6**(7):1057-1067

[45] Davidson MH. Rosuvastatin: A highly efficacious statin for the treatment of dyslipidaemia. Expert Opinion on Investigational Drugs. 2002; **11**(1):125-141

[46] Lee HW, Bok YK, Joong BA, Sung KK, Lee JH, Jae SS, et al. Molecular design, synthesis, and hypoglycemic and hypolipidemic activities of novel pyrimidine derivatives having thiazolidinedione. European Journal of Medicinal Chemistry. 2005;**40**(9):862-874

[47] Serby MD, Zhao H, Szczepankiewicz BG, Kosogof C, Xin Z, Liu B, et al. 2,4-Diaminopyrimidine derivatives as potent growth hormone secretagogue receptor antagonists. Journal of Medicinal Chemistry. 2006;**49**(8):2568-2578. DOI: 10.1021/jm0510934

[48] Hu G, Wang C, Xin X, Li S, Li Z, Zhao Y, et al. Design, synthesis and biological evaluation of novel 2,4-diaminopyrimidine derivatives as potent antitumor agents. New Journal of Chemistry. 2019;**43**(25):10190-10202. DOI: 10.1039/C9NJ02154J

[49] Patel S, Modi P, Ranjan V, Chhabria M. Structure-based design, synthesis and evaluation of 2,4-diaminopyrimidine derivatives as novel caspase-1 inhibitors. Bioorganic Chemistry. 2018;**78**:258-268

[50] Manley PJ, Balitza AE, Bilodeau MT, Coll KE, Hartman GD, McFall RC, et al. 2,4-Disubstituted pyrimidines: A novel class of KDR kinase inhibitors. Bioorganic Medicinal Chemistry Letters. 2003;**13**(10):1673-1677

[51] Xu Y, Hao SY, Zhang XJ, Li WB, Qiao XP, Wang ZX, et al. Discovery of novel 2,4-disubstituted pyrimidines as Aurora kinase inhibitors. Bioorganic Medicinal Chemistry Letters. 2020;**30**(3):126885

[52] Luo G, Tang Z, Lao K, Li X, You Q, Xiang H. Structure-activity relationships of 2, 4-disubstituted pyrimidines as dual ERα/VEGFR-2 ligands with anti-breast cancer activity. European Journal of Medicinal Chemistry. 2018;**150**:783-795

[53] Patel H, Ansari A, Pawara R, Ansari I, Jadhav H, Surana S. Design and synthesis of novel 2,4-disubstituted aminopyrimidines: Reversible non-covalent T790M EGFR inhibitors. Journal of Receptors and Signal Transduction. 2018;**38**(5–6):393-412

[54] Pawara R, Ahmad I, Surana S, Patel H. Computational identification of 2,4-disubstituted amino-pyrimidines as L858R/T790M-EGFR double mutant

inhibitors using pharmacophore mapping, molecular docking, binding free energy calculation, DFT study and molecular dynamic simulation. Silico Pharmacology. 2021;**9**(1):54

[55] Fang Z, Zheng S, Chan KF, Yuan W, Guo Q, Wu W, et al. Design, synthesis and antibacterial evaluation of 2,4-disubstituted-6-thiophenyl-pyrimidines. European Journal of Medicinal Chemistry. 2019;**161**:141-153

[56] Mukherjee P, Li H, Sevrioukova I, Chreifi G, Martásek P, Roman LJ, et al. Novel 2,4-disubstituted pyrimidines as potent, selective, and cell-permeable inhibitors of neuronal nitric oxide synthase. Journal of Medicinal Chemistry. 2015;**58**(3):1067-1088

[57] Mohamed T, Rao PPN. Design, synthesis and evaluation of 2,4-disubstituted pyrimidines as cholinesterase inhibitors. Bioorganic Medicinal Chemistry Letters. 2010;**20** (12):3606-3609

[58] Josef S, Thomas B, Thomas M, Thomas B, Thomas C, Armin Z, et al. 4,6-Disubstituted pyrimidines and their use as protein kinase inhibitors. Berlin, DE: EPO, Bayer Schering Pharma AG; 2011. p. 7879853

[59] Nguyen DT, Shayahi S. Pazopanib: approval for soft-tissue sarcoma. Journal of the Advanced Practitioner in Oncology. 2013;**4**(1):53-57

[60] Qi H, Chen L, Liu B, Wang X, Long L, Liu D. Synthesis and biological evaluation of novel pazopanib derivatives as antitumor agents. Bioorganic Medicinal Chemistry Letters. 2014;**24**(4):1108-1110. DOI: 10.1016/j.bmcl.2014.01.003

[61] Mei YC, Yang BW, Chen W, Huang DD, Li Y, Deng X, et al. A novel practical synthesis of Pazopanib: An anticancer drug. Letters in Organic Chemistry. 2012;**9**(4):276-279

[62] Cope, Arthur C Burrows, W Dickinson. Clarke-Eschweiler Cyclization. Scope and Mechanism. The Journal of Organic Chemistry. 1966;**31** (10):3099-3103

[63] Kumar R, GIRI P, Barman D, Nath A, Prasad M, et al. Process for the preparation of pazopanib or salts thereof. United States, O2014097152A1, WIPO/PCT. 2014. pp. 1-12

[64] Schafer PH, Kivitz AJ, Ma J, Korish S, Sutherland D, Li L, et al. Spebrutinib (CC-292) affects markers of B cell activation, chemotaxis, and osteoclasts in patients with rheumatoid arthritis: Results from a mechanistic study. Rheumatology Therapy. 2020;**7**(1):101-119. DOI: 10.6084/

[65] Liang C, Tian D, Ren X, Ding S, Jia M, Xin M, et al. The development of Bruton's tyrosine kinase (BTK) inhibitors from 2012 to 2017: A mini-review. European Journal of Medicinal Chemistry. 2018;**151**:315-326

[66] Tasso B, Spallarossa A, Russo E, Brullo C. The development of btk inhibitors: A five-year update. Molecules. 2021;**26**(23):7411

[67] Liu XJ, Xu-Liu PXJ, Ying Yuan X, Yu GX, Li YR, et al. Progress in the development of small molecular inhibitors of the Bruton's tyrosine kinase (BTK) as a promising cancer therapy. Bioorganic & Medicinal Chemistry. 2021;**47**:116358-116388. DOI: 10.1016/j.bmc.2021.116358

[68] Zhang D, Gong H, Meng F. Recent advances in btk inhibitors for the treatment of inflammatory and

autoimmune diseases. Molecules. 2021; **26**(16):4907. DOI: 10.3390/molecules26164907

[69] Li Y, Huang Y, Cheng H, Xu F, Qi R, Dai B, et al. Discovery of BRAF/HDAC dual inhibitors suppressing proliferation of human colorectal cancer cells. Frontiers in Chemistry. 2022;**10**(July):1-13

[70] Rheault TR, Stellwagen JC, Adjabeng GM, Hornberger KR, Petrov KG, Waterson AG, et al. Discovery of dabrafenib: A selective inhibitor of Raf kinases with antitumor activity against B-Raf-driven tumors. ACS Medicinal Chemistry Letters. 2013;**4**(3):358-362

[71] Schafer JJ, Short WR. Rilpivirine, a novel non-nucleoside reverse transcriptase inhibitor for the management of HIV-1 infection: A systematic review. Antiviral Therapy. 2012;**17**(8):1495-1502. DOI: 10.3851/IMP2254

[72] Li S-L, Xu P, Zhang L, Sun G-X, Lu Z-J. Effectiveness and safety of rilpivirine, a non-nucleoside reverse transcriptase inhibitor, in treatment-naive adults infected with HIV-1: A meta-analysis. HIV Clinical Trials. 2015; **16**(1):22-29

[73] Ding L, Pannecouque C, De Clercq E, Zhuang C, Chen FE. Improving Druggability of novel Diarylpyrimidine NNRTIs by a fragment-based replacement strategy: From biphenyl-DAPYs to Heteroaromatic-biphenyl-DAPYs. Journal of Medicinal Chemistry. 2021;**64**(14):10297-10311

[74] Jin KJ, Yin H, De Clercq E, Pannecouque C, Meng G, Chen FE. Discovery of biphenyl-substituted diarylpyrimidines as non-nucleoside reverse transcriptase inhibitors with high potency against wild-type and mutant HIV-1. European Journal of Medicinal Chemistry. 2018;**145**:726-734

[75] Ludovici DW, De Corte BL, Kukla MJ, Ye H, Ho CY, Lichtenstein MA, et al. Evolution of anti-HIV drug candidates. Part 3: Diarylpyrimidine (DAPY) analogues. Bioorganic Medicinal Chemistry Letters. 2001;**11**(17):2235-2239

[76] Farag AK, Hassan AHE, Chung KS, Lee JH, Gil HS, Lee KT, et al. Diarylurea derivatives comprising 2,4-diarylpyrimidines: Discovery of novel potential anticancer agents via combined failed-ligands repurposing and molecular hybridization approaches. Bioorganic Chemistry. 2020;**103**:104121

[77] Application F, Data P, Agnes PT, Herman A, Peeters J, Elisabeth A, et al. SALT OF 4-4-4-(2-CYANOETHENYL)-2,6- DIMETHYLPHENYLAMINO-2-PYRIMIDINYLAMINO BENZONITRILE. United States 20110008434 (12).; 2006;**1**(10). Available from: https://www.freepatentsonline.com/y2011/0008434.html.

[78] Guillemont JEG, Stevens PTA, Copmans AH, Peeters J, Stappers AE, Vande Cruys RPG, Stoffels P (2006) Salt of 4-[[4-[[4-(2-cyanoethenyl)-2,6-dimethylphenyl]amino]-2-pyrimidinyl]amino]benzonitrile, US 20060111379A1. Available from: https://www.freepatentsonline.com

[79] Zhang T, Yang J, Zhou Z, Fu Z, Cherukupalli S, Kang D, et al. The development of an effective synthetic route of rilpivirine. BMC Chemistry. 2021;**15**(1):1-9. DOI: 10.1186/s13065-021-00749-y

[80] Huo Z, Zhang H, Kang D, Zhou Z, Wu G, Desta S, et al. Discovery of novel Diarylpyrimidine derivatives as potent

HIV-1 NNRTIs targeting the "nNRTI adjacent" binding site. ACS Medicinal Chemistry Letters. 2018;**9**(4):334-338

[81] Ma XD, Yang SQ, Gu SX, He QQ, Chen FE, DeClercq E, et al. Synthesis and anti-HIV activity of Aryl-2-[(4-cyanophenyl)amino]-4-pyrimidinone hydrazones as potent non-nucleoside reverse transcriptase inhibitors. Chem Med Chem. 2011;**6**(12):2225-2232

[82] Angst D, Gessier F, Janser P, Vulpetti A, Wälchli R, Beerli C, et al. Discovery of LOU064 (Remibrutinib), a potent and highly selective covalent inhibitor of Bruton's tyrosine kinase. Journal of Medicinal Chemistry. 2020;**63**(10):5102-5118

[83] Maurer M, Giménez-Arnau A, Jain V, Tillinghast J, Tolcachier A, Nigen S, et al. Remibrutinib treatment improves quality of life in patients with chronic spontaneous Urticaria. The Journal of Allergy and Clinical Immunology. 2022;**149**(2):AB179. DOI: 10.1016/j.jaci.2021.12.589

[84] Wang X, Sathunuru R, Melendez V, Kozar MP, Lin AJ. Facile synthesis of 2,4-diamino-6-alkyl- or 6-aryl-pyrimidine derivatives. Journal of Heterocyclic Chemistry. 2010;**47**(5):1056-1061

[85] Garrido-Castro AC, Saura C, Barroso-Sousa R, Guo H, Ciruelos E, Bermejo B, et al. Phase 2 study of buparlisib (BKM120), a pan-class I PI3K inhibitor, in patients with metastatic triple-negative breast cancer. Breast Cancer Research. 2020;**22**(1):1-13

[86] Bohnacker T, Prota AE, Beaufils F, Burke JE, Melone A, Inglis AJ, et al. Deconvolution of Buparlisib's mechanism of action defines specific PI3K and tubulin inhibitors for therapeutic intervention. Nature Communications. 2017;**8**:1-13

[87] De Gooijer MC, Zhang P, Buil LCM, Çitirikkaya CH, Thota N, Beijnen JH, et al. Buparlisib is a brain penetrable pan-PI3K inhibitor. Scientific Reports. 2018;**8**(1):1-8

[88] Burger MT, Pecchi S, Wagman A, Ni ZJ, Knapp M, Hendrickson T, et al. Identification of NVP-BKM120 as a potent, selective, orally bioavailable class i PI3 kinase inhibitor for treating cancer. ACS Medicinal Chemistry Letters. 2011;**2**(10):774-779

[89] Goswami S, Ghosh K, Mukherjee R, Adak AK, Mahapatra AK. N-bromosuccinimide reactions of some heterocycles in the presence or absence of water: An overview of ring versus side chain bromination for the synthesis of important brominated heterocyclic synthons. Journal of Heterocyclic Chemistry. 2001;**38**(1):173-178

[90] Xu Y. Process for preparing PI3K inhibitor buparsilib. United States 9481665 B2. 2016;**2**(12):3-8. Available from: https://www.freepatentsonline.com/9481665.html

[91] Luo Y, Deng YQ, Wang J, Long ZJ, Tu ZC, Peng W, et al. Design, synthesis and bioevaluation of N-trisubstituted pyrimidine derivatives as potent aurora a kinase inhibitors. European Journal of Medicinal Chemistry. 2014;**78**:65-71

[92] Gao C, Si X, Chi L, Wang H, Dai H, Liu L, et al. Synthesis and Antiproliferative activity of 2,4,5,6-Tetrasubstituted pyrimidine derivatives containing anisole. Chinese Journal of Organic Chemistry. 2022;**42**(6):1677-1686

[93] Rembold H, Schramm HJ. Kondensation des 2.4-Diamino-6-

hydroxy-pyrimidins mit Aldosen. Chemische Berichte. 1963;**96**(10):2786-2797

[94] Altenbach RJ, Adair RM, Bettencourt BM, Black LA, Fix-Stenzel SR, Gopalakrishnan SM, et al. Structure-activity studies on a series of a 2-aminopyrimidine-containing histamine H4 receptor ligands. Journal of Medicinal Chemistry. 2008;**51**(20):6571-6580

[95] Oliver KC. 100 years of the Biginelli Dihydropyrimidine synthesis. Tetrahedron. 1993;**49**(32):6937-6963

[96] Joshi S, Maikap GC, Titirmare S, Chaudhari A, Gurjar MK. An improved synthesis of etravirine. Organic Process Research and Development. 2010;**14**(3):657-660

[97] Diering M, Mitchell F. Clinical pharmacokinetics and pharmacodynamics of Etravirine: An updated review. Physiology & Behavior. 2018;**176**(1):139-148

[98] Havens JP, Podany AT, Scarsi KK, Fletcher CV. Clinical pharmacokinetics and pharmacodynamics of Etravirine: An updated review. Clinical Pharmacokinetics. 2020;**59**(2):137-154

[99] Elsayed RK, Caldwell DJ. Etravirine: A novel nonnucleoside reverse transcriptase inhibitor for managing human immunodeficiency virus infection. American Journal of Health Pharmacy. 2010;**67**(3):193-205

[100] Feng D, Wei F, Wang Z, Kang D, Zhan P, Liu X. Development of a practical synthesis of etravirine via a microwave-promoted amination. Chemistry Central Journal. 2018;**12**(1):4-9. DOI: 10.1186/s13065-018-0504-4

[101] Sica DA. Minoxidil: An underused vasodilator for resistant or severe hypertension. Journal of Clinical Hypertension (Greenwich, Conn.). 2004;**6**(5):283-287

[102] Schier O, Marxer A. Antihypertensive agents 1969-1980. Progress in Drug Research. 1981;**25**:9-132

[103] Goren A, Naccarato T. Minoxidil in the treatment of androgenetic alopecia. Dermatologic Therapy. 2018;**31**(5):e12686

[104] Kelly Y, Blanco A, Tosti A. Androgenetic alopecia: An update of treatment options. Drugs. 2016;**76**(14):1349-1364

[105] Eisavi R, Ahmadi F, Zeynizadeh B, Kouhkan M. The promoted synthesis of minoxidil by magnetic nanoparticles of cobalt ferrite ($CoFe_2O_4$) as a heterogeneous reusable catalyst. Turkish Journal of Chemistry. 2019;**43**(5):1425-1435

[106] Rewcastle GW. Pyrimidines and their benzo derivatives. In: Katritzky AR, Ramsden CA, Scriven EFV, Taylor RJK, editors. Comprehensive Heterocyclic Chemistry III. Oxford: Elsevier; 2008. pp. 117-272

Chapter 5

Selected Imidazole Derivatives: Synthesis and X-Ray Crystal Structure – A Review

Aravazhi Amalan Thiruvalluvar

Abstract

This chapter accounts for the synthesis and X-ray structure of selected imidazole derivatives: DMDPIMH: 4,5-dimethyl-1,2-diphenyl-1*H*-imidazole monohydrate, DMPPTI: 4,5-dimethyl-2-phenyl-1-(p-tolyl)-1*H*-imidazole, FPDMMPI: 2-(4-fluorophenyl)-4,5-dimethyl-1-(4-methylphenyl)-1*H*-imidazole, DMPDMPIHH: 1-(3,5-dimethylphenyl)-4,5-dimethyl-2-phenyl-1*H*-imidazole hemihydrate, DMPFPDMI: 1-(3,5-Dimethylphenyl)-2-(4-fluorophenyl)-4,5-dimethyl-1*H*-imidazole, FPMOPDMI: 2-(4-fluorophenyl)-1-(4-methoxyphenyl)-4,5-dimethyl-1*H*-imidazole, DMOPFPDMI: 1-(3,5-dimethoxyphenyl)-2-(4-fluorophenyl)-4,5-dimethyl-1*H*-imidazole, and FPTPI: 2-(4-fluorophenyl)-1,4,5-triphenyl-1*H*-imidazole. The molecular crystal structure was determined by this investigation, along with the impact of substitutions on the geometry of the imidazole units. The uses and applications of the imidazole compounds are described in the introduction. All eight imidazole derivatives' syntheses were discussed. The essential characteristics of the imidazole derivatives are provided by the structural analyses of the eight derivatives and a portion of a database survey. All the derivatives have a planar imidazole unit. The crystal structures are stabilized by intermolecular O—H ... N, O—H ... O, C—H ... N, hydrogen bonds, and C—H ... π interactions.

Keywords: synthesis of imidazole, single crystal, X-ray structural analysis, intermolecular hydrogen bonds, C—H ... π interactions

1. Introduction

An online search reveals that $C_3N_2H_4$ is the formula for the chemical molecule imidazole. A solid that is either colorless or white and water-soluble only slightly produces an alkaline solution. It is a diazole and an aromatic heterocycle in a chemical sense since it has meta-substituted nitrogen atoms that are not contiguous. Imidazole was named in 1887 by the German chemist Arthur Rudolf Hantzsch (1857–1935) [1].

Because of their ability to fluoresce and glow under certain conditions, imidazoles are widely used in analytical processes [2]. The noticeable alteration in fluorescence following metal binding is a crucial characteristic that makes imidazole derivatives more desirable as a chelator [3]. As a result, extremely sensitive fluorescent

IntechOpen

chemisensors for sensing and imaging metal ions have been built using imidazole derivatives [4]. Its chelates, especially those that include Ir3^{+}, are crucial parts of organic light-emitting diodes and are excellent candidates for fluorescent chemisensors for metal ions [5].

Chemical compounds with an imidazole ring system have many pharmacological properties and play indispensable roles in biochemical processes [6]. Lots of the substituted imidazoles are known as inhibitors of fungicides and herbicides, plant growth regulators, and therapeutic agents. Modern advancements in green chemistry and organometallic chemistry have broadened the limit of imidazoles to the synthesis and application of a prominent category of imidazoles as ionic liquids and imidazole-associated N-heterocyclic carbenes. Imidazole derivatives are also applied as possible anticancer agents. Lately, heterocyclic imidazole derivatives have drawn attention due to their unique optical properties. These compounds play a particularly substantial part in chemistry as mediators for synthetic reactions and the development of functionalized materials. The amino acid histidine, vitamin B12, a component of DNA base structure, purines, histamine, and biotin are just a few examples of well-known components of human bodies that contain the imidazole core. Other examples include azomycin, cimetidine, and metronidazole [7]. The use of imidazole-based compounds was reported for multidrug-resistant tuberculosis, antifungal and antimycobacterial activity, and bactericidal effects [8].

The most essential technique for determining the relative atomic locations in a molecule structure is X-ray diffraction. Additionally, it can offer objective proof of the dimensions and geometry of molecules. The stability of a structure is aided by hydrogen bonds. As a result, it is a component of molecular conformation in that symmetry, and the subsequent arrangement of the molecules should result in the production of as many hydrogen bonds as feasible [9]. This chapter describes the author's work on the synthesis and crystal structure determination of a few significant novel imidazole derivatives.

2. Imidazole derivatives: synthetic process

2.1 Synthesis of 4,5-dimethyl-1,2-diphenyl-1*H*-imidazole monohydrate ($C_{17}H_{16}N_2.H_2O$): DMDPIMH

In 10 ml of ethanol, at a constant temperature of 333 K, 15 mmol of benzaldehyde, 15 mmol of pure butane-2,3-dione, 15 mmol of aniline, and 15 mmol of ammonium acetate were added over the course of an hour. The reaction mixture was refluxed for seven days before dichloromethane extraction was used to complete the process. Using column chromatography (CC), the solid is separated and purified as the eluent (hexane: ethyl acetate). The yield was 1.79 g or 48% (**Figure 1**). A suitable single crystal was used to collect the X-ray diffraction data [10].

2.2 Synthesis of 4,5-dimethyl-2-phenyl-1-(p-tolyl)-1*H*-imidazole ($C_{18}H_{18}N_2$): DMPPTI

Over about an hour, at a constant temperature of 333 K, 15 mmol of pure butane-2,3-dione, 15 mmol of p-toluidine, 15 mmol of ammonium acetate, and 15 mmol of benzaldehyde were added to 10 ml of ethanol. The actions that follow are completed

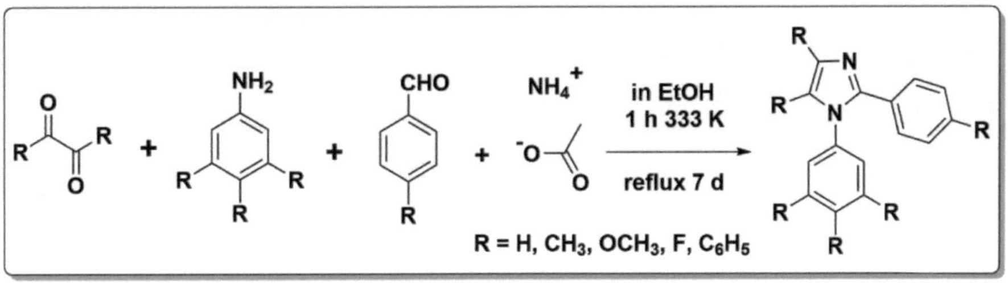

Figure 1.
Schematic diagram for the synthesis of 2.1–2.8.

in the manner outlined in DMDPIMH. The yield was 1.93 g or 46% (**Figure 1**). An appropriate single crystal was used to acquire the X-ray diffraction data [11].

2.3 Synthesis of 2-(4-fluorophenyl)-4,5-dimethyl-1-(4-methylphenyl)-1*H*-imidazole ($C_{18}H_{17}FN_2$): FPDMMPI

For an hour, at a steady temperature of 333 K, 15 mmol of pure biacetyl in 10 ml of ethanol, 15 mmol of p-toluidine, 15 mmol of ammonium acetate, and 15 mmol of p-fluorobenzaldehyde were added. The actions that follow are completed in the manner outlined in DMDPIMH. The yield was 1.93 g or 46% (**Figure 1**). A suitable single crystal was used to collect the X-ray diffraction data [12].

2.4 Synthesis of 1-(3,5-dimethylphenyl)-4,5-dimethyl-2-phenyl-1*H*-imidazole hemihydrate ($C_{19}H_{20}N_2.0.5H_2O$): DMPDMPIHH

15 mmol of pure butane-2,3-dione in 10 ml of ethanol was combined with 15 mmol of 3,5-xylidine, 15 mmol of ammonium acetate, and 15 mmol of benzaldehyde for an hour at a constant temperature of 333 K. The subsequent steps are conducted in the manner specified in DMDPIMH. The yield was 1.91 g or 46% (**Figure 1**). The X-ray diffraction data were gathered with a suitable single crystal [13].

2.5 Synthesis of 1-(3,5-dimethylphenyl)-2-(4-fluorophenyl)-4,5-dimethyl-1*H*-imidazole ($C_{19}H_{19}FN_2$): DMPFPDMI

15 mmol of 3,5-xylidine, 15 mmol of ammonium acetate, and 15 mmol of 4-fluorobenzaldehyde were added to 15 mmol of pure butane-2,3-dione in 10 ml of ethanol for about 1 hour at a constant temperature of 333 K. The remaining procedure is followed as given in DMDPIMH. The yield was 2.1 g or 48% (**Figure 1**). Crystals suitable for X-ray diffraction studies were developed by slow solvent evaporation using dichloromethane [14].

2.6 Synthesis of 2-(4-fluorophenyl)-1-(4-methoxyphenyl)-4,5-dimethyl-1*H*-imidazole ($C_{18}H_{17}FN_2O$): FPMOPDMI

15 mmol of p-anisidine, 15 mmol of ammonium acetate, and 15 mmol of 4-fluorobenzaldehyde were added to 15 mmol of pure biacetyl in 10 ml of ethanol for

about 1 hour at a constant temperature of 333 K. The subsequent steps are specified in DMDPIMH. The yield was 1.77 g or 40% (**Figure 1**). The X-ray diffraction data were gathered using an appropriate single crystal [15].

2.7 Synthesis of 1-(3,5-dimethoxyphenyl)-2-(4-fluorophenyl)-4,5-dimethyl-1*H*-imidazole ($C_{19}H_{19}FN_2O_2$): DMOPFPDMI

Pure butane-2,3-dione in 10 ml of ethanol, 3,5-dimethoxyaniline, ammonium acetate, and 4-fluorobenzaldehyde were added in 15 mmol increments over an hour at a constant temperature of 333 K. The remaining actions were completed by the guidelines in DMDPIMH. 2.20 g or 45% of the total were yielded (**Figure 1**). Crystals suitable for X-ray diffraction measurements were made using dichloromethane as a slow solvent to evaporate [16].

2.8 Synthesis of 2-(4-fluorophenyl)-1,4,5-triphenyl-1*H*-imidazole ($C_{27}H_{19}FN_2$): FPTPI

Over an hour at a constant temperature of 333 K, 15 mmol of aniline, 15 mmol of ammonium acetate, and 15 mmol of p-fluorobenzaldehyde were added to 15 mmol of benzil in 10 ml of ethanol. The actions that follow are completed in the manner outlined in DMDPIMH. The yield was 3.51 g or 60% (**Figure 1**). The X-ray diffraction data were gathered using a suitable single crystal [17].

3. Eight imidazole derivatives: structural analyses

3.1 Structural analysis of 4,5-dimethyl-1,2-diphenyl-1*H*-imidazole monohydrate (DMDPIMH)

This section shows the structure details of DMDPIMH [10]. SIR2004 [18] solved the structure by using direct methods. SHELXL97 [19] refined the final crystal structure model.

This compound belongs to the tetragonal crystal system with the space group $I4_1/a$. Molecular formula: $C_{17}H_{16}N_2.H_2O$; molecular weight: 266.33; crystal data: $a = b = 25.5498$ Å; $c = 9.3792$ (1) Å; $\alpha = \beta = \gamma = 90°$; $V = 6122.67$ (9) Å^3; $Z = 16$; $D_x = 1.156$ Mg m^{-3}; $F(000) = 2272$; final $R[F^2 > 2\sigma(F^2)] = 0.043$ and $wR(F^2) = 0.134$ for 2610 reflections observed with $I > 2\sigma(I)$.

In multiple Fourier maps, the water atoms H1W and H2W coupled to the oxygen atom O1W were found and freely refined. With C—H = 0.93–0.96 Å; $U_{iso}(H) = kU_{eq}(C)$, where k = 1.5 for —CH_3 and 1.2 for the other H atoms, the other hydrogen atoms were positioned geometrically and supported by their parent atoms. The —CH_3 groupings are disorganized over two locations. Each was enhanced as an idealized disordered —CH_3 group.

In this compound (**Figure 2**), with a maximum deviation of 0.0037 (7) Å for C5, the imidazole ring is planar. The imidazole ring makes dihedral angles of 80.74 (7) and 41.62 (7) ° with the ring (C11—C16) and the ring (C21—C26), respectively. 75.83 (8) ° is the dihedral angle among the phenyl rings. Intermolecular hydrogen bonds O1W—H1W···N3 (y + 1/4, −x + 1/4, z + 1/4) and O1W—H2W···O1W (y + 1/4, −x + 1/4, z + 1/4) are found in the crystal structure (**Figure 3**).

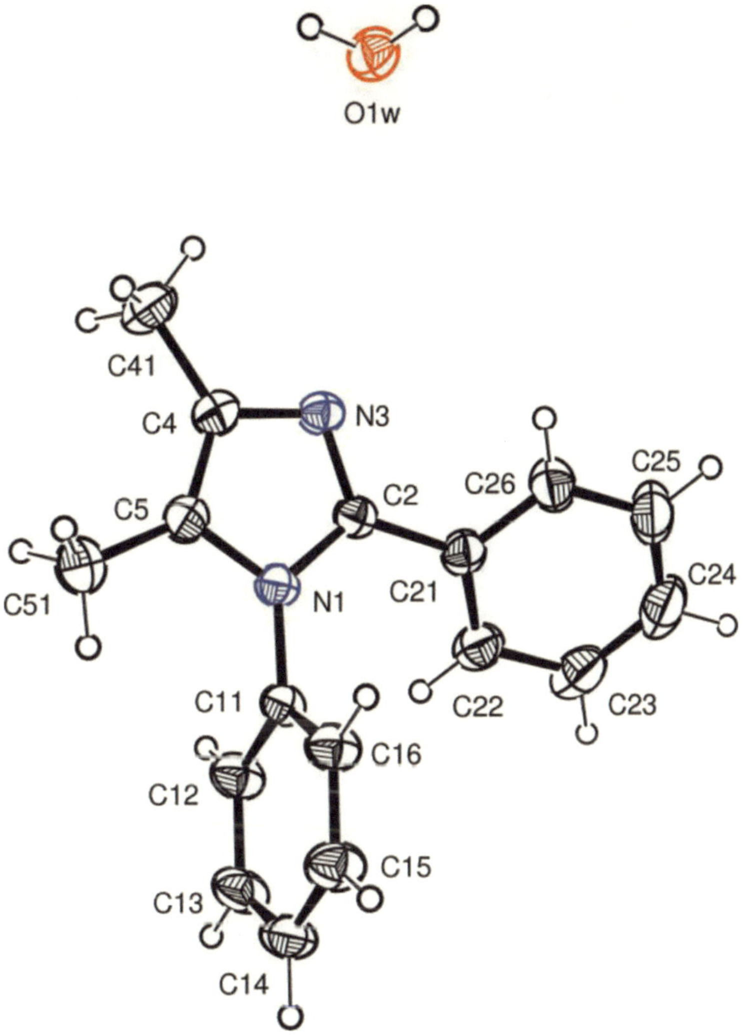

Figure 2.
The thermal displacement ellipsoid plot [20] is shown at a 30% probability level.

The ORTEP-3 for Windows drew the thermal displacement ellipsoid plot with a 30% probability level [20] (**Figure 2**). The crystal packing using the PLATON [21] was viewed along the c-axis (**Figure 3**).

3.2 Structural analysis of 4,5-dimethyl-2-phenyl-1-(p-tolyl)-1*H*-imidazole (DMPPTI)

This section depicts the structure details of DMPPTI [11]. The crystal structure was solved by direct methods using the program SIR2002 [22]. The program SHELXL97 [19] refines the final crystal structure model.

This compound belongs to the monoclinic crystal system with the space group $P2_1/n$. Molecular formula: $C_{18}H_{18}N_2$; molecular weight: 262.34; crystal data: a = 9.6971 (3) Å; b = 7.5458 (2) Å; c = 19.8407 (7) Å; β = 96.604 (3)°; V = 1442.16 (8) $Å^3$; Z = 4;

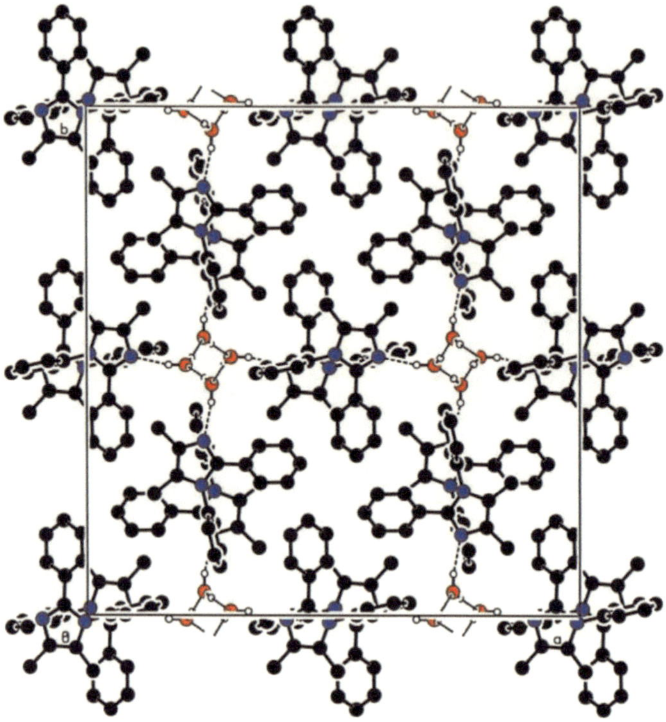

Figure 3.
The crystal packing with hydrogen bonds [21], viewed along the c-axis.

D_x = 1.208 Mg m^{-3}; F(000) = 560; final $R[F^2 > 2\sigma(F^2)]$ = 0.049 and $wR(F^2)$ = 0.138 for 2529 reflections observed with $I > 2\sigma(I)$.

With C—H = 0.95 to 0.98 Å; $U_{iso}(H) = kU_{eq}(C)$, k = 1.5 for $—CH_3$ and 1.2 for the other H atoms, all the hydrogen atoms were arranged geometrically and allowed to drive on their parent atoms.

In (**Figure 4**), with the maximum deviation = 0.004 (1) Å for N3, the imidazole ring is planar. This ring makes dihedral angles of 68.91 (8)°, and 20.43 (9)° with the (C11—C16) and (C21—C26) rings, respectively. 73.62 (8)° is the dihedral angle among the (C11—C16) and (C21—C26) rings. The C12—H12···$N3^{i}$ and C16—H16···$N3^{ii}$ intermolecular nonclassical hydrogen bonds stabilize the packing (**Figure 5**). Symmetry codes are: (i) −x + 2, −y + 1, −z and (ii) −x + 2, −y + 2, −z.

The ORTEP-3 for Windows drew the thermal displacement ellipsoid plot with a 30% probability level [20] (**Figure 4**). The crystal packing using the PLATON [21] was viewed along the a-axis (**Figure 5**).

3.3 Structural analysis of 2-(4-fluorophenyl)-4,5-dimethyl-1-(4-methylphenyl)-1*H*-imidazole (FPDMMPI)

The FPDMMPI's detailed structure is illustrated in this section [12]. The structure was solved by SHELXS97 [19]. The final crystal structure model is enhanced using the tool SHELXL97 [19].

This compound belongs to the monoclinic crystal system with the space group $P2_1/n$. Molecular formula: $C_{18}H_{17}N_2$; molecular weight: 280.34; crystal data: a = 9.8888 (2) Å; b = 7.6693 (1) Å; c = 20.1017 (5) Å; β = 95.915 (1)°; V = 1516.40 (4) $Å^3$; Z = 4;

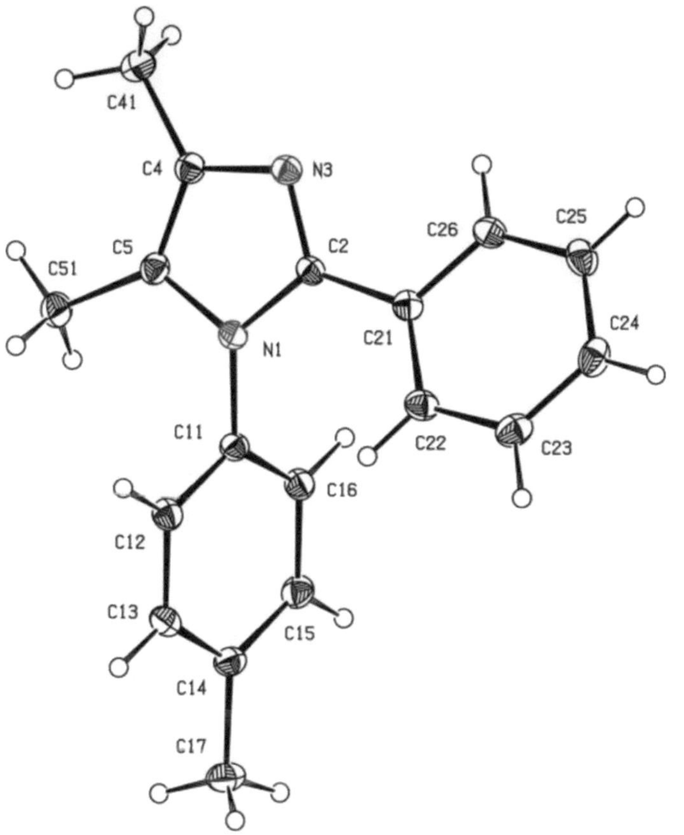

Figure 4.
The thermal displacement ellipsoid plot [20] is shown at a 30% probability level.

D_x = 1.228 Mg m^{-3}; F(000) = 592; final $R[F^2 > 2\sigma(F^2)]$ = 0.049 and $wR(F^2)$ = 0.153 for 2617 reflections observed with $I > 2\sigma(I)$.

With C—H = 0.93–0.96 Å; $U_{iso}(H) = kU_{eq}(C)$, where k = 1.5 for $-CH_3$ and 1.2 for the other H atoms, all the hydrogen atoms were geometrically arranged and admitted to riding on their parent atoms.

In this molecule (**Figure 6**), the dihedral angles made by (C11—C16) and (C21—C26) rings are 72.33 (8)° and 18.71 (8)° with the planar imidazole ring. Benzene rings makes a 75.05 (7)° dihedral angle. The C12—H12···N3 (2 − x, 1 − y, −z) and C16—H16···N3 (2 − x, 2 − y, −z) intermolecular hydrogen bonds stabilize the crystal (**Figure 7**).

The ORTEP-3 for Windows drew the thermal displacement ellipsoid plot with a 30% probability level [20] (**Figure 6**). The crystal packing using the PLATON [21] was viewed along the a-axis (**Figure 7**).

3.4 Structural analysis of 1-(3,5-dimethylphenyl)-4,5-dimethyl-2-phenyl-1*H*-imidazole hemihydrate (DMPDMPIHH)

The specifics of DMPDMPIHH's structure are shown in this section [13]. The crystal structure was solved by SHELXS97 [19]. The final crystal structure model is enhanced using the tool SHELXL97 [19].

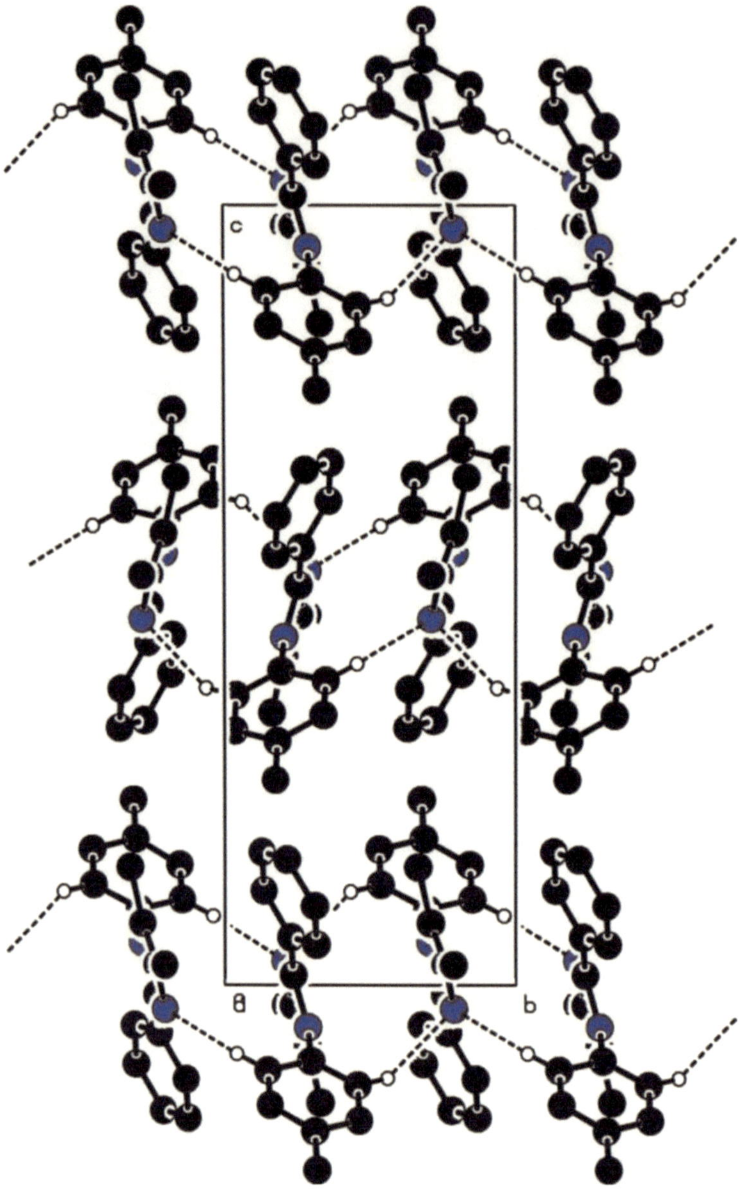

Figure 5.
The crystal packing with hydrogen bonds [21], viewed along the a-axis.

This compound belongs to the orthorhombic crystal system with the space group Pbcn. Molecular formula: $C_{19}H_{20}N_2.0.5H_2O$; molecular weight: 285.38; crystal data: a = 16.7611 (2) Å; b = 11.5467 (2) Å; c = 16.6563 (2) Å; $\alpha = \beta = \gamma = 90°$; V = 3223.58 (8) $Å^3$; Z = 8; D_X = 1.176 Mg m^{-3}; F(000) = 1224; final $R[F^2 > 2\sigma(F^2)] = 0.062$ and wR $(F^2) = 0.171$ for 2630 reflections observed with $I > 2\sigma(I)$.

The atom H1W was introduced alongside the atom O1W, found in a differential Fourier map and refined without restriction. O1W is home to an additional hydrogen atom that follows the (−x, y, 1/2 − z) symmetry. With

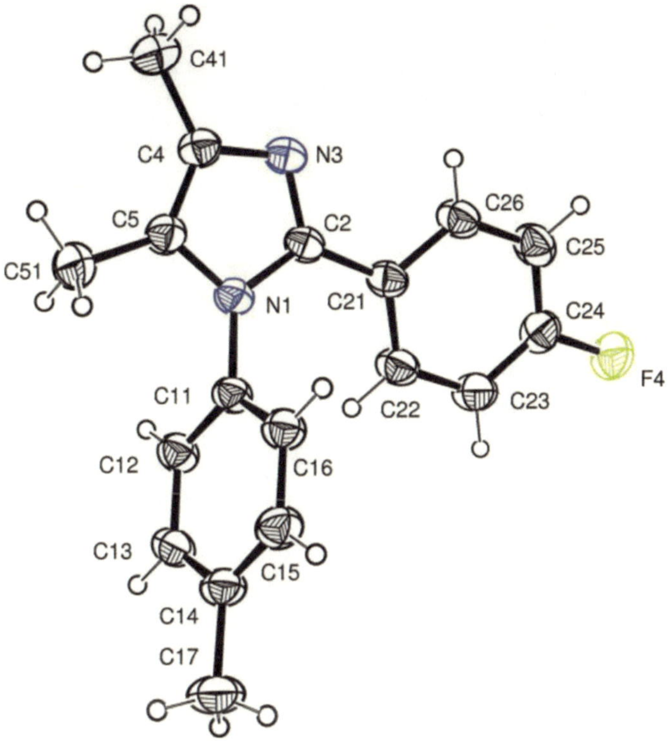

Figure 6.
The thermal displacement ellipsoid plot [20] is shown at a 30% probability level.

C—H - 0.95–0.98 Å; $U_{iso}(H)$ = $kU_{eq}(C)$, where k = 1.5 for $—CH_3$ and 1.2 for the other H atoms, the residual H atoms were oriented geometrically and acknowledged riding on their parent atoms.

In this compound (**Figure 8**), with a maximum deviation of 0.005 (1) Å for N3, the imidazole ring is planar. The imidazole ring makes dihedral angles of 67.46 (10) and 23.10 (11)° with the (C11—C16) and (C21—C26), respectively. 68.22 (10)° is the dihedral angle between the benzene and phenyl rings. Intermolecular O1W—H1W···N3 (−x, y, −z + 1/2) and C12—H12···N3 (−x, −y + 1, −z) hydrogen bonds are found in the crystal structure (**Figure 9**).

The ORTEP-3 for Windows drew the thermal displacement ellipsoid plot with a 30% probability level [20] (**Figure 8**). The crystal packing using the PLATON [21] was viewed along the b-axis (**Figure 9**).

3.5 Structural analysis of 1-(3,5-dimethylphenyl)-2-(4-fluorophenyl)-4,5-dimethyl-1*H*-imidazole (DMPFPDMI)

This section depicts the structure details of DMPFPDMI [14]. SHELXS97 [19] solved the crystal structure. The program SHELXL97 [19] refines the final crystal structure model.

This compound belongs to the triclinic crystal system with the space group P$\bar{1}$. Molecular formula: $C_{19}H_{19}FN_2$; molecular weight: 294.36; crystal data: a = 8.4226 (10) Å; b = 9.5572 (10) Å; c = 11.0351 (11) Å; α = 105.423 (9)°; β = 105.677 (9)°; γ = 95.781

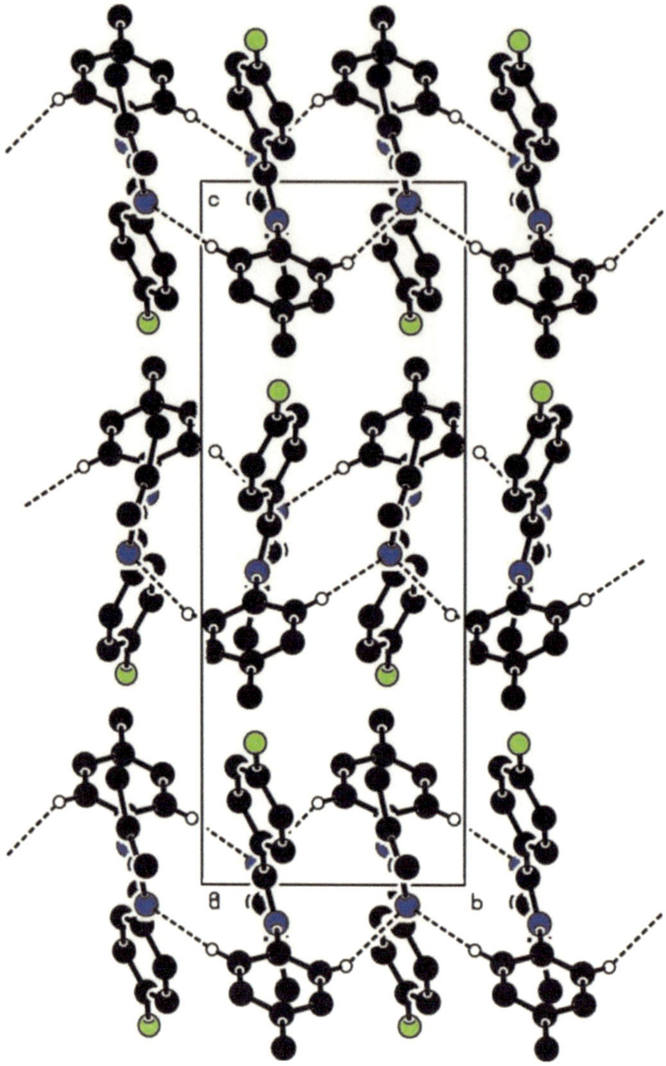

Figure 7.
The crystal packing with hydrogen bonds [21], viewed along the a-axis.

(9)°; V = 810.07 (17) Å^3; Z = 2; D_x = 1.207 Mg m^{-3}; F(000) = 312; final R [$F^2 > 2\sigma(F^2)$] = 0.052 & wR(F^2) = 0.159 for 2771 reflections observed with I > 2σ(I).

All of the H atoms admitted to riding on their parent atoms and were directed geometrically, with C—H = 0.95–0.98 Å; $U_{iso}(H) = kU_{eq}(C)$, where k = 1.5 for —CH_3 and 1.2 for the other H atoms.

In this compound (**Figure 10**), with a maximum deviation of 0.0015 (9) Å for C4, the imidazole ring is planar. The imidazole ring makes dihedral angles of 77.61 (9) and 26.93 (10)° with the (C11—C16) and (C21—C26) rings, respectively. 78.84 (8)° is the dihedral angle among the (C11—C16) and (C21—C26) rings. A C12—H12···π (−x, −y, −z + 1) interaction involving the (C21—C26) ring stabilizes the crystal.

The ORTEP-3 for Windows drew the thermal displacement ellipsoid plot with a 30% probability level [20] (**Figure 10**).

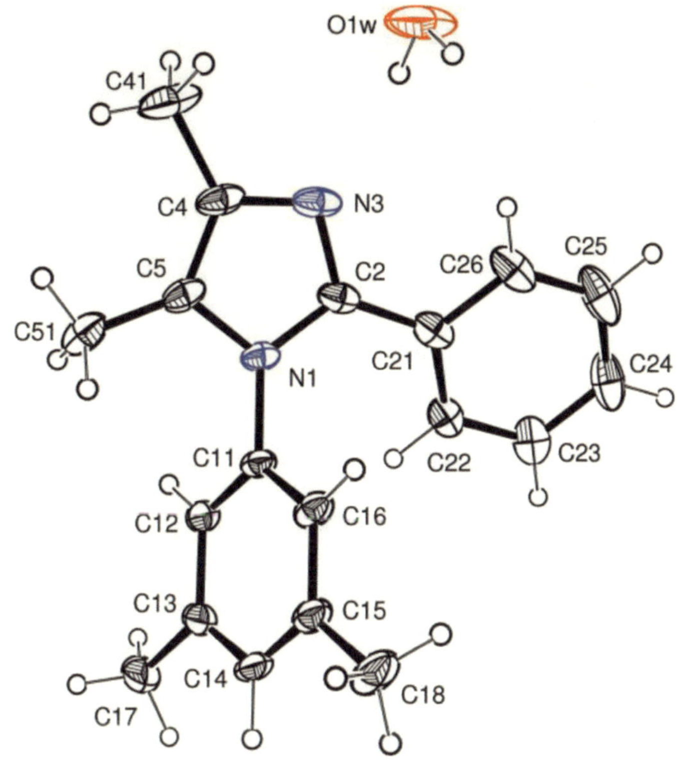

Figure 8.
The thermal displacement ellipsoid plot [20] is shown at a 30% probability level.

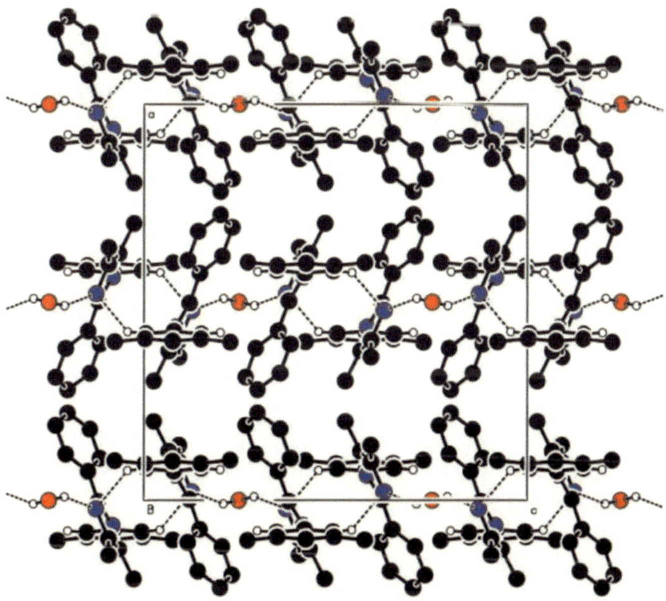

Figure 9.
The crystal packing with hydrogen bonds [21] is viewed along the b-axis.

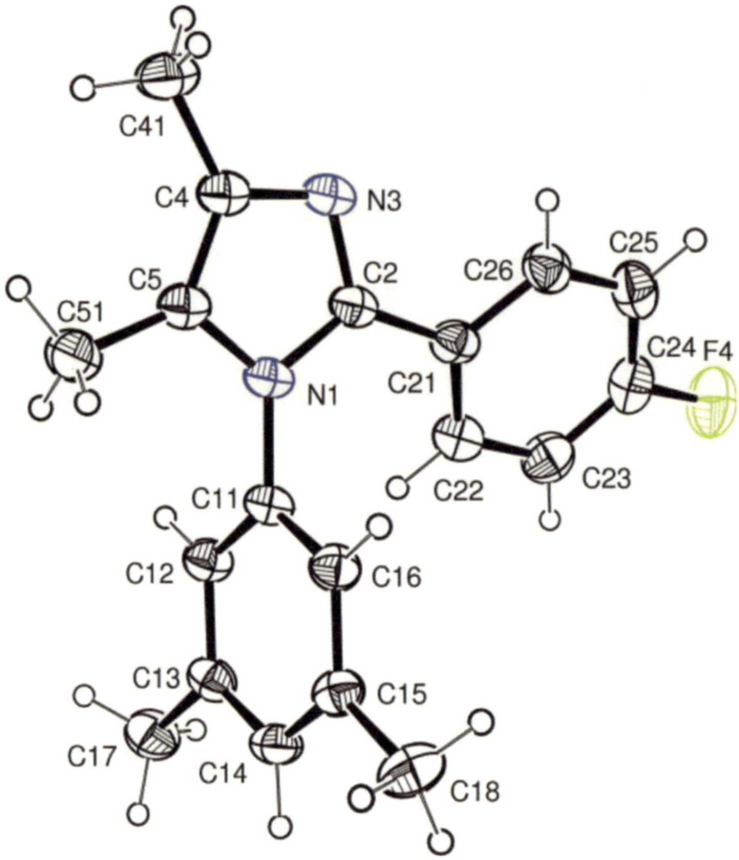

Figure 10.
The thermal displacement ellipsoid plot [20] is shown at a 30% probability level.

3.6 Structural analysis of 2-(4-fluorophenyl)-1-(4-methoxyphenyl)-4,5-dimethyl-1*H*-imidazole (FPMOPDMI)

This section depicts the structure details of FPMOPDMI [15]. SHELXS97 [19] solved the crystal structure. The program SHELXL97 [19] refines the final crystal structure model.

This compound belongs to the monoclinic crystal system with the space group $P2_1/c$. Molecular formula: $C_{18}H_{17}FN_2O$; molecular weight: 296.34; crystal data: a = 8.5132 (1) Å; b = 9.5128 (2) Å; c = 19.2610 (3) Å and β = 96.798 (2)°; V = 1548.87 (4) Å^3; Z = 4; D_x = 1.271 Mg m^{-3}; F(000) = 624; final $R[F^2 > 2\sigma(F^2)] = 0.041$ and $wR(F^2) = 0.130$ for 2744 reflections observed with $I > 2\sigma(I)$.

With C—H = 0.93 Å for Csp^2 and 0.96 Å for Csp^3, and $U_{iso}(H) = kU_{eq}(C)$, where k = 1.5 for —CH_3 and 1.2 for the other H atoms, all the hydrogen atoms were guided geometrically and admitted to riding on their parent atoms.

In this molecule (**Figure 11**), with a maximum deviation of 0.005 (1) Å for C2, the imidazole ring is planar. The imidazole ring makes dihedral angles of 76.46 (7)° and 40.68 (7)° with the (C11—C16) and (C21—C26) rings, respectively. 71.25 (6)° is the dihedral angle among benzene rings. No classical hydrogen bonds are found in the crystal.

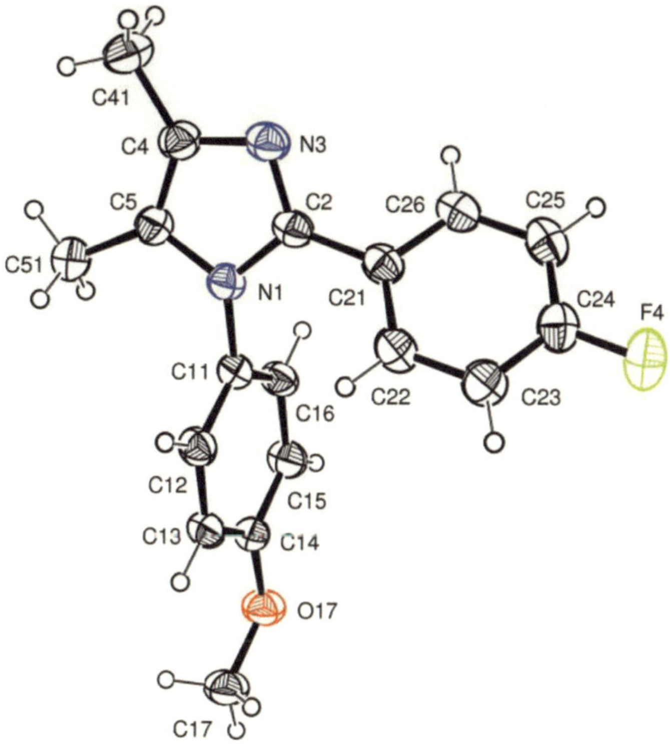

Figure 11.
The thermal displacement ellipsoid plot [20] is shown at a 30% probability level.

The ORTEP-3 for Windows drew the thermal displacement ellipsoid plot with a 30% probability level (**Figure 11**) [20].

3.7 Structural analysis of 1-(3,5-dimethoxyphenyl)-2-(4-fluorophenyl)-4,5-dimethyl-1*H*-imidazole (DMOPFPDMI)

This section depicts the structure details of DMOPFPDMI [16]. SHELXS97 [19] solved the crystal structure. The program SHELXL97 [19] refines the final crystal structure model.

This compound belongs to the monoclinic crystal system with the space group $P2_1/n$. Molecular formula: $C_{19}H_{19}FN_2O_2$; molecular weight: 326.36; crystal data: a = 6.9654 (1) Å; b = 17.8520 (3) Å; c = 13.7121 (3) Å; β = 97.833 (2)°; V = 1689.14 (5) Å^3; Z = 4; D_x = 1.283 Mg m^{-3}; F(000) = 688; final $R[F^2 > 2\sigma(F^2)] = 0.046$ and $wR(F^2) = 0.138$ for 2723 reflections observed with $I > 2\sigma(I)$.

The H atoms were all geometrically directed, with C—H = 0.93–0.96 Å; $U_{iso}(H) = kU_{eq}(C)$, where k = 1.5 for $—CH_3$ and 1.2 for the remaining H atoms. All of the H atoms are acknowledged to ride on their parent atoms.

In this compound, in **Figure 12**, with a maximum deviation of 0.0030 (8) Å for C4, the imidazole ring is planar. The imidazole ring makes dihedral angles of 66.45 (7) and 29.98 (7)° with the (C11—C16) and (C21—C26) rings, respectively. The dihedral angle among the benzene rings is 64.79 (7)°. A C23—H23···π (x − 1/2, −y + 1/2, z − 1/2) interaction involving the imidazole (N1/C2/N3/C4/C5) ring is

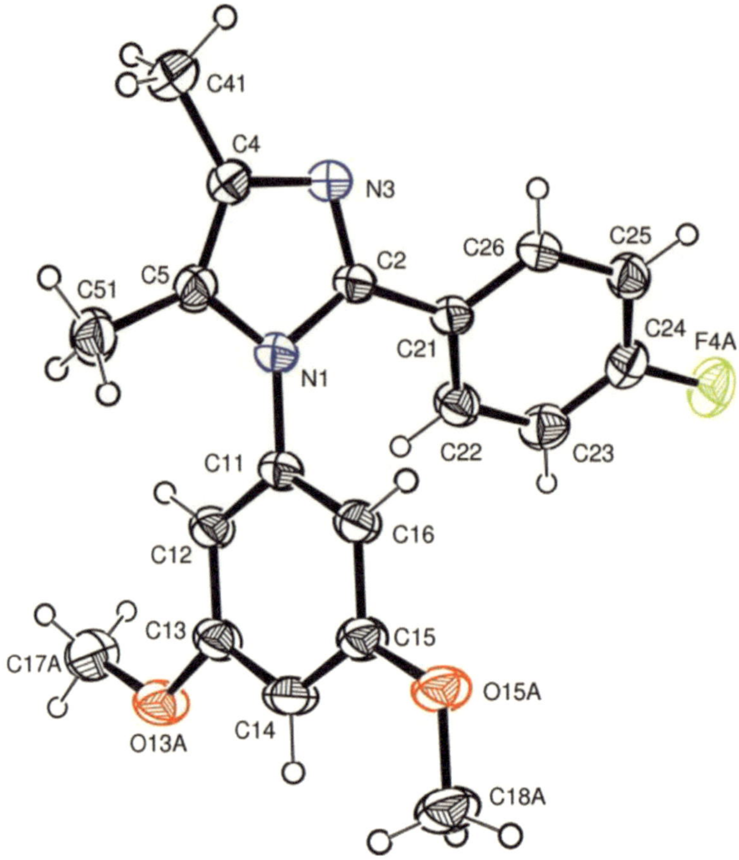

Figure 12.
The thermal displacement ellipsoid plot [20] is shown at a 30% probability level. The minor component of —O—CH$_3$ and F atoms was omitted for clarity.

found in the crystal structure. The —O—CH$_3$ groups were disordered [occupancy factors: 0.803 (4) & 0.197 (4)]. The F atom is disordered [occupancy factors: 0.929 (4) & 0.071 (4)].

The ORTEP-3 for Windows drew the thermal displacement ellipsoid plot with a 30% probability level (**Figure 12**) [20].

3.8 Structural analysis of 2-(4-fluorophenyl)-1,4,5-triphenyl-1*H*-imidazole (FPTPI)

This section depicts the structure details of FPTPI [17]. SHELXS86 [19] solved the crystal structure. The program SHELXL97 [19] refines the final crystal structure model.

This compound belongs to the triclinic crystal system with the space group P$\bar{1}$. Molecular formula: $C_{27}H_{19}FN_2$; molecular weight: 390.44; crystal data: a = 10.1794 (5) Å; b = 10.5239 (6) Å; c = 10.6175 (6) Å; α = 80.750 (5)°; β = 85.776 (4)°; γ = 67.348 (5)°; V = 1035.95 (11) Å^3; Z = 2; D_x = 1.252 Mg m^{-3}; F(000) = 408; final R [$F^2 > 2\sigma(F^2)$] = 0.049 and wR(F^2) = 0.128 for 3489 reflections observed with I > 2σ(I).

All the H atoms were directed geometrically and admitted to riding on their parent atoms with C—H = 0.93 Å and $U_{iso}(H) = 1.2U_{eq}$ (parent atom).

DOI: http://dx.doi.org/10.5772/intechopen.108488

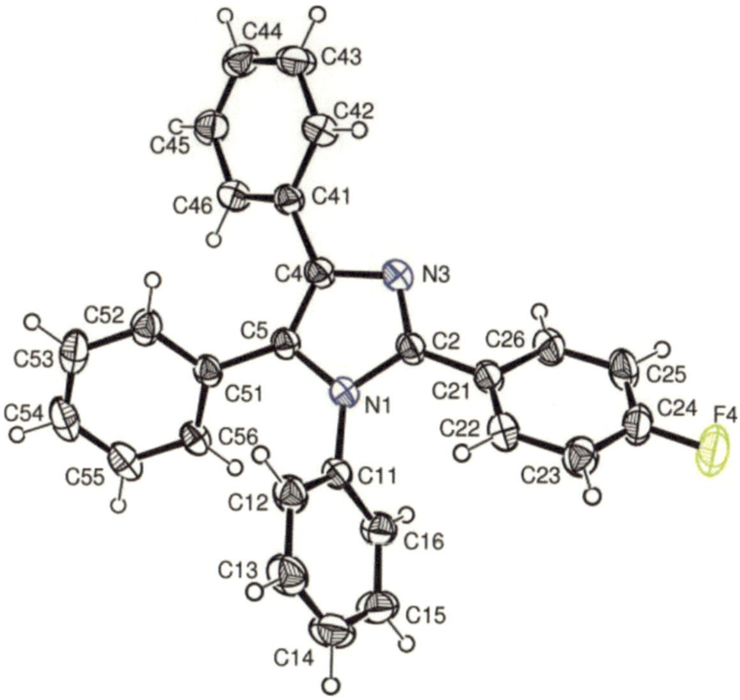

Figure 13.
The thermal displacement ellipsoid plot [20] is shown at a 30% probability level.

In **Figure 13**, the imidazole ring is planar with a maximum deviation = 0.004 (1) Å for N1. The dihedral angles made by the imidazole ring with the (C11—C16), (C21—C26), (C41—C46), and (C51—C56) rings are 62.80 (6), 36.98 (6), 33.16 (6) and 46.24 (6)°, respectively. The dihedral angles made by the (C11—C16) ring are 54.26 (6), 85.21 (7), and 65.02 (6) ° with the fluorophenyl, (C41—C46) and (C51—C56) rings attached to C4 and C5, respectively. The (C41—C46) ring makes the dihedral angle of 51.10 (6)° with the (C51—C56) ring. The crystal exhibits no conventional hydrogen bond interactions.

The ORTEP-3 for Windows drew the thermal displacement ellipsoid plot with a 30% probability level [20] (**Figure 13**).

4. Comparative structural study of the eight imidazole derivatives

Section 3 presented the single crystal X-ray structure analyses of eight closely related organic imidazole compounds. All the eight structures have the imidazole core essentially as the basic skeleton, with different groups (DMDPIMH: —CH_3, —CH_3, —C_6H_5, —C_6H_5, with H_2O, DMPPTI: —CH_3, —CH_3, —C_6H_5, —C_6H_4—CH_3, FPDMMPI: —CH_3, —CH_3, —C_6H_4—F, —C_6H_4—CH_3, DMPDMPIHH: —CH_3, —CH_3, —C_6H_5, —C_6H_3—2(CH_3), with 0.5H_2O, DMPFPDMI: —CH_3, —CH_3, —C_6H_4—F, —C_6H_3—2(CH_3), FPMOPDMI: —CH_3, —CH_3, —C_6H_4—F, —C_6H_4—O—CH_3, DMOPFPDMI: —CH_3, —CH_3, —C_6H_4—F, —C_6H_3—2(O—CH_3), FPTPI: —C_6H_5, —C_6H_5, —C_6H_4—F, —C_6H_5) as different substituents. The structure elucidation of the compounds has revealed a few features, such as (1) the hydrogen bonds: O—H ... N, O—H ... O, C—H ... N; and (2) the interactions C—H ... π.

C—H ... π interactions: Weak acids and soft bases interact. A particular hydrogen bond between two such systems is called a C—H ... π interaction. The interactions between aromatic acceptors and C—H aliphatic donors and those between aromatic acceptors and C—H donors are the two most prominent. The non-covalent connections that surround these systems are referred to in chemistry as "π interactions" or "π effects" [9].

Related structures: On September 11, 2022, the terms **imidazole** and **1*H*-imidazole** on IUCr Journals Crystallography Journals Online (https://journals.iucr.org/) returned 6276 and 290 articles, respectively, matching the searches. The IUCr Journals' paper reference codes: tk2693 (4,5-dimethyl-1,2-diphenyl-1*H*-imidazole monohydrate), rk2234 (4,5-dimethyl-2-phenyl-1-(p-tolyl)-1*H*-imidazole), si2269 (2-(4-fluorophenyl)-4,5-dimethyl-1-(4- methylphenyl)-1*H*-imidazole), hg2722 (1-(3,5-dimethylphenyl)-4,5-dimethyl-2-phenyl-1*H*-imidazole hemihydrate), hg5018 (1-(3,5-dimethylphenyl)-2-(4-fluorophenyl)-4,5-dimethyl-1*H*-imidazole), wn2396 (2-(4-fluorophenyl)-1-(4-methoxyphenyl)-4,5-dimethyl-1*H*-imidazole), hg5156 (1-(3,5-dimethoxyphenyl)-2-(4-fluorophenyl)-4,5-dimethyl-1*H*-imidazole), and tk2707 (2-(4-fluorophenyl)-1,4,5-triphenyl-1H-imidazole), affirm that the geometry of the five-membered imidazole cores are very like in all the described structures.

Database survey: The search using the CCDC ConQuest (Version 2022.2.0) in the Cambridge Structural Database (CSD, Version 5.43, update November 2021) [23] for the compound name **imidazole** gave 21,489 hits, and for the compound name **1*H*-imidazole** gave 9198 hits. A single query on 2D diagram search using the CCDC ConQuest (Version 2022.2.0) gave 142 hits for DMDPIMH (including the CSD Refcode: SUYRUA), 45 hits for DMPPTI (including the CSD Refcode: VABBUX), 5 hits for FPDMMPI (including the CSD Refcode: QUZLED), 25 hits for DMPDMPIHH (including the CSD Refcode: VUZSOZ), one hit for DMPFPDMI (with the CSD Refcode: URALEF), two hits for FPMOPDMI (including the CSD Refcode: YUQBUI), three hits for DMOPFPDMI (including the CSD Refcode: GAPGEL), and six hits for FPTPI (including the CSD Refcode: KUZFUH).

5. Conclusions

The synthesis and single-crystal X-ray structure characterization of new, carefully chosen imidazole derivatives of biological, pharmacological, and technological relevance are covered in the preceding sections. The geometry of these eight imidazole derivatives (bond lengths, bond angles, torsion angles, and dihedral angles between the least-squares planes) shows that the imidazole is essentially planar in the compounds DMDPIMH, DMPPTI, FPDMMPI, DMPDMPIHH, DMPFPDMI, FPMOPDMI, DMOPFPDMI, and FPTPI as expected and as shown by the most recent literature survey ((https://journals.iucr.org/) (CSD, Version 5.43, update November 2021). The imidazole skeleton has a planer five-membered heterocyclic ring, according to the current X-ray analysis. The fundamental geometrical measurements of the imidazole core in the DMDPIMH molecule agree with those found in other closely related imidazole derivatives. Around the imidazole rings, all of the substituent groups are at their predicted locations. The molecular structure and atom connectivity of the aforementioned compounds, as seen in **Figures 2, 4, 6, 8, 10–13**, are confirmed by the X-ray investigation. Finally, the stabilization of the crystal structures is accomplished by the O-H ... N, O—H ... O, C—H ... N, and C—H ... π contacts.

Acknowledgements

Aravazhi Amalan Thiruvalluvar acknowledges Professor R. J. Butcher and the late Professor J. P. Jasinski for their help collecting the single-crystal X-ray diffraction data and Professor J. Jayabharathi for her help in synthesizing the compounds.

Conflict of interest

There is no conflict of interest in writing this chapter.

Thanks

Aravazhi Amalan Thiruvalluvar thanks his wife Lilly for all her emotional support and help at various stages of this review writing, his son Uthaya Raj, his daughter-in-law Amudha, and his daughter Manju Princy for their loving support.

A partial crystallographic information file (CIF) for the compound DMDPIMH (only as an example)

data_I
_chemical_name_systematic
;
4,5-Dimethyl-1,2-diphenyl-1<i>H</i>-imidazole monohydrate
;
_chemical_name_common 1H-imidazole
_chemical_formula_moiety 'C17 H16 N2, H2 O'
_chemical_formula_sum 'C17 H18 N2 O'
_chemical_formula_structural 'C17 H16 N2, H2 O'
_chemical_formula_iupac 'C17 H16 N2, H2 O'
_chemical_formula_weight 266.33
_chemical_melting_point 375
_symmetry_cell_setting tetragonal
_symmetry_space_group_name_H-M 'I 41/a'
_symmetry_space_group_name_Hall '-I 4ad'
_cell_length_a 25.5498(2)
_cell_length_b 25.5498(2)
_cell_length_c 9.37920(10)
_cell_angle_alpha 90
_cell_angle_beta 90
_cell_angle_gamma 90
_cell_volume 6122.67(9)
_cell_formula_units_Z 16
_cell_measurement_reflns_used 4027
_cell_measurement_theta_min 4.8917
_cell_measurement_theta_max 77.3788
_cell_measurement_temperature 295(2)
_cell_special_details ?

_exptl_crystal_description 'irregular plate'
_exptl_crystal_colour 'colourless'
_exptl_crystal_size_max 0.53
_exptl_crystal_size_mid 0.42
_exptl_crystal_size_min 0.18
_exptl_crystal_size_rad ?
_exptl_crystal_density_diffrn 1.156
_exptl_crystal_density_meas ?
_exptl_crystal_density_method 'Not Measured'
_exptl_crystal_F_000 2272
_exptl_absorpt_coefficient_mu 0.572
_exptl_absorpt_correction_type multi-scan
_exptl_absorpt_process_details
;
(CrysAlis Pro; Oxford Diffraction, 2010)
;
_exptl_absorpt_correction_T_min 0.80515
_exptl_absorpt_correction_T_max 1.00000
_diffrn_ambient_temperature 295(2)
_diffrn_radiation_type 'CuK\a'
_diffrn_radiation_wavelength 1.54184
_diffrn_radiation_source 'Enhance (Cu) X-ray Source'
_diffrn_radiation_monochromator graphite
_diffrn_measurement_device_type 'Oxford Diffraction Xcalibur Ruby Gemini'
_diffrn_measurement_method '\w scans'
_diffrn_detector_area_resol_mean 10.5081
_diffrn_reflns_number 8078
_diffrn_reflns_av_R_equivalents 0.0190
_diffrn_reflns_av_sigmaI/netI 0.0162
_diffrn_reflns_theta_min 4.90
_diffrn_reflns_theta_max 77.60
_diffrn_reflns_theta_full 67.50
_diffrn_measured_fraction_theta_max 0.953
_diffrn_measured_fraction_theta_full 0.971

Author details

Aravazhi Amalan Thiruvalluvar
Kunthavai Naacchiyaar Government Arts College for Women (Autonomous), Thanjavur, Tamil Nadu, India

*Address all correspondence to: thiruvalluvar.a@gmail.com

References

[1] Wikipedia Contributors. Imidazole. Wikipedia, The Free Encyclopedia. Available from: https://en.wikipedia.org/w/index.php?title=Imidazole&oldid=1154797754 [Accessed June 20, 2023]

[2] Jayabharathi J, Thanikachalam V, Srinivasan N, Perumal MV. Fluorescence spectral studies of some imidazole derivatives. Spectrochimica Acta Part A: Molecular and Biomolecular Spectroscopy. 2012;**90**:125-130. ISSN 1386-1425. DOI: 10.1016/j.saa.2012.01.030. Available from: https://www.sciencedirect.com/science/article/pii/S1386142512000431

[3] Jayabharathi J, Thanikachalam V, Perumal MV, Srinivasan N. Fluorescence resonance energy transfer from a bio-active imidazole derivative 2-(1-phenyl-1H-imidazo[4,5-f][1,10]phenanthrolin-2-yl)phenol to a bioactive indoloquinolizine system. Spectrochimica Acta Part A: Molecular and Biomolecular Spectroscopy. 2011;**79**(1):236-244. ISSN 1386-1425. DOI: 10.1016/j.saa.2011.02.049. Available from: https://www.sciencedirect.com/science/article/pii/S1386142511001429

[4] Saravanan K, Srinivasan N, Thanikachalam V, et al. Synthesis and photophysics of some novel imidazole derivatives used as sensitive fluorescent chemisensors. Journal of Fluorescence. 2011;**21**:65-80. DOI: 10.1007/s10895-010-0690-5

[5] Erdogdu Y, Güllüoğlu MT, Yurdakul, S. et al. DFT simulations, FT-IR, FT-raman, and FT-NMR spectra of 4-(4-chlorophenyl)-1H-imidazole molecules. Optics and Spectroscopy. 2012;**113**:23-32. DOI: 10.1134/S0030400X12070089

[6] Kanakaraju S, Prasanna B, Venkatanarayana P, et al. An efficient one-pot three-component synthesis of imidazol-2-ones using [Bmim]BF4 ionic liquid. Journal of the Iranian Chemical Society. 2012;**9**:933-937. DOI: 10.1007/s13738-012-0110-7

[7] Verma A, Joshi s, Singh D. Imidazole: Having versatile biological activities. Journal of Chemistry. 2013;**2013**:329412. DOI: 10.1155/2013/329412

[8] Rosepriya S, Thiruvalluvar A, Saravanan K, Jayabharathi J, Butcher R.J. 1-(3,5-Dimethoxyphenyl)-2-(4-fluorophenyl)-4,5-dimethyl-1H-imidazole. Acta Cryst. 2012;**E68**:o256. DOI: 10.1107/S1600536811055012

[9] Amalan Thiruvalluvar A, Vasuki G, Jayabharathi J, Rosepriya S. X-Ray crystal structure analysis of selected benzimidazole derivatives [Internet]. Chemistry and applications of benzimidazole and its derivatives. IntechOpen. 2019. DOI: 10.5772/intechopen.85291

[10] Gayathri P, Thiruvalluvar A, Saravanan K, Jayabharathi J, Butcher RJ. 4,5-Dimethyl-1,2-diphenyl-1*H*-imidazole monohydrate. Acta Crystallographica Section E: Structure Reports Online. 2010;**E66**:o2219

[11] Gayathri P, Thiruvalluvar A, Srinivasan N, Jayabharathi J, Butcher RJ. 4,5-Dimethyl-2-phenyl-1-(p-tolyl)-1*H*-imidazole. Acta Crystallographica Section E: Structure Reports Online. 2010;**E66**:o2826

[12] Gayathri P, Jayabharathi J, Srinivasan N, Thiruvalluvar A, Butcher RJ. 2-(4-Fluorophenyl)-4,5-dimethyl-1-(4- methylphenyl)-1*H*-imidazole. Acta Crystallographica Section E: Structure Reports Online. 2010;**E66**:o1703

[13] Gayathri P, Thiruvalluvar A, Srinivasan N, Jayabharathi J, Butcher RJ. 1-(3,5-Dimethylphenyl)-4,5-dimethyl-2-phenyl-1*H*-imidazole hemihydrate. Acta Crystallographica Section E: Structure Reports Online. 2010;**E66**:o2776

[14] Rosepriya S, Thiruvalluvar A, Jayabharathi J, Srinivasan N, Butcher RJ, Jasinski JP, et al. 1-(3,5-Dimethylphenyl)-2-(4-fluorophenyl)-4,5-dimethyl-1*H*-imidazole. Acta Crystallographica Section E: Structure Reports Online. 2011;**E67**:o1065

[15] Gayathri P, Jayabharathi J, Saravanan K, Thiruvalluvar A, Butcher RJ. 2-(4-Fluorophenyl)-1-(4-methoxyphenyl)-4,5-dimethyl-1*H*-imidazole. Acta Crystallographica Section E: Structure Reports Online. 2010;**E66**:o1791

[16] Rosepriya S, Thiruvalluvar A, Saravanan K, Jayabharathi J, Butcher RJ. 1-(3,5-Dimethoxyphenyl)-2-(4-fluorophenyl)-4,5-dimethyl-1*H*-imidazole. Acta Crystallographica Section E: Structure Reports Online. 2012;**E68**:o256

[17] Gayathri P, Thiruvalluvar A, Srinivasan N, Jayabharathi J, Butcher RJ. 2-(4-Fluorophenyl)-1,4,5-triphenyl-1*H*-imidazole. Acta Crystallographica Section E: Structure Reports Online. 2010;**E66**:o2519

[18] Burla MC, Caliandro R, Camalli M, Carrozzini B, Cascarano GL, De Caro L, et al. SIR2004: An improved tool for crystal structure determination and refinement. Journal of Applied Crystallography. 2005;**38**:381-388

[19] Sheldrick GM. A short history of SHELX. Acta Crystallographica Section A Foundations and Advances. 2008;**A64**:112-122

[20] Farrugia LJ. ORTEP-3 for windows - a version of ORTEP-III with a graphical user Interface (GUI). Journal of Applied Crystallography. 1997;**30**:565

[21] Spek AL. Structure validation in chemical crystallography. Acta Crystallographica Section D Structural Biology. 2009;**D65**:148-155

[22] Burla MC, Camalli M, Carrozzini B, Cascarano GL, Giacovazzo C, Polidori G, et al. SIR2002: The program. Journal of Applied Crystallography. 2003;**36**:1103

[23] Groom CR, Bruno IJ, Lightfoot MP, Ward SC. The Cambridge structural database. Acta Crystallographica Section B: Structural Science, Crystal Engineering and Materials. 2016;**B72**:171-179

Chapter 6

Synthesis of Five-Membered Heterocycles Using Water as a Solvent

Abha Kathuria, Anwar Jahan and Poonam Dwivedi

Abstract

Water owing to its modest cost, easy accessibility, and non-toxic and non-flammable attributes has been considered one of the most ideal and promising solvents in organic synthesis, especially from the green and sustainable point of view. Furthermore, considering the immense enzyme-mediated biosynthesis in nature, water serves as a favourable medium for the versatile synthesis of a wide variety of complicated molecules and compounds. Over the past decades, considerable efforts have been deployed in conducting organic reactions using water as a solvent. In recent years, more and more general organic reactions were successfully exploited to perform in water instead of organic solvents to achieve sustainable and environmental benefits.

Keywords: water, furan, pyrrole, thiophene, green synthesis, five-membered heterocycles

1. Introduction

Among pharmaceutically relevant natural and synthetic compounds, heterocyclic compounds hold a unique place. They have an incredible ability to serve as biomimetics and potential pharmacophores and are standard key components of several drugs [1, 2]. There is a critical need for new small organic molecules both in lead identification and lead optimization processes. Conventional methods of organic synthesis are often too slow to satisfy the demand for generation of such compounds. The synthetic chemists worldwide have been under tremendous pressure to produce them in good numbers and that too in an environmentally benign fashion. A typical reaction involves a reaction medium apart from reactants and reagents to form products. There are several issues that influence the choice of solvent, especially in context of green chemistry. It should be relatively non-reactive, non-toxic and non-hazardous. The solvent should also be contained, that is, it should not be released into the environment [3].

Water is one of the most intriguing media among all the alternative solvents available, owing to its peculiar properties. It is the most abundant and available molecule on the planet and many biochemical processes occur in aqueous medium. Despite its accessibility, it has not been a favourite reaction medium for organic chemists because its presence causes the decomposition of organometallic reagents, which are used preferably in dry organic solvents. In fact, water has been generally used to work-up organic reactions and, therefore, it has been associated with a waste-production step and to the consequent obvious problems of cleaning-up water from reactants' residues arises. The low solubility of most organic compounds in water (hampering their reactivity), and the instability of many intermediates and catalysts in water are perhaps the two main causes of this problem. Today, it is a different scenario altogether. This very aqueous medium has captured the interest of organic chemists and many others, so much so that many surprising discoveries have been attributed to its use. Many organic reactions such as pericyclic, condensation, oxidation and reduction reactions can be conducted efficiently in aqueous medium and, in some cases, water is necessary to enhance the reaction rate and increase selectivity. Further, water-tolerant catalysts have been prepared that allow organometallic reactions to be carried out in aqueous medium. Reactions that were earlier thought impossible in water have today become a certainty [3].

1.1 Evolution of water as a solvent

Although Diels and Alder themselves used water as reaction medium in the cycloaddition of furan with maleic anhydride in 1931, but it was Breslow's observations (in 1980) confirming the acceleration of some Diels-Alder reaction when carried out in water with respect to other organic solvents that played a pivotal role in the development of organic synthesis in aqueous media [4, 5]. Prior to this, in 1948, Woodward and Baer also employed aqueous maleic acid as dienophile, and in 1973 the beneficial effect of aqueous medium on the reaction was also successfully investigated by Koning and Carlson. In fact, Breslow's kinetic work was the first that quantitatively showed the valuable effects of water on the reactivity and selectivity of an organic reaction. Herein, the reaction of butanone with cyclopentadiene in water was enhanced by 740 times as compared to that carried out in isooctane [6–8]. Though, there are numerous advantages of using an aqueous reaction media. Among them, the possibility of recovery and reuse of the aqueous medium containing all the soluble species dissolved in it (catalysts and reactants), and the possibility of controlling the pH is most prominent.

1.2 Water as a solvent in the synthesis of heterocycles

Water plays a dominant role as a solvent for the development of numerous heterocycles by replacing the use of hazardous organic solvents. Despite its overwhelming advantages as reaction solvent for organic synthesis, poor solubility of many organic compounds in water limits its use under standard conditions. However, when used in high temperature and pressure conditions, it undergoes a change in its properties and can act as a potent solvent for good number of organic reactions [9]. Owing to tremendous potential of water as a medium for production of heterocyclic moieties, in this chapter, we have tried to compile the 'in water' synthesis of five-membered heterocycles.

2. Synthesis of nitrogen-containing heterocycles

Nitrogen heterocycles are most abundant in nature and are of immense significance. Their structural subunits exist in many natural products, such as vitamins, hormones, antibiotics and alkaloids, as well as pharmaceuticals, herbicides, dyes and many more compounds.

2.1 Pyrroles

Tetra substituted pyrroles were synthesised by the three-component condensation reaction of acid chlorides, amino acids and dialkyl acetylenedicarboxylates, in the presence of ionic liquids, as catalysts in water at room temperature (**Figure 1**) [3].

Polysubstituted pyrroles are obtained in good yields from α-azido chalcones and 1,3-dicarbonyl compounds in water. $InCl_3$ was used as a catalyst for this regioselective transformation (**Figure 2**) [10].

Another reaction involving water as a solvent employed a catalysed amination-annulation strategy for the synthesis of 2-aminopyrrole-4-carboxylates. This work developed a $Zn(ClO_4)_2$ catalysed approach to generate pyrrole ring-formation in high yields (**Figure 3**) [11].

An efficient approach for the synthesis of polysubstituted pyrrolidin-1,2-diones was achieved by a one-pot three-component reaction of nitroarenes, formaldehyde and dialkyl acetylenedicarboxylates using indium in dilute aqueous HCl at room temperature (**Figure 4**) [12].

Pyrrolo[2,1-*a*]isoquinolines and pyrrolo[1,2-*a*]quinolines have been obtained in good to excellent yields using quinoline or isoquinoline, phenacyl bromide derivatives and activated alkynes in aqueous medium (**Figure 5**) [13].

2.2 Azines

A convenient and fast procedure for the synthesis of cycl[3.2.2]azines through a three-component reaction of 2-picoline, α-bromoacetophenone and alkyne in aqueous medium under microwave irradiation (**Figure 6**) [14].

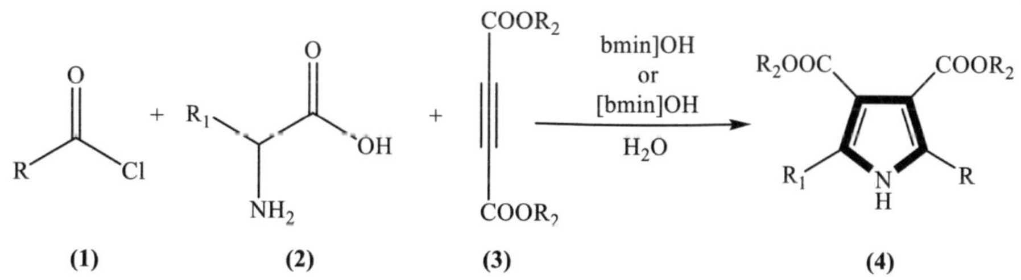

R = Me, Ph, 4-MeC$_6$H$_4$, 4-ClC$_6$H$_4$, 4-NO$_2$C$_6$H$_4$
R$_1$ = *i*-Bu, *sec*-Bu, Bu, Ph
R$_2$ = Me, Et

Figure 1.
Synthesis of tetra substituted pyrroles.

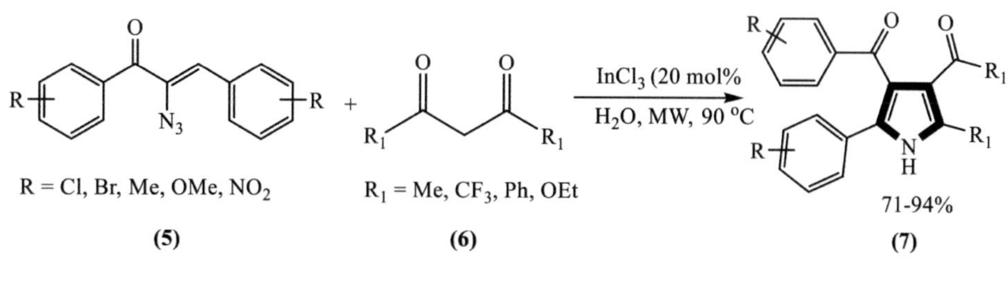

R = Cl, Br, Me, OMe, NO$_2$

R$_1$ = Me, CF$_3$, Ph, OEt

Figure 2.
Synthesis of polysubstituted pyrroles using $InCl_3$ in presence of water as a solvent.

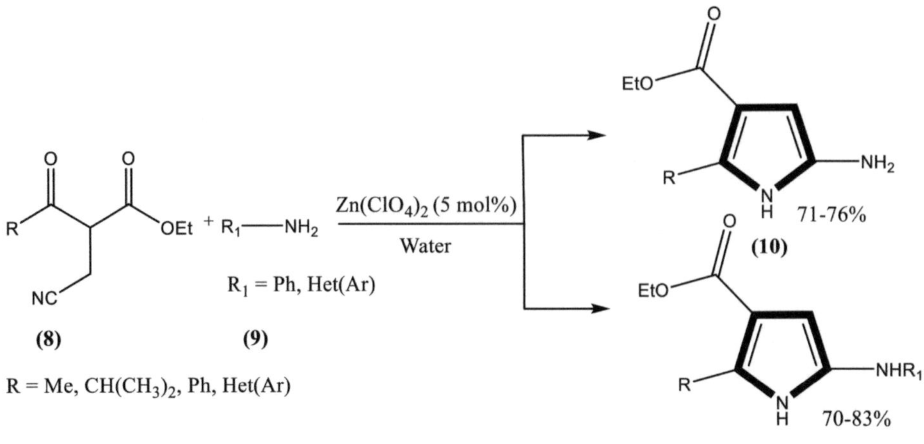

(8) (9) (10) (11)

R = Me, CH(CH$_3$)$_2$, Ph, Het(Ar)

Figure 3.
Zn(ClO$_4$)$_2$ catalysed synthesis of pyrrole in water.

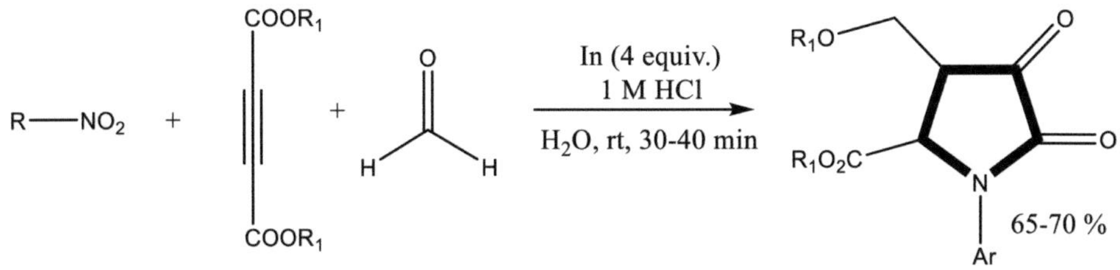

(12) (13) (14) (15)

R = Ph, Het(Ph)
R$_1$ = Me, Et

Figure 4.
Synthesis of poly-substituted pyrrolidin-1,2-diones in water.

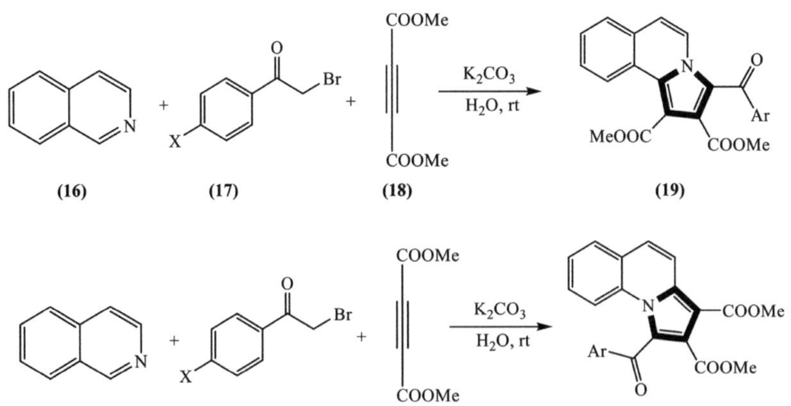

(16) (17) (18) (19)

(20) (21) (22) (23)

X = H, OMe, Br, Ph

Figure 5.
Synthesis of pyrrolo[2,1-a]isoquinolines and pyrrolo[1,2-a]quinolines.

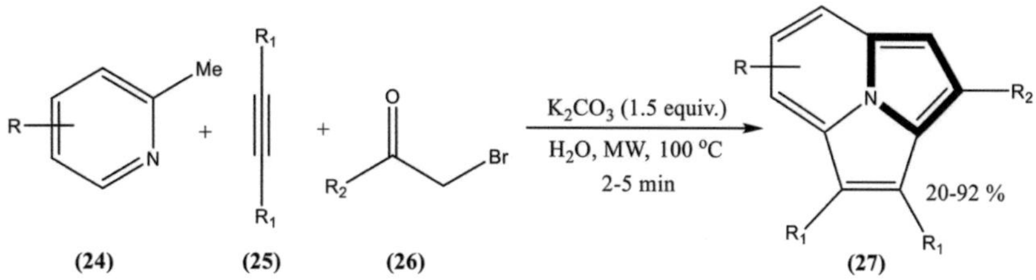

(24) (25) (26) (27)

R = H, Me
R_1 = COOMe, COOEt
R_2 = Ph, Het(Ar)

Figure 6.
Synthesis of cycl[3.2.2]azines in aqueous medium.

2.3 Indoles

2-aryl indoles were obtained by alkynylation coupling of aryl iodides with terminal alkynes catalysed by a water-soluble copper complex (sulfonato–Cu(salen)). This reaction was stirred using 2-iodoaniline and aryl acetylene to obtain 2-arylindoles in excellent yields (**Figure 7**) [15].

A greener *N*-heterocyclization with water as solvent and no transition metal catalysts showed good yields in a fraction of the time to produce isoindolines (**Figure 8**) [16].

2.4 Pyrazoles

A variety of substituted pyrazoles have been synthesised by the condensation of hydrazine with diketones and β-keto esters, respectively. One such method involves polystyrene sulfonic acid (PSSA) catalysed assembly of the above in water as a solvent (**Figure 9**) [17].

2.5 Pyrrolo pyrazines

Pyrrolo[2–*b*]pyrazines were synthesised by treating phenylacetylene with 5-(alkyl-arylamino)-6-chloropyrazine-2,3-dicarbonitriles **39**. This coupling reaction is a variant of the Larock indole synthesis and was performed in water using $Pd(Ph_3P)_2Cl_2$, CuI as a catalyst, sodium lauryl sulfate and K_2CO_3 at 70°C for 24 h (**Figure 10**) [18].

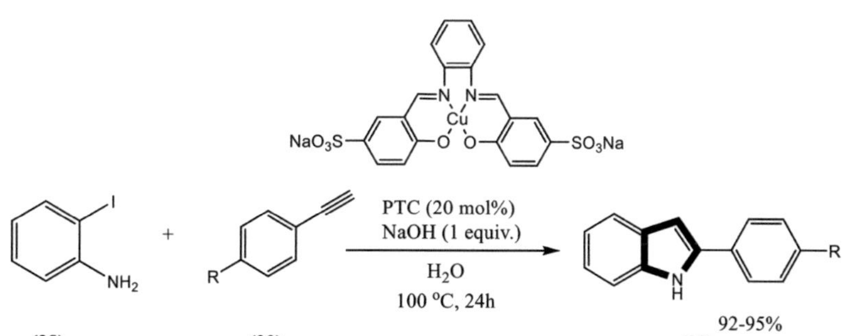

(28) (29) (30)

R = H, Me, NO_2

Figure 7.
Synthesis of 2-aryl indoles catalysed by a water-soluble copper complex (sulfonato–Cu(salen).

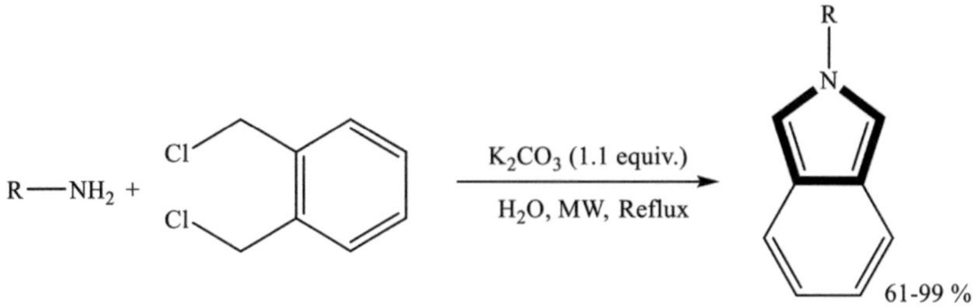

(31) (32) (33)

R = Ph, Het(Ar)

Figure 8.
Synthesis of isoindolines in aqueous medium.

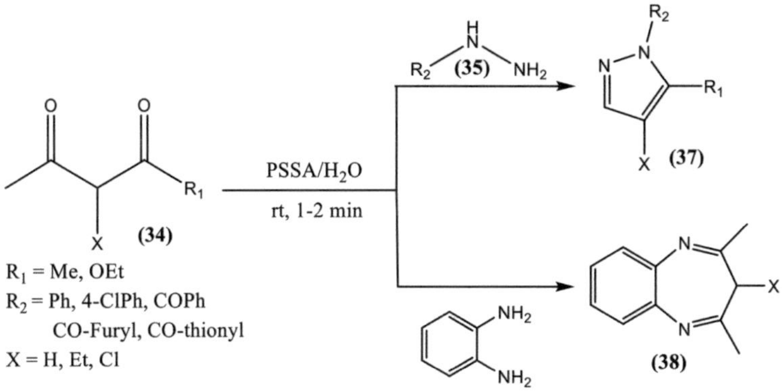

Figure 9.
Polystyrenesulphonicacid (PSSA) catalysed assembly of nitrogen-containing heterocycles in water.

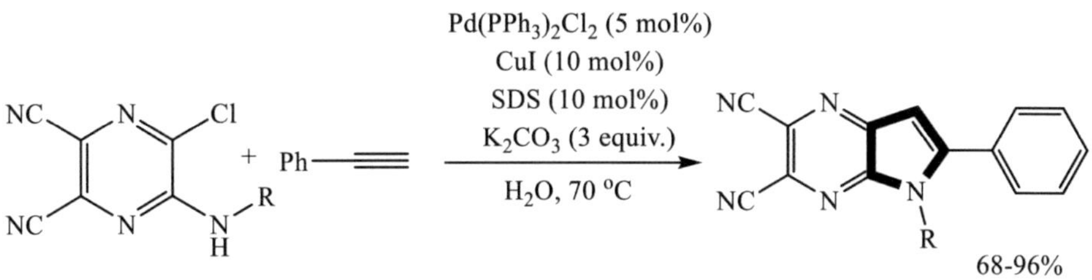

(39) (40) (41)

R = Aryl, Het(Ar)

Figure 10.
Synthesis of pyrrolo[2–b]pyrazines in water using $Pd(Ph_3P)_2Cl_2$ and CuI.

3. Synthesis of oxygen-containing heterocycles

3.1 Furans

Furan moieties are important substructures that have been found in numerous natural products, such as kailolides and combranolides. These heterocycles are

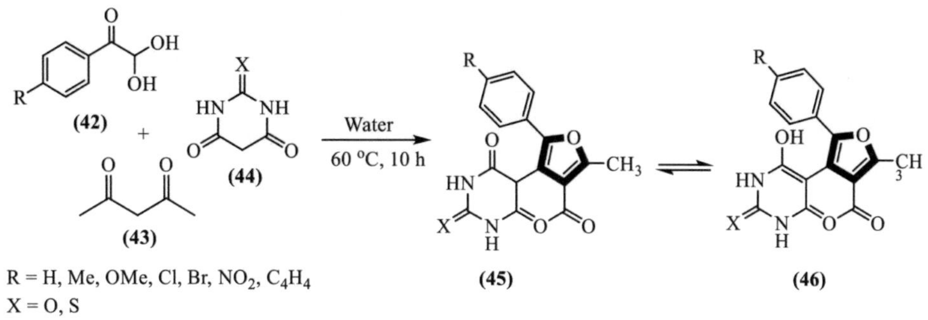

Figure 11.
Synthesis of 5-(furan-3-yl)barbiturate/thiobarbiturate derivatives.

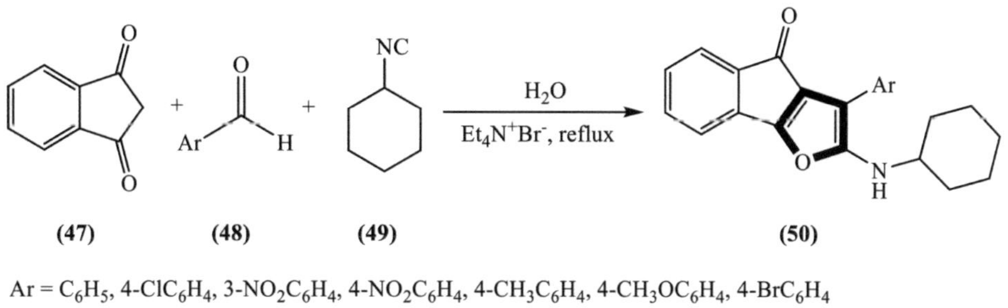

Figure 12.
Synthesis of 2-(cyclohexylamino)-3-aryl-indeno [1,2-b] furan-4-ones in water.

also found in a variety of commercial products such as pharmaceuticals, fragrances and dyes.

A simple three-component reaction between arylglyoxal monohydrates, acetylacetone and barbituric or thiobarbituric acid resulted in the synthesis of polyfunctionalized 5-(furan-3- yl)barbiturates and 5-(furan-3-yl)thiobarbiturate derivatives in good yields (**Figure 11**). The method employs readily available starting materials, neutral reaction conditions and water as an environmentally green solvent [19].

An efficient and a simple synthesis of 2-(cyclohexylamino)-3-aryl- indeno[1,2-b] furan-4-ones was attained via a one-pot three-component reaction of aldehydes, cyclohexylisocyanide and 1,3-indandione in water for 5 h in excellent yields (**Figure 12**). Water was used as a solvent to avoid the use of other highly toxic and environmentally unfavourable solvents for this synthesis [20].

4. Synthesis of Sulphur containing heterocycles

Sulphur heterocycles are also important classes of heterocycles in pharmaceuticals and organic synthesis, which are known to possess important biological properties and have attracted much attention from medicinal chemists over the years.

4.1 Thiophenes

A simple and convenient synthesis of 2-aminothiophenes was investigated by the reaction of ketones with malononitrile using water as a solvent in a reaction ignited

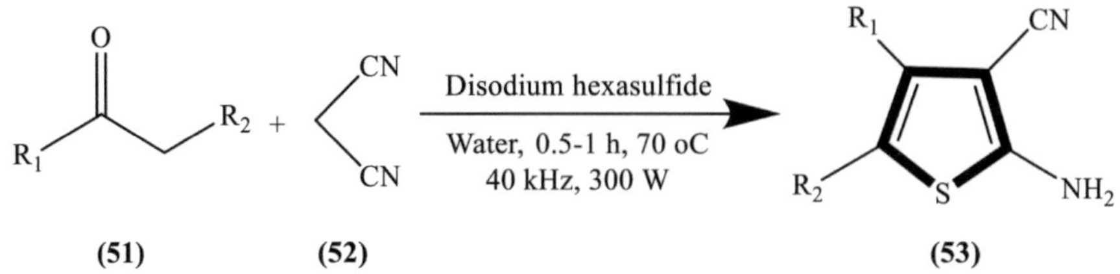

R_1 = -$(CH_2)_3$, -$(CH_2)_4$, Ph, Tolyl
R_2 = -$(CH_2)_3$-, -$(CH_2)_4$-, Me, Et, *i*-Bu, *i*-Pr, *n*-Pr, *n*-Bu, Ph. Bn

Figure 13.
Synthesis of 2-aminothiophenes from ketones and malonodinitrile in water.

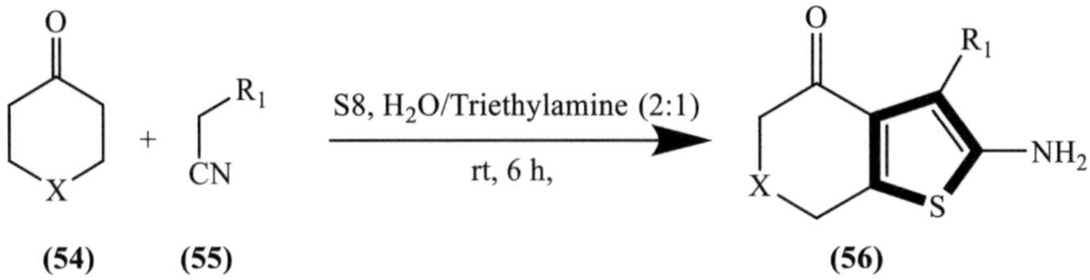

R_1 = CN, COOEt
X = S, O, NH, H, CH_2, -$(CH_2)_2$-, -$(CH_2)_3$-

Figure 14.
Synthesis of 2-aminothiophenes from ketones and nitriles.

by sodium polysulfides in catalyst-free conditions under ultrasound activation with 42–90% product yields (**Figure 13**). The use of water as a solvent as well as sonication activation is part of green chemistry principles for this protocol [21].

A simple and effective reaction of ketones with nitriles and elemental sulphur (S8) in a mixture of triethylamine and water at room temperature led to the desired 2-aminothiophenes in 75–98% yields after recrystallization from a mixture of ethyl acetate and hexanes (**Figure 14**) [22]. The use of water as solvent at room temperature with a good atom economy makes the reaction green compatible.

5. Miscellaneous

Substituted 2-aminothiazoles were prepared in water through a highly efficient and facile synthesis, without the use of a catalyst or co-organic solvent. The reaction was carried out at ambient temperature and the products were obtained in excellent yields. The developed protocol is successfully utilised for the preparation of an anti-inflammatory drug, fanetizole (**Figure 15**) [23].

A new and facile protocol for the synthesis of 2-amino-1,3,4-thiadiazoles in water by the reaction of acid hydrazides with dithiocarbamates in moderate to excellent yields (**Figure 16**) [24].

The reaction between TosMIC **63** and benzaldehyde **64** was carried out in the presence of various amounts of Et_3N and β-CD (β-Cyclodextrin) at 50°C in water (**Figure 17**). The simple procedure, relatively short reaction times, easy workup and high yields are

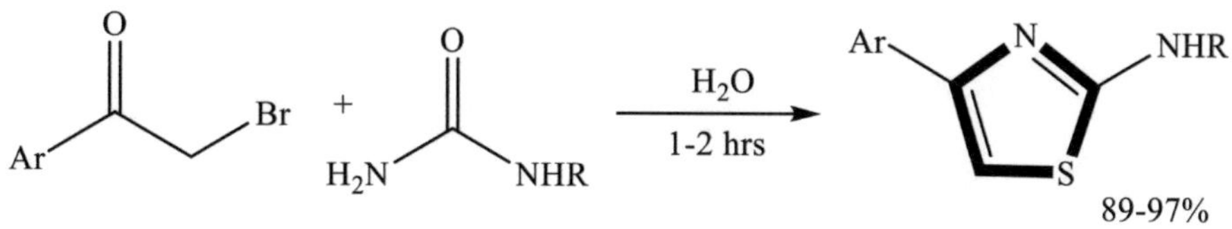

(57) (58) (59)

Ar = Ph; R = Me, Et

Figure 15.
Synthesis of substituted 2-aminothiazoles in water.

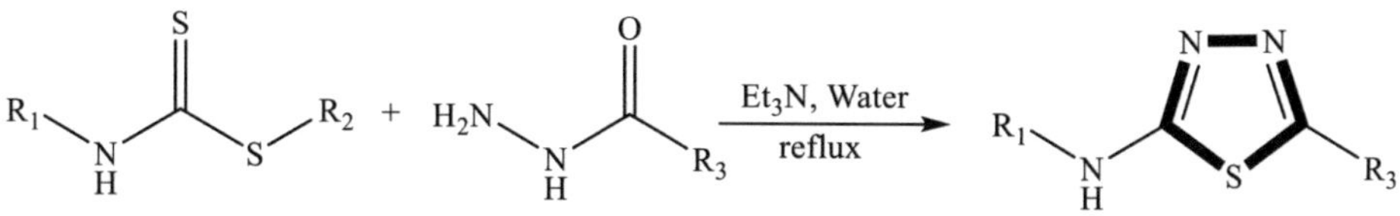

(60) (61) (62)

Figure 16.
Synthesis of 2-amino-1, 3, 4-thiadiazoles.

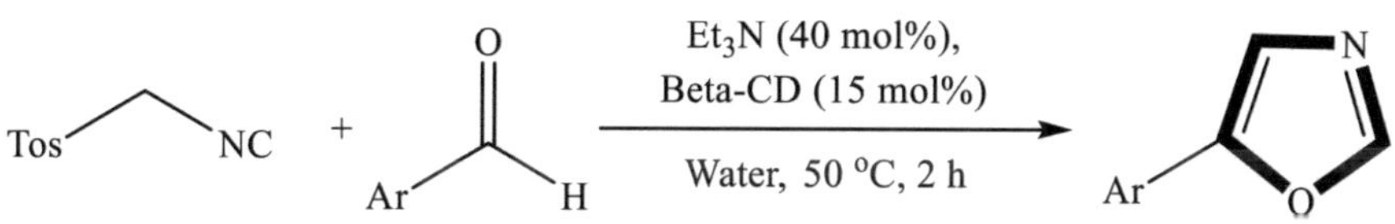

(63) (64) (65)

Ar = Ph, 4-FC_6H_4, 4-MeOC_6H_4, 3-MeOC_6H_4

Figure 17.
Green method for the Van Leusen synthesis of oxazoles.

the reasons that promote the use of β-CD/water as an inexpensive and environmentally benign catalyst for the synthesis of differently substituted oxazoles [25, 26].

6. Recent literature

Ring-closing metathesis of oxygen-containing dienes with ruthenium catalyst, in H_2O as a solvent, provides a good platform for the synthesis of dihydro furan (**Figure 18**) [27].

The hydrosilylation and olefin ring-closing metathesis reaction in the presence of Grubbs-2 catalyst in an aqueous medium is another alternative route (**Figure 19**) [27].

Candida rugosa lipase (CRL) mediated reaction of 1,3-diketone with fumaronitrile resulted in the formation of poly cyano substituted indoles in excellent yields (**Figure 20**) [28].

Polyhydroxyalkyl furans were also prepared from lipase-catalysed Knoevenagel reaction in water using malononitrile and reducing sugars as substrates. Out of the various lipases deployed in the reaction Novozyme-435 showed the highest catalytic activity (**Figure 21**) [28].

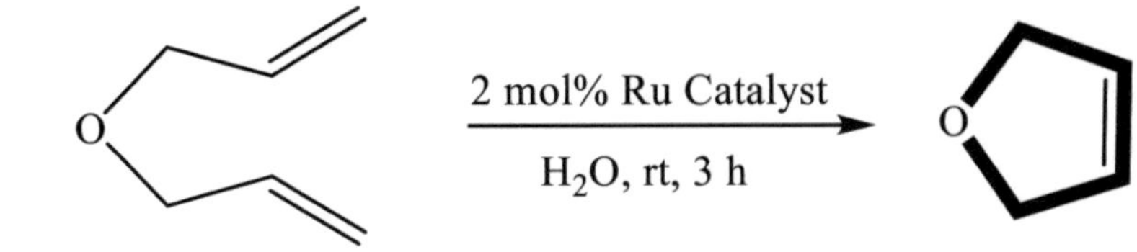

(66) **(67)**

Figure 18.
Synthesis of dihydrofuran.

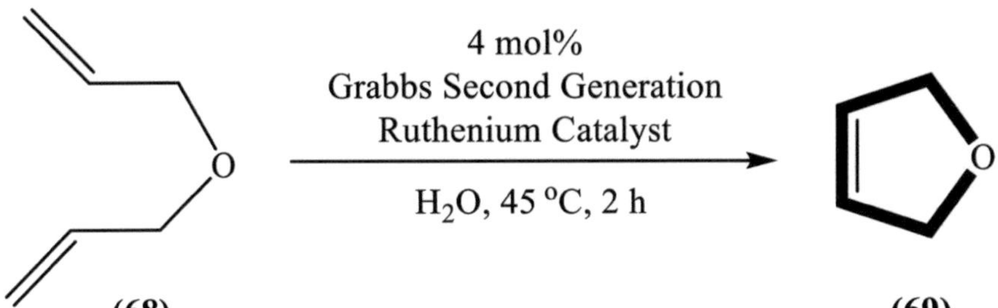

(68) **(69)**

Figure 19.
Synthesis of 2,5-dihydrofuran using Grubbs second Generation catalyst.

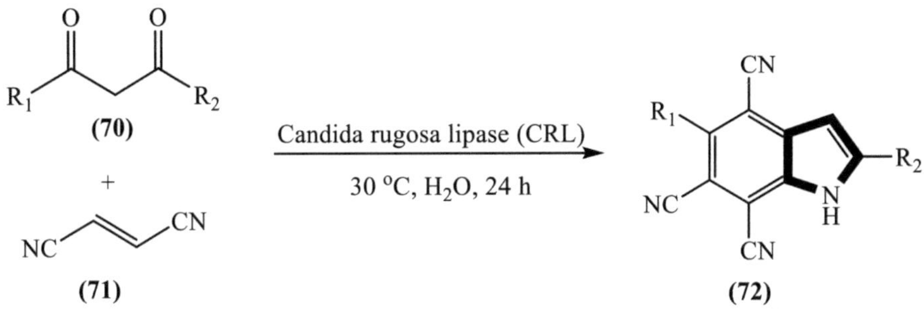

$R_1 = R_2 =$ Me, Et, *i*-Pr, *t*-Bu, *i*-Bu, 57-88% yield
$R_1 \neq R_2$, R_1 or $R_2 =$ Me, Et, *i*-Pr, *t*-Bu, *i*-Bu, $CH(CH_3)_2$, $(CH_2)_4CH_3$, 63-85 % yield

Figure 20.
Lipase-catalysed synthesis of cyano-containing multisubstituted indoles.

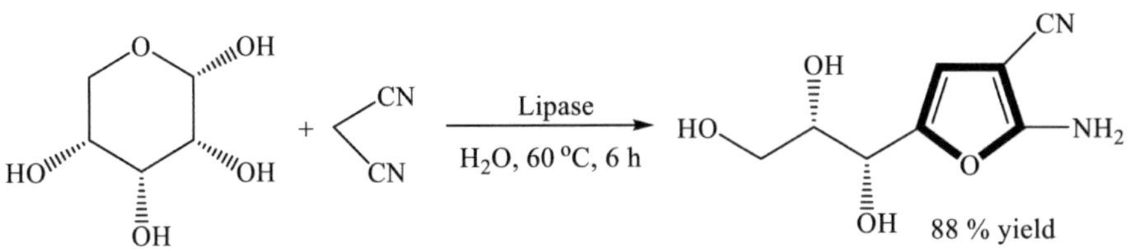

(73) (74) (75)

Figure 21.
Lipase-catalysed synthesis of polyhydroxyalkyl furans.

7. Conclusion

The chapter has attempted to recapitulate some of the most promising new organic reactions in aqueous medium. It has been observed that organic synthesis when carried out in water as a solvent can significantly reduce the number of steps. This shortening of synthetic routes, increasing product selectivity and reducing volatile organic consumption will certainly provide economic, health and environmental benefits to the mankind.

Author details

Abha Kathuria*, Anwar Jahan and Poonam Dwivedi
Department of Chemistry, Ramjas College, University of Delhi, Delhi, India

*Address all correspondence to: kathuriaabha@gmail.com

References

[1] Gupta RR, Kumar M, Gupta V. Vol II – Five Membered Heterocycles. Heterocycles. Berlin, Heidelberg, New York: Springer, Springer-Verlag; 1999

[2] Eicher T, Hauptmann S. The Chemistry of Heterocycles: Structure, Reactions, Syntheses, and Applications. 2nd ed. Weinheim: Wiley-VCH; 2003

[3] Panda SS, Jain SC. "In water" syntheses of heterocyclic compounds. Mini-Reviews in Organic Chemistry. 2011;**8**:455-464

[4] Rideout DC, Breslow R. Hydrophobic acceleration of Diels-Alder reactions. Journal of the American Chemical Society. 1980;**102**:7816-7817

[5] Breslow R, Maitra U, Rideout D. Selective Diels-Alder reactions in aqueous solutions and suspensions. Tetrahedron Letters. 1983;**24**:1901-1904

[6] Otto D, Alder K, Beckmann S. Syntheses in the hydroaromatic series. VIII. Diene syntheses of anthracene. Anthracene formula. Justus Liebigs Annalen der Chemie. 1931;**486**:191-202

[7] Woodward RB, Baer H. The reaction of furan with maleic anhydride. Journal of the American Chemical Society. 1948;**70**:1161-1166

[8] Carlson BA, Sheppard WA, Webster OW. Reaction of diazonium salts with dienes. Route to pyridazines and pyridazinium salts. Journal of the American Chemical Society. 1975;**97**:5291-5293

[9] Bryson TA, Gibson JM, Stewart JJ, Voegtle H, Tiwari A, Dawson JH, et al. Synthesis of quinolines, pyridine ligands and biological probes in green media. Green Chemistry. 2003;**5**:177-180

[10] Suresh R, Muthusubramanian S, Nagaraj M, Manickam G. Indium trichloride catalyzed regioselective synthesis of substituted pyrroles in water. Tetrahedron Letters. 2013;**54**:1779-1784

[11] Demir AS, Emrullahoglu M. Zinc perchlorate catalyzed one-pot amination–annulation of a-cyanomethyl-b-ketoesters in water. Regioselective synthesis of 2-aminopyrrole-4-carboxylates. Tetrahedron. 2006;**62**:1452-1458

[12] Das B, Shinde DB, Kanth BS, Satyalakshmi G. An efficient multicomponent synthesis of Polysubstituted Pyrrolidines and Tetrahydropyrimidines starting directly from nitro compounds in water. Synthesis. 2010;**16**:2823-2827

[13] Kianmehr E, Estiri H, Bahreman A. Efficient synthesis of pyrrolo[2,1-a] isoquinoline and pyrrolo[1,2-a] quinoline derivatives in aqueous media. Journal of Heterocyclic Chemistry. 2009;**46**(6):1203-1207

[14] Gogoi S, Dutta M, Gogoi J, Boruah RC. Microwave promoted synthesis of cycl[3.2.2]azines in water via a new three-component reaction. Tetrahedron Letters. 2011;**52**(7):813-816

[15] Yu L, Jiang X, Wang L, Li Z, Wu D, Zhou X. Catalytic Alkynylation coupling reactions by copper (II) complex in water and its applications to domino synthesis of 2-Arylindoles. European Journal of Organic Chemistry. 2010;**29**:5560-5562

[16] Barnard TM, Vanier GS, Collins MJ. Scale-up of the green synthesis of Azacycloalkanes and Isoindolines under microwave irradiation. Organic Process Research and Development. 2006;**10**:1233-1237

[17] Polshettiwar V, Varma RS. Greener and expeditious synthesis of bioactive heterocycles using microwave irradiation. Pure and Applied Chemistry. 2008;**80**(4):777-790

[18] Keivanloo A, Bakherad M, Nasr-Isfahani H, Esmaily S. Highly efficient synthesis of 5,6-disubstituted-5*H*-pyrrolo[2,3-b]pyrazine-2,3-dicarbonitriles through a one-pot palladium-catalyzed coupling reaction/cyclization in water. Tetrahedron Letters. 2012;**53**:3126-3130

[19] Dehghanzadeh F, Shahrokhabadi F, Anary-Abbasinejad M. A simple route for synthesis of 5-(furan-3-yl) barbiturate/thiobarbiturate derivatives via a multi-component reaction between arylglyoxals, acetylacetone and barbituric/thiobarbituric acid. ARKIVOC. 2019;**part v**:133-141

[20] Heravi MM, Baghernejad B, Oskooie HA. Organic synthesis in water: A green protocol for the synthesis of 2-(cyclohexylamino)-3-aryl-indeno[1,2-b]furan-4-ones. Molecular Diversity. 2009;**1**:385-387

[21] Duvauchellel V, Meffre P, Benfodda Z. Green methodologies for the synthesis of 2-aminothiophene. Environmental Chemistry Letters. 2022:1-25. DOI: 10.1007/s10311-022-01482-1

[22] Puterová Z, Krutošíková A, Véghc D. Gewald reaction: Synthesis, properties and applications of substituted 2-aminothiophenes. ARKIVOC. 2010;**2010**:209-246

[23] Taterao MP, Ingale SA, Srinivasan KV. Catalyst–free efficient synthesis of 2-aminothiazoles in water at ambient temperature. Tetrahedron. 2008;**64**(22):5019-5022

[24] Aryanasab F, Halimehjani AZ, Saidi MR. Dithiocarbamate as an efficient intermediate for the synthesis of 2-amino-1,3,4-thiadiazoles in water. Tetrahedron Letters. 2010;**51**(5):790-792

[25] Reddy GR, Reddy TR, Chary RG, Joseph SC, Mukherjee S, Pal M. β-Cyclodextrin mediated MCR in water: Synthesis of dihydroisoindolo[2,1-a]quinazoline-5,11-dione derivatives under microwave irradiation. Tetrahedron Letters. 2013;**54**:6744-6746

[26] Rahimzadeh G, Kianmehr E, Mahdavi M. Improvement of the Van Leusen reaction in the presence of β-cyclodextrin: A green and efficient synthesis of oxazoles in water. Zeitschrift für Naturforschung. 2017;**72**(12):923-926

[27] Kaur N. Ruthenium catalyzed synthesis of five – Membered O – Heterocycles. Inorganic Chemistry Communications. 2019;**99**:82-107

[28] Ma X-L, Wang Y-H, Shen J-H, Hu Y. Progress in the synthesis of heterocyclic compounds catalyzed by lipases. Pharmaceutical Fronts. 2021;**3**(1):e87-e97

Chapter 7

Current Development in the Synthesis of Benzimidazole-Quinoline Hybrid Analogues and Their Biological Applications

Zebabanu Khalifa, Rachana Upadhyay, Premlata Kumari and Amit B. Patel

Abstract

Heterocyclic compounds have an essential role in many domains of medicinal chemistry. Many pharmaceutical industries use and investigate nitrogen-containing heterocycles because they are crucial in discovering and developing novel therapeutically active compounds. The benzimidazole moiety is a fundamental component of many heterocyclic scaffolds, which play an important role in producing a wide range of biological activities. Similarly, quinoline is also a versatile bicyclic heterocyclic scaffold with many medicinal applications. It is an essential scaffold for drug discovery leads, and it plays a significant role in medicinal chemistry and has biological activities similar to benzimidazole scaffolds. The present chapter discusses the quinoline-benzimidazole hybrids scaffolds and their potential pharmacological activities.

Keywords: quinoline, benzimidazole, hybrid molecules, synthesis, pharmacological activity

1. Introduction

Heterocycles are common elements found in the majority of commercial drugs and are a target of medicinal chemistry in the drug delivery process [1]. In recent years, nitrogen-containing five- and six-member heterocycles have received a significant amount of interest because of their major pharmacological and synthetic implications [2]. Structure elucidation, identification of novel biologically active compounds, development of efficient therapies and discovering the mode of action of newer molecules are all critical aspects of medicinal chemistry that nitrogen-containing heterocyclic compounds perform. They are commonly employed in drug development because of the discovery of robust synthetic pathways that can rapidly produce large amounts of desired chemicals, which helps to accelerate the drug development process. Their pharmacophores have a wide range of pharmacological activities against disease or disorder [3, 4].

The "one drug, one target, one disease" strategy is no longer appropriate in today's complicated and infectious diseases [5]. Drug resistance is a problem that cannot be solved using usual disease treatment methods. The hybridisation of physiologically active compounds is a potent drug discovery approach that can be utilised to treat a wide range of diseases. It opens up the possibility of better medications [6]. Benzimidazole is the building block of various synthetic medicinal and biochemical compounds with important biological activities like anticancer [7, 8], antimicrobial [9, 10], antibacterial [11, 12], antimycobacterial [13, 14], anti-inflammatory [15, 16], etc. Similarly, quinoline is found in various natural products. It is frequently utilised to explore a range of bioactive compounds with varying pharmacological properties like antifungal [17, 18], anti-inflammatory [19, 20], anticancer [21, 22], antimicrobial [23, 24], anticoagulant [25, 26], antiviral [27, 28], antimalarial [29, 30], antitrypanosomal [30, 31] etc. Due to the therapeutic efficacy of benzimidazole and quinoline scaffolds, medicinal chemists have been interested in developing hybrid analogues with enhanced potency by incorporating them, and molecular hybridisation technology has been employed to do so. With a virtual planning and developing procedure, finding the enhanced pharmacological activity of new or modified present drugs is quite challenging. In such cases, these benzimidazole-quinoline hybrid compounds play a crucial role since they can reduce time and expenses by employing techniques including X-ray screening, molecular docking, NMR skeletons of biomolecules and computer-aided drug design [32, 33].

The fundamental purpose of benzimidazole-quinoline hybrid compounds is to strengthen their potential to interact with several biological targets. Due to their excellent potential in drug discovery and development, we have introduced this study of the pharmacological activities of benzimidazole-incorporated quinoline compounds.

2. Biological applications of benzimidazole-quinoline hybrids

2.1 Anticancer activity

Hranjec and co-workers developed the synthesis and biological characterisation of novel benzimidazo[1,2-*a*]quinoline-6-carbonitriles derivatives (**Figure 1**). The reaction was initiated by condensing 2-cyanomethylbenzimidazole with heteroaromatic aldehyde yielded acrylic substituted acrylonitriles. The photochemical dehydrocyclisation procedure was utilised to produce benzimidazo[1,2-*a*]quinoline-6-carbonitriles and heteroaromatic fluorenes. All the synthesised compounds were investigated for antiproliferative efficacy against five tumour cell lines such as HeLa, MiaPaCa-2, SW

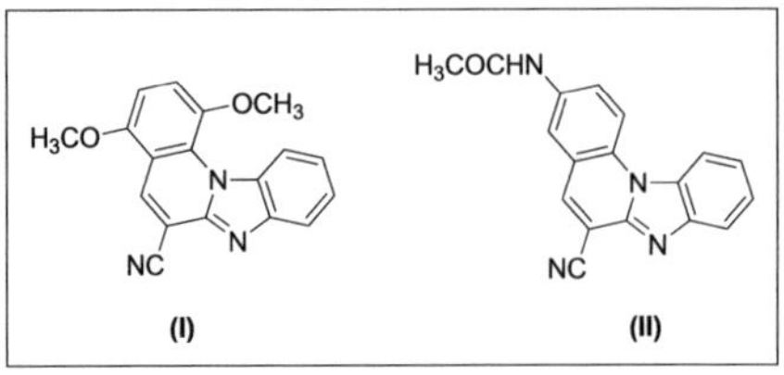

Figure 1.
Benzo[4,5]imidazo[1,2-a]quinoline-6-carbonitrile methoxy/anilide derivatives.

620, MCF-7 and H460. Among them, compounds **(I)** and **(II)** showed good antiproliferative activity against all the cell lines with an IC_{50} range of 2–90 μM [34].

Perin *et al.* synthesised benzimidazo[1,2-*a*]quinolines and benzimidazo[1,2-*a*] quinolines-6-carbonitriles derivatives by the thermal reaction using sulfolane for dehydrohalogenation cyclisation at 280°C followed by UV/Vis spectroscopy (**Figure 2**). Out of all the synthesised compounds **(III)**, **(IV)**, **(V)**, **(VI)**, **(VII)** and **(VIII)** showed strong antiproliferative activity with IC_{50} range of 2–19 μM [35].

Further, Perin and co-workers have also synthesised amino-substituted cyclic benzimidazo[1,2-*a*]quinoline derivatives and tested *in vitro* for their antiproliferative, antibacterial and antiviral activities (**Figure 3**). Low to moderate yields of amide substituted cyclic derivatives were obtained by condensing acyl halides with an excess quantity of the corresponding amines in absolute dichloromethane. The antiproliferative activity of **(IX)**, **(X)** and **(XI)** was found to be most promising with an IC_{50} value ranging between 0.48–4.1 μM against HeLa cell line and 0.24–0.69 μM against MCF7 cell line, and it also caused apoptosis in human cervical carcinoma HeLa cells at micromolar doses. Antiviral activity was observed in compounds **(XII)**, **(XIII)** and **(XIV)** having a side chain at position-2 against the herpes simplex virus (HSV) with an EC_{50} value of 1.8–6.8 μM and human coronavirus with EC_{50} value of 4–12 μM. In addition, **(IX)** and **(X)** substituting an amide side chain position-6 of the tetracyclic skeleton were efficacious against *S. epidermidis* and *C. albicans* [36].

Perin *et al.* synthesised 2-substituted benzimidazo[1,2-*a*]-quinoline derivatives by microwave-assisted amination of 2-chloro/fluorobenzo[4,5]imidazo[1,2-*a*]quinoline-6-carbonitrile precursors, which were prepared by photochemical dehydrogenation of non-fused *E*-2-(2-benzimidazolyl)-3-(4-halophenyl)acrylonitriles with 4-halobenzaldehyde (**Figure 4**). Furthermore, the antiproliferative efficacy of amino-substituted derivatives was investigated. Compound **(XV)** showed prominent activity against HCT116 and H460 cells with IC_{50} values of 0.5 and 0.3 μM, respectively. The compounds **(XVI)** and **(XVII)** with the presence of cyclic tertiary amino substituent displayed excellent antiproliferative action against HCT116 and H460 cell lines with IC_{50} range of 0.06–0.3 μM [37].

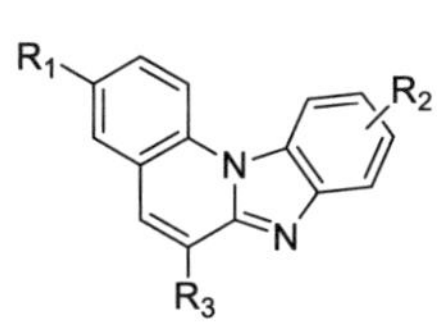

(III): R_1= $NH_3^+Cl^-$, R_2= $NH_3^+Cl^-$, R_3= H
(IV): R_1= NH_2, R_2= CH_3, R_3= CN
(V): R_1= $NH_3^+Cl^-$, R_2= H, R_3= CN
(VI): R_1= H, R_2= $NH_3^+Cl^-$, R_3= CN
(VII): R_1= R_2= $NH_3^+Cl^-$, R_3= CN
(VIII): R_1= $NH_3^+Cl^-$, R_2= CH_3, R_3= CN

Figure 2.
Amino substituted benzo[4,5]imidazo[1,2-a]quinoline derivatives.

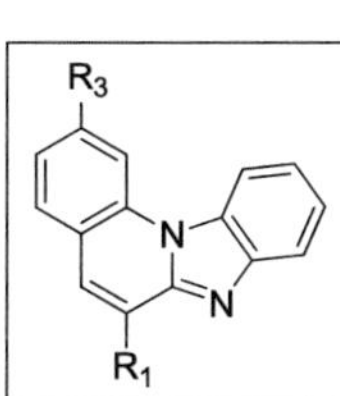

(IX): R_3= H, R_1= $CONHCH_2CH_2CH_2N(CH_3)_2$
(X): R_3= H, R_1= $CONHCH_2CH_2N(CH_2CH_3)_2$
(XI): R_1= R_3= $CONHCH_2CH_2CH_2N(CH_3)_2$
(XII): R_1= H, R_3= $CONHCH_2CH(CH_3)_2$
(XIII): R_1= H, R_3= $CONHCH_2CH_2CH_2N(CH_3)_2$
(XIV): R_1= H, R_3= $CONHCH_2CH_2N(CH_2CH_3)_2$

Figure 3.
Benzo[4,5]imidazo[1,2-a]quinoline amide derivatives.

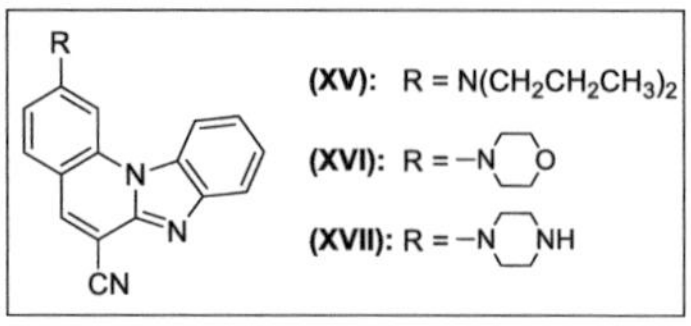

Figure 4.
Benzo[4,5]imidazo[1,2-a]quinoline-6-carbonitrile derivatives.

Tantawy *et al.* synthesised novel 2-(1*H*-benzo[*d*]imidazol-2-yl)quinoline-4-carboxylic acid derivatives by the reaction of 2-acetylimidazoles with isatin (**Figure 5**). Compounds **(XVIII), (XIX)** and **(XX)** displayed excellent *in vitro* antitumour activity with IC_{50} in the range of 12.7–16.13 μg/mL (percentage inhibition of 71.9–74.0%) against MCF7 breast cancer cell lines [38].

Shi and co-workers have designed and synthesised *N*-(2-phenyl-1*H*-benzo[*d*]imidazol-5-yl) quinolin-4-amine derivatives as potent inhibitors of VEGFR-2 (KDR) kinase (**Figure 6**). These derivatives were prepared by the condensation of 4-chloroquinoline with benzimidazole quinoline-4-amines. The synthesised compound **(XXI)** showed the most potent inhibitory activity against VEGFR-2 with IC_{50} value of 0.03 μM. Moreover, this compound also showed the highest antiproliferative activity against MCF-7 and Hep-G2 cell lines with IC_{50} values of 1.2 and 13.3 μM, respectively [39].

Brajsa *et al.* prepared 2-Imidazolinyl-substituted triaza-benzo[*c*]fluorenes derivatives from triaza-benzo[*c*]fluorenes using the Pinner reaction and tested for antitumour efficacy in 2D and 3D cell culture (**Figure 7**). The compound **(XXII)** showed the most prominent activity against PANC-1 and MDA-MB-231 cell lines with IC_{50} values of 1 and 0.2 μM, respectively [40].

Kuang *et al.* prepared a series of 3-(1*H*-benzimidazol-2-yl)quinoline-2(1*H*)-one analogues by coupling reaction of 2-oxo-quinoline-3-carbaldehyde with various *o*-phenylenediamines (**Figure 8**). The HepG2 tumour growth was effectively inhibited by **(XXIII)** with IC_{50} value of 8.45 μM in an *in vitro* antitumour experiment.

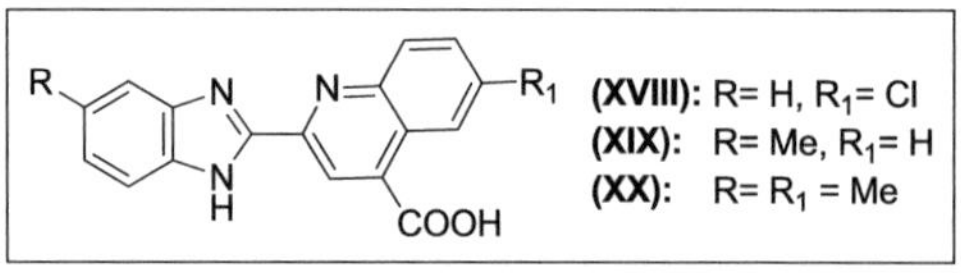

Figure 5.
*2-(1*H*-benzo[*d*]imidazol-2-yl)quinoline-4-carboxylic acid derivatives.*

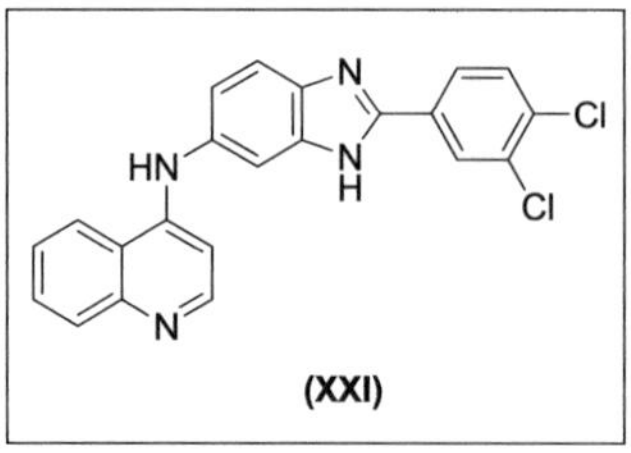

Figure 6.
N-(2-(3,4-dichlorophenyl)-1H-benzo[d]imidazol-6-yl)quinolin-4-amine derivatives.

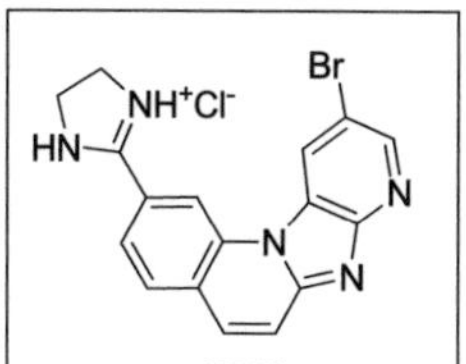

(XXII)

Figure 7.
*2-(10-Bromopyrido[2',3'4,5]imidazo[1,2-a]quinoline-2yl)-4,5-dihydro-1*H*-imidazol-3-ium chloride derivatives.*

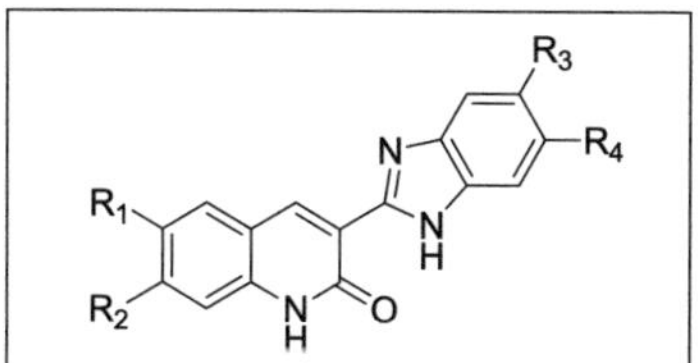

(XXIII): R_1= R_2= R_4 = H, R_3= Cl
(XXIV): R_1= R_2= O-CH_2-O, R_3=Cl, R_4= H

Figure 8.
*3-(1*H*-benzo[*d*]imidazol-2-yl)quinolin-2(1*H*)-one derivatives.*

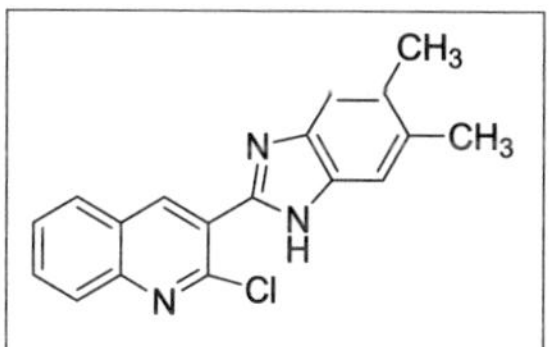

(XXV)

Figure 9.
*2-Chloro-3-(5,6-dimethyl-1*H*-benzo[*d*]imidazol-2-yl)quinoline derivatives.*

In addition, the compound **(XXIV)** showed excellent antiproliferative activity against BEL-7402 with IC_{50} value of 9.06 μM [41].

Similarly, Kuang *et al.* further synthesised 2-chloro-3-(1*H*-benzo[*d*]imidazol-2-yl) quinoline derivatives as potent antitumour agents (**Figure 9**). *In vitro* antitumour assay findings revealed that some compounds had moderate to high inhibitory effects against HepG2, SK-OV-3, NCI-H460 and BEL-7404 tumour cells, as compared to 5-FU and cisplatin. In the HepG2 xenograft model, compound **(XXV)** effectively inhibited tumour growth with IC_{50} value 7.54 μM [42].

Macan and co-workers prepared 2-Fluoro-5-(1*H*-1,2,3-triazol-1-yl)benzo[4,5] imidazo[1,2-*a*]quinoline-6-carbonitrile derivatives *via* Huisgen 1,3-dipolar cycloaddition of the azide derivatives and the corresponding alkynes using Cu(OAc)2 as a catalyst (**Figure 10**). The presence of 3-chloropropyl functional group in compound **(XXVI)** and 2-hydroxyethyl side-chain in compound **(XXVII)** exhibited the most pronounced growth-inhibitory effect on colon cancer (HCT116) cells with IC_{50} value of 0.5 μM and 0.6 μM, respectively [43].

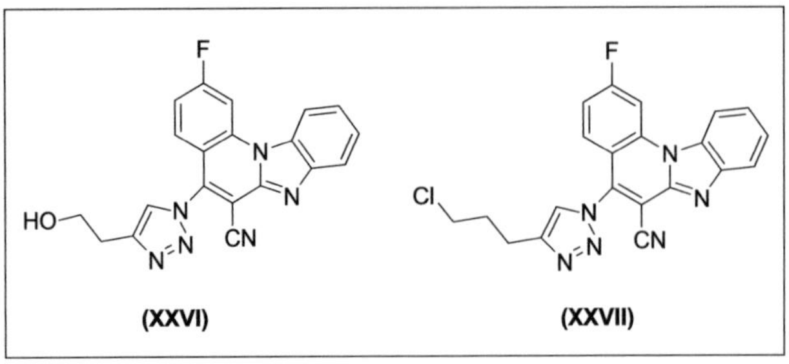

Figure 10.
*2-Fluoro-5-(1*H*-1,2,3-triazol-1-yl)benzo[4,5]imidazo[1,2-*a*]quinoline-6-carbonitrile derivatives.*

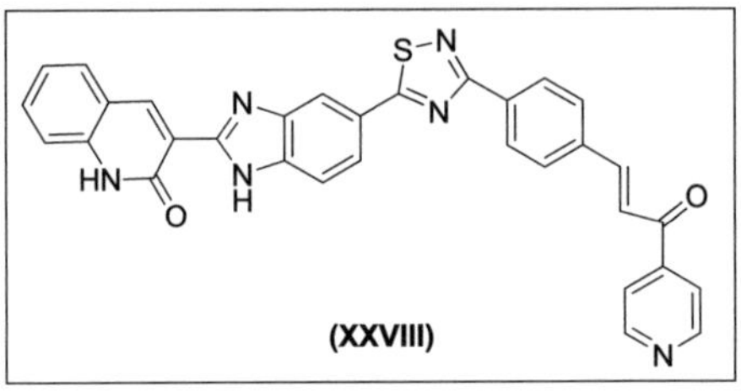

Figure 11.
*(*E*)-3-(5-(3-(4-(3-oxo-3-(pyridin-4-yl)prop-1-en-1-yl)phenyl)-1,2,4-thiadiazol-5-yl)-1*H*-benzo[*d*] imidazol-2-yl)quinoline-2(1*H*)-one derivatives.*

Pragathi and co-workers have synthesised a series of chalcone incorporated 2-quinoline-benzimidazole-1,2,4-thiadiazoles by aldol condensation (**Figure 11**). Synthesised compounds were examined for their anticancer activities against different cancer cell lines such as MCF-7, A549, Colo-205 and A2780. From all the investigated compounds, **(XXVIII)** showed excellent anticancer activities with IC_{50} range of 0.012–1.46 μM [44].

Gaikwad *et al.* developed a series of 1,2,3-triazole-based quinoline-benzimidazole hybrid scaffolds by a click reaction using the [3+2] azide-alkyne cycloaddition reaction using copper(I)Iodide in DMF (**Figure 12**). Synthesised compounds were screened against a panel of NCI-60 human cancer cell lines for in vitro cytotoxicity evaluation. The compound **(XXIX)** showed an excellent antiproliferative effect with a significant IC_{50} value of 0.59 μM against the human breast cancer BT-474 cell line [45].

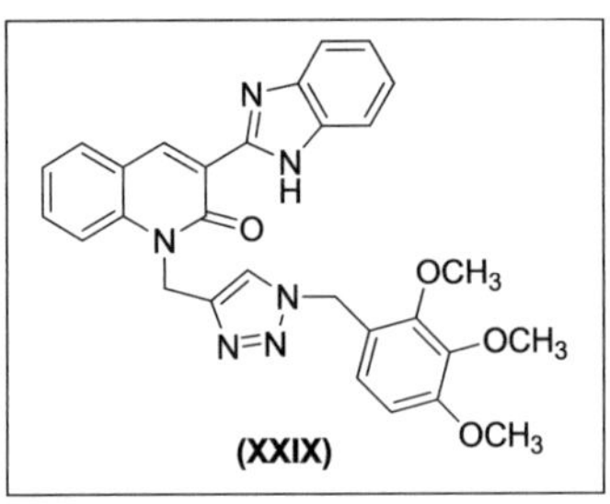

Figure 12.
*3-(1*H*-benzo[*d*]imidazol-2-yl)-1-((1-(2,3,4-trimethoxybenzyl)-1H-1,2,3-triazol-4-yl)methyl)quinoin-2 (1*H*)-one derivatives.*

2.2 Antimicrobial activity

Sonar *et al.* synthesised 4-((5-difluoromethoxy)-1*H*-benzo[*d*]imidazol-2-ylthio) methyl)tetrazolo[1,5-*a*]quinoines derivatives (**Figure 13**). These targeted molecules were synthesised by the conversion of 2-chloroquinoline-3-carbaldehyde to tetrazolo [1,5-*a*]quinoline-4-carbaldehyde after treatment with sodium azide, followed by reduction to the corresponding alcohol derivatives and conversion to chloride with thionyl chloride and coupling with 5-(difluoromethoxy)-1*H*-benzo[*d*]-imidazol-2-thiol. The compounds **(XXX)**, **(XXXI)**, **(XXXII)**, **(XXXIII)** and **(XXXIV)** were evaluated for their antibacterial activity against Gram-positive (*B. subtilis* and *S. aureus)* and Gram-negative (*E. Coli* and *S. aboney*) microorganisms. The synthesised compounds demonstrated significant activity against all the bacteria with zone of inhibition ranging from 7 to 17 mm at MIC level of 10 mg/mL [46].

Chaudhari *et al.* synthesised 8-(1-alkyl/alkysulsulphonyl/alkoxycarbonyl-benzimidazol-2ylmethoxy)-5-chloroquine derivatives (**Figure 14**). The first step was the reaction of 5-chloro-8-hydroxyquinoline with monochloroacetic acid methyl ester followed by hydrolysis to form 5-chloroquinolin-8-yl)oxy]acetic acid. Which was further reacted with *o*-phenylenediamine to obtain 8-[(1*H*-benzimidazol-2-yl) methoxy]-5-chloroquinoline. The final analogues were synthesised by the reaction of 8-[(1*H*-benzimidazol-2-yl)methoxy]-5-chloroquinoline with different electrophiles in the presence of an appropriate base. Synthesised compounds were tested for their antifungal activity, among them compound **(XXXV)**, **(XXXVI)**, **(XXXVII)** and **(XXXVIII)** showed good antifungal activity against *A. niger* MTCC282 with 18–19 mm zone of inhibition at 50 μg/mL MIC. In addition, compound **(XXXIX)** showed potent antibacterial activity with 15 mm zone of inhibition (MIC, 25 μg/mL) against *S. aureus* MTCC-96 and *S. typhimurium* MTCC-98 [47].

Gowda *et al.* used one-step method to synthesise 2-(1*H*-benzimidazol-2-yl)-6-substituted thieno[2,3-*b*]quinoline derivatives (**Figure 15**). They introduced nucleophilic substitution reaction of 2-(chloromethyl)-1*H*-benzimidazole 2/2-(mercaptomethyl)-1*H*-benzimidazoles followed by the cyclisation with 2-mercapto-quinoline-3-carbaldehyde 1/2-chloroquinoline-3-carbaldehyde. The antibacterial activity of the synthesised compounds was tested *in vitro* against Gram-positive (*E. coli* and *S. aureus*) and Gram-negative (*P. aeruginosa* and *K. pneumoniae)* pathogens.

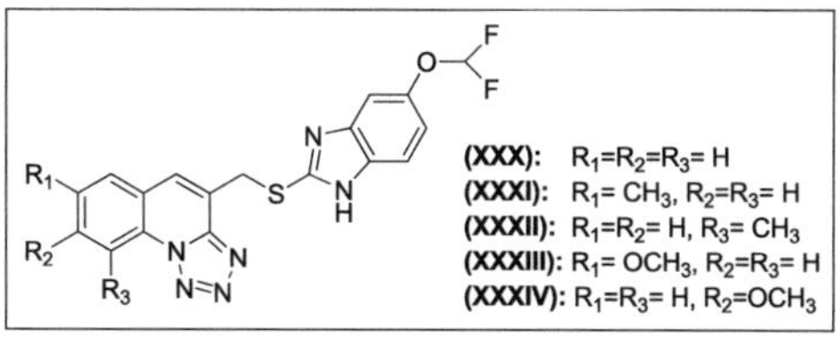

Figure 13.
*4-(((5-(Difluoromethoxy)-1*H*-benzo[*d*]imidazol-2-yl)thiomethyl)tetrazolo[1,5-*a*]quinoline derivatives.*

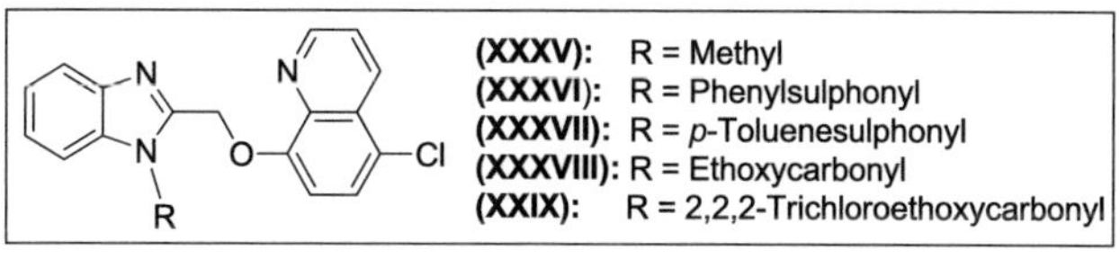

Figure 14.
*8-((1*H*-benzo[*d*]imidazol-2-yl)methoxy)-5-chloroquinoline derivatives.*

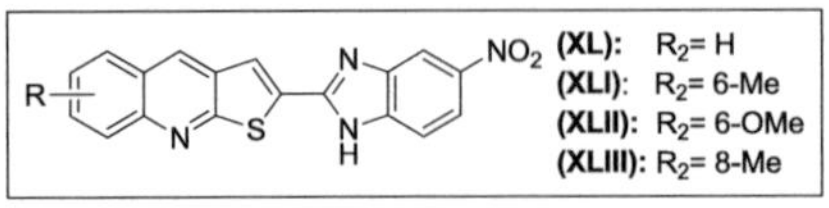

Figure 15.
2-(1H-benzo[d]imidazo-2-yl)thieno[2,3-b]quinolines derivatives.

Compounds **(XL)**, **(XLI)**, **(XLII)** and **(XLIII)** with the substitution of nitro group at the position-5 of benzimidazole ring showed highest MIC at 12.5 μg/mL [48].

Mungra *et al.* synthesised a new class of potent quinoline fused benzimidazole derivatives (**Figure 16**). The final compounds were prepared by the reaction of tetrazolo[1,5-*a*]quinoline-4-carbaldehyde with *o*-phenylenediamine by microwave irradiation. Compounds were further tested for their antimicrobial activity. Among all compounds, **(XLIV)** displayed excellent activity against Gram-positive bacteria *B. subtills* with MIC value of 100 μg/mL [49].

Garudachari *et al.* synthesised quinoline incorporated benzimidazole from aniline and isatin via the multistep process (**Figure 17**). These derivatives were synthesised by multi-component one-pot Doebner and Pfitzinger reaction. Newly synthesised compounds were tested *in vitro* for their antibacterial and antifungal activities. From them, compounds **(XLV)**, **(XLVI)**, **(XLVII)** and **(XLVIII)** showed significant antibacterial activity against *S. aureus* and *E. coli* with a zone of inhibition in the range of 10–16 mm at MIC value of 6.25 μg/mL. Compound **(XLIX)** was showed potent antifungal activity against *A. niger*, and *Penicillium sp.* with 12–15 mm zone of inhibition, and **(L)**, **(LI)** and **(LII)** showed good antifungal activity against *A. niger* and *Penicillium sp.* at 6.25 μg/mL with 12–16 mm zone of inhibition [50].

De Souza *et al.* introduced well-known fluorescent benzoimidazo[1,2-*a*]quinoline derivatives with DNA binding and anticancer properties. Compounds were tested as bifunctional agents that could detect yeast biofilms on stainless steel surfaces and act as biocidal agents (**Figure 18**). The biocidal activity of benzimidazole compound

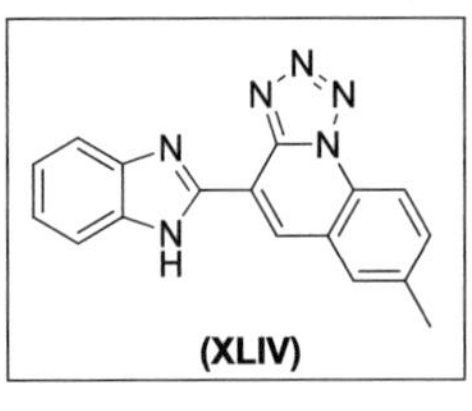

Figure 16.
4-(1H-benzo[d]imidazol-2-yl)-7-methyltetrazolo[1,5-a]quinoline derivatives.

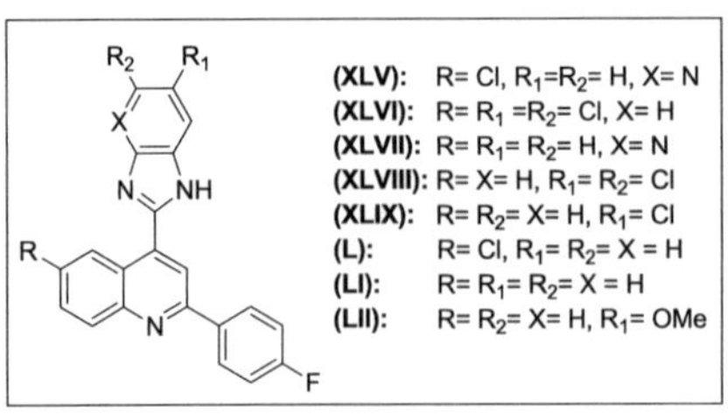

Figure 17.
4-(1H-benzo[d]imidazol-2-yl)-2-(4-fluorophenyl)quinoline derivatives.

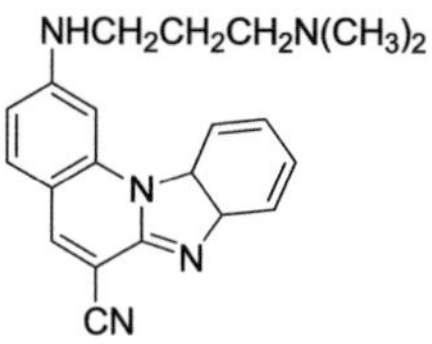

(LIII)

Figure 18.
2-((3-(Dimethylamino)propyl)amino)-7a,11a-dihydrobenzo[4,5]imidazo[1,2-a]quinoline-6-carbonitrile derivatives.

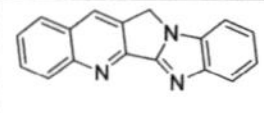

(LIV)

Figure 19.
*12*H*-benzo[4′,5′]imidazo[1′,2′:1,2]pyrrolo[3,4-*b*]quinoline derivatives.*

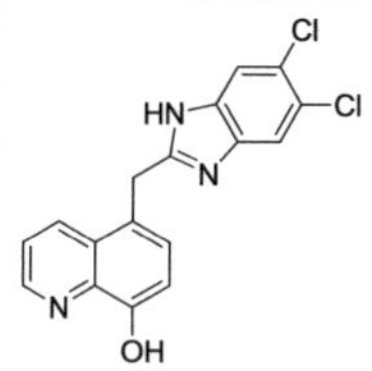

(LV)

Figure 20.
*5-((5,6-Dichoro-1*H*-benzo[*d*]imidazol-2-yl)methyl)quinolin-8-ol derivatives.*

(LIII) against yeast cells and biofilms seemed promising. The compound showed significant activity against *C. Albicans* CA01 with MIC value of 4 μg/mL [51].

Villa *et al.* synthesised 12*H*-benzo[4′,5′]imidazo[1′,2′:1,2]pyrrolo[3,4-*b*]quinoline derivatives using two different pathways (**Figure 19**). In the first path, titled analogues were prepared by the reaction of arylimines and 20 mol% of $BF_3O(C_2H_5)_2$ in dichloroethane. The arylimines were obtained by the reaction of benzimidazole carbaldehyde with arylamine in the presence of $InCl_3$. However, another path was accomplished by a one-pot intramolecular Povarov reaction without isolating the intermediates. Among them, compound **(LIV)** significantly reduced the metabolic activity of *C. albicans* biofilms fungal cells at MIC value of 0.5 μg/mL [52].

El Faydy and co-workers synthesised 8-quinolinol analogues bearing a benzimidazole moiety by condensing 5-(carboxymethyl)-8-quinolinol with substituted *o*-phenylenediamines in an acidic medium (**Figure 20**). The synthesised compounds were tested in vitro against two Gram-positive (*B. subtilis and S. aureus)* and two Gram-negative (*E. Ludwigii and E. coli)* microorganisms. The compound **(LV)** showed good antibacterial activity against all the microorganisms, with 28–42 mm zone of inhibition at MIC in the range of 10–20 μg/mL [53].

2.3 Miscellaneous activity

El-Feky *et al.* investigated the anti-inflammatory potential of several novel *N*-(substituted-phenyl)-2-(2-(6-fluoro-2(4-fluorophenyl)-quinoline-4-yl)-1*H*-

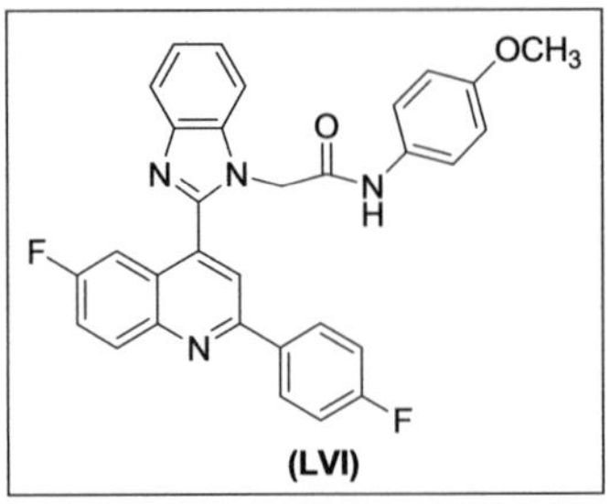

Figure 21.
*2-(2-(6-Fluoro-2-(4-fluorophenyl)quinolin-4-yl)-1*H*-benzo[*d*]imidazol-1-yl)-*N*-(4-methoxyphenyl) acetamide derivatives.*

benzimidazol-1-yl)acetamides derivatives by the reaction with 4-(1*H*-benzo[*d*]imidazol-2-yl)-6-fluoro-2-(4-fluorophenyl)quinoline with 2-chloro-*N*-arylacetamide (**Figure 21**). Synthesised compounds were investigated for their anti-inflammatory activity. Among them, the compound **(LVI)** showed the highest activity with 55 mg/kg dose (%) [54].

Madawali *et al.* synthesised novel 2-chloro-3-[3-(6-nitro-1*H*-benzimidazol-2-yl)-1*H*-pyrazol-5-yl]quinolines derivatives from 2-chloro-3-[3-(6-nitro-1H-benzimidazol-2-yl)-4,5-dihydro-1H-pyrazol-5-yl]quinolines by dehydrogenation reaction (**Figure 22**). Among them, compounds **(LVII)**, **(LVIII)** and **(LIX)** showed excellent anti-inflammatory activity with excellent % inhibition of denaturation 82.81, 82.35 and 82.81, respectively [55].

Ukrainets and co-workers synthesised benzimidazo-2-ylamides of 4-hydroxy-2-oxo-1,2-dihydroquinoline-3-carboxylic acids (**Figure 23**). The targeted compounds were prepared by the well-known reaction, which involved heating ethyl esters with 2-aminobenzimidazole. They were tested for their antituberculosis and antithyroid activities. Compounds **(LX)** and **(LXI)** showed the highest activity against *M. tuberculosis H37Rv* ATCC 27294 with MIC value of 1.56 μg/mL. Moreover, the compound **(LXII)** (1.46 nmol/l) showed higher antithyroid activity as compared to the standard drug Mercazolyl (1.63 nmol/l) against the triiodothyronine (T3) hormone [56].

Mantu *et al.* designed and synthesised 2-(1*H*-benzo[*d*]imidazol-1-yl)-*N*-(quinoline-8-yl)acetamide analogues from 2-chloro-*N*-(quinolin-8-yl)acetamide and imidazole

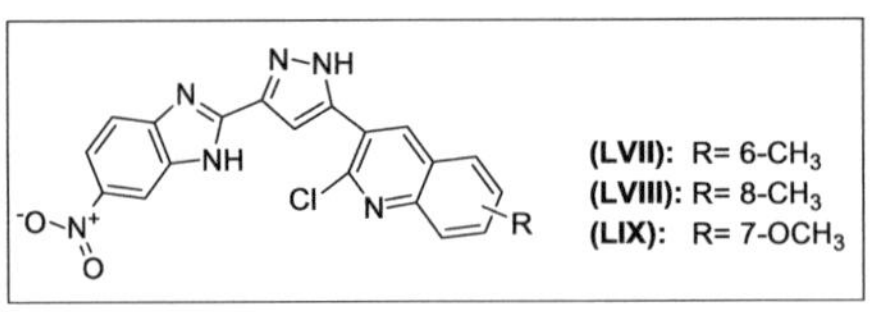

Figure 22.
*2-Chloro-3-[3-(6-nitro-1*H*-benzimidazol-2-yl)-4,5-dihydro-1*H*-pyrazol-5-yl]quinoline derivatives.*

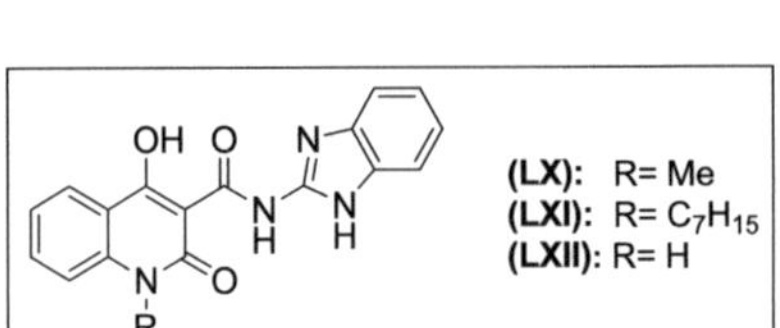

Figure 23.
N-(1*H*-benzo[*d*]imidazol-yl-)-4-hydroxy-2-oxo-1,2-dihydroquinoline-3-carboxamide derivatives.*

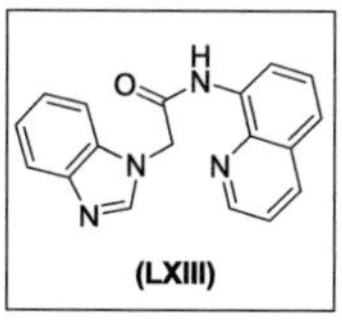

Figure 24.
*2-(1*H*-benzo[*d*]imidazol-1-yl)-*N*-(quinoline-8-yl)acetamide derivatives.*

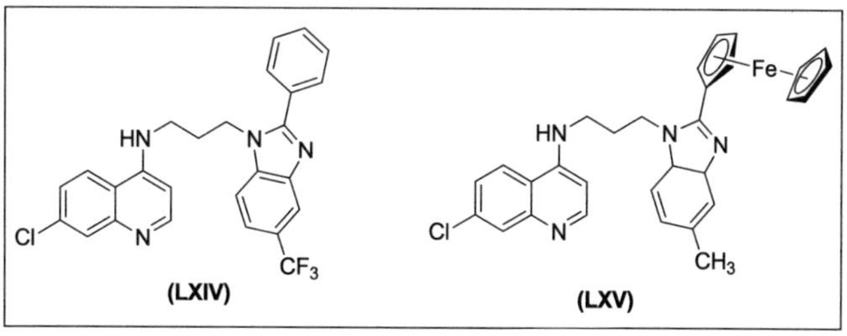

Figure 25.
*7-Chloro-*N*-(3-(2-phenyl/ferrocenyl-5-(trifuoromethyl)-1*H*-benzo[*d*]imidazol-1-yl)propyl)quinolin-4-amine derivatives.*

derivatives (**Figure 24**). The synthesised derivatives were tested *in vitro* for anticancer and antimycobacterial activities. Among all hybrid compounds, **(LXIII)** showed good antitumour activity against Renal Cancer A498 (52.92 μM) and Breast Cancer MDA-MB-468 (56.54 μM). This compound also showed antimycobacterial activity against *M. tuberculosis H37Rv* with MIC >100 μM and IC_{90} value of 77 μg/mL [57].

Baartzes and co-workers synthesised phenyl-benzimidazole and ferrocenyl-benzimidazoles fused aminoquinoline derivatives and evaluated their antiplasmodial activity (**Figure 25**). Out of all the synthesised hybrids analogues, phenyl **(LXIV)** and ferrocenyl **(LXV)** showed excellent antiplasmodial activity against multi-drug resistant K1 strains of the human malaria parasite at IC_{50} values of 0.151 and 0.283 μM, respectively. They were further assessed *in vivo* against *Plasmodium berghei* infected mice. Treatment with either compound decreased parasitemia, with **(LXV)** demonstrating superior activity [58].

Pomel *et al.* synthesised 7-chloro-4-(4-(5-fluoro-1*H*-benzo[*d*]imidazol-2-yl) phenoxy)quinoline derivatives by the reaction of 2-phenol-1*H*-6-fluorobenzimidazole and 4,7-dichloroquinoleine chloride. *T. b. gambiense*, a human parasite, was used *in vitro* to test antitrypanosomal activity (**Figure 26**). The compound **(LXVI)** showed the best *in vitro* activity with IC_{50} value of 1.98 μM, cytotoxicity CC_{50} value 10.56 μM and selective index (SI) was 5.3 [59].

Bharadwaj *et al.* synthesised benzimidazole-containing quinolinyl oxadiazoles from quinoline-4-carboxylic acid and benzimidazole hydrazide (**Figure 27**). These

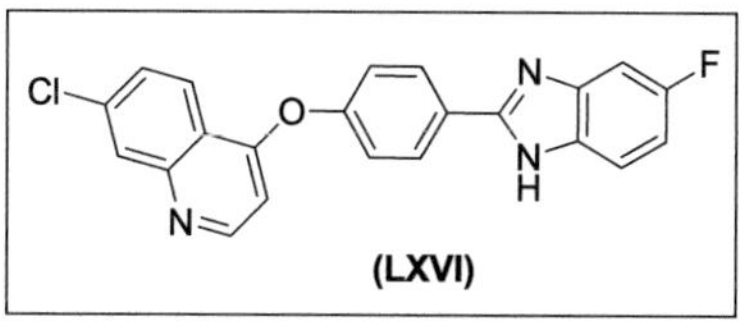

Figure 26.
*7-Chloro-4-(4-(5-fluoro-1*H*-benzo[*d*]imidazol-2-yl)phenoxy)quinoline derivatives.*

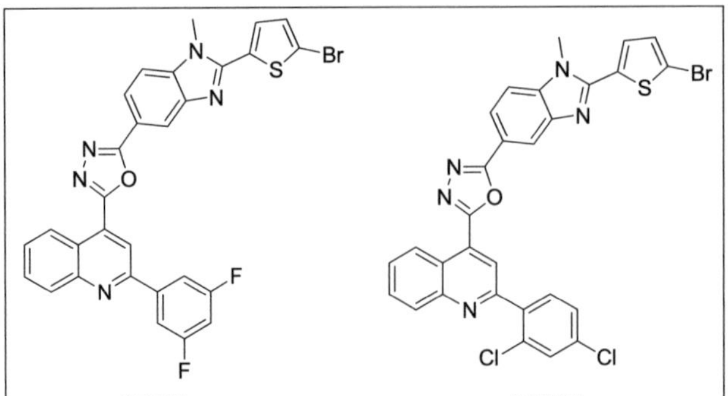

Figure 27.
2-(2-(5-Bromothiophen-2-yl)-1-methyl-1H-benzo[d]imidazol-5-yl)-5-(2-phenylquinolin-4-yl)-1,3,4-oxadiazole derivatives.

derivatives were investigated for their antidiabetic, anticoagulant and antiplatelet activity. The anticoagulant activity showed by compound **(LXVII)** and **(LXVIII)** with IC_{50} values of 0.66 and 0.68 μg/mL, respectively. The compound **(LXVII)** also showed antiplatelet activity [60].

2.4 Summary

The study summarises the detailed literature review on recent advancements in synthesis and the pharmacological properties of benzimidazole-integrated quinoline hybrids. The most common and efficient approach for the synthesis of benzimidazole derivatives is the condensation reaction between *o*-Phenylenediamine and various carbonyl compounds. However, quinoline derivatives are basically synthesised by the Skraup reaction. It involves a reaction with aniline or a substituted aniline with glycerol in the presence of sulphuric acid, ferrous sulfate, and nitrobenzene. The result and discussion showed that benzimidazole-quinoline hybrids could display various biological activities such as anticancer, antimicrobial, anti-inflammatory, antimycobacterial, antiplasmodial, antitrypanosomal, anticoagulant and antithyroid activities. This established a new level of revolution for the synthesis of hybrid

Entry	Type	Activity
(XVII)	Anticancer activity	IC_{50}= 0.06–0.2 μM
(LIV)	Antimicrobial activity	MIC = 0.5 μg/mL
(LVI)	Anti-inflammatory activity	Protection dose (%) = 55
(LX) and (LXI)	Antimycobacterial activity	MIC = 1.56 μg/mL
(LXIV)	Antiplasmodial activity	IC_{50} = 0.151 μM
(LXVI)	Antitrypanosomal activity	IC_{50} = 1.98 μM
(LXVII)	Anticoagulant activity	IC_{50} = 0.66 μg/mL
(LXII)	Antithyroid activity	1.46 nmol/l

Table 1.
Biological activity of benzimidazole-quinoline hybrids.

molecules. The benzimidazole-quinoline hybrids with the highest reported activities are listed in **Table 1**.

3. Conclusions

The present chapter narrates the distinct pharmacological actions of benzimidazole integrated quinoline hybrid scaffolds. The prime objective of this study is to investigate the pharmacological potential of hybrid compounds consisting of benzimidazole and quinoline moieties in a single molecule. This can help improve their potential to interact with a wide range of biological targets and enhance their pharmacological properties. The efforts are rationalised to develop such scaffolds that could impart numerous pharmacological effects to accomplish the desired aim. Further, the structure–activity relationship, pharmacokinetic, pharmacodynamic, and toxicological studies of benzimidazole-quinoline hybrid derivatives will be highly beneficial for designing and developing novel bioactive agents in the future.

Acknowledgements

The authors are thankful to Veer Narmad South Gujarat University, Surat, for financial assistance (Grant No: IPR/UGC/18441/2021, dated: 26/11/2021). We are also thankful to the U.T. Administration of Dadra & Nagar Haveli and Daman & Diu, and the Principal, Government College, Daman, for encouragement and facilities.

Conflict of interest

The authors declare no conflict of interest.

Author details

Zebabanu Khalifa[1], Rachana Upadhyay[1], Premlata Kumari[2] and Amit B. Patel[1*]

1 Department of Chemistry, Government College, Daman (Affiliated to Veer Narmad South Gujarat University, Surat), Daman (U.T.), India

2 Department of Chemistry, S.V. National Institute of Technology, Surat, India

*Address all correspondence to: amit.svnit10@gmail.com; amit.patel10@ddd.gov.in

References

[1] Jampilek J. Heterocycles in medicinal chemistry. Molecules. 2019;**24**:3839. DOI: 10.3390/molecules24213839

[2] Jangir N, Poonam DS, Jangid DK. Recent advances in the synthesis of five- and six-membered heterocycles as bioactive skeleton: A concise overview. ChemistrySelect. 2022;7:e202103139. DOI: 10.1002/slct.202103139

[3] Kerru N, Gummidi L, Maddila S, Gangu KK, Jonnalagadda SB. A review on recent advances in nitrogen-containing molecules and their biological applications. Molecules. 2020;**25**:1909. DOI: 10.3390/molecules25081909

[4] Patel AB, Kumari P, Chikhalia KH. One-Pot synthesis of novel quinoline-fused Azeto[1,2-a]benzimidazole analogs via intramolecular Pd-catalyzed C–N coupling. Catalysis Letters. 2014; **144**:1332-1338. DOI: 10.1007/s10562-014-1266-9

[5] Wermuth CG. Multitargeted drugs: The end of the 'one-target-one-disease' philosophy? Drug Discovery Today. 2004;**9**:826-827. DOI: 10.1016/S1359-6446(04)03213-1

[6] Bérubé G. An overview of molecular hybrids in drug discovery. Expert Opinion on Drug Discovery. 2016;**11**: 281-305. DOI: 10.1517/17460441.2016.1135125

[7] Wang Z, Deng X, Xiong S, Xiong R, Liu J, Zou L, et al. Design, synthesis and biological evaluation of chrysin benzimidazole derivatives as potential anticancer agents. Natural Product Research. 2018;**32**:2900-2909. DOI: 10.1080/14786419.2017.1389940

[8] Kanwal A, Saddique FA, Aslam S, Ahmad M, Zahoor AF, Mohsin N-A. Benzimidazole ring system as a privileged template for anticancer agents. Pharmaceutical Chemistry Journal. 2018;**51**:1068-1077. DOI: 10.1007/s11094-018-1742-4

[9] Alasmary F, Snelling A, Zain M, Alafeefy A, Awaad A, Karodia N. Synthesis and evaluation of selected benzimidazole derivatives as potential antimicrobial agents. Molecules. 2015;**20**: 15206-15223. DOI: 10.3390/molecules 200815206

[10] Özkay Y, Tunalı Y, Karaca H, Işıkdağ İ. Antimicrobial activity of a new series of benzimidazole derivatives. Archives of Pharmacal Research. 2011; **34**:1427-1435. DOI: 10.1007/s12272-011-0903-8

[11] Srivastava R, Gupta SK, Naaz F, Singh A, Singh VK, Verma R, et al. Synthesis, antibacterial activity, synergistic effect, cytotoxicity, docking and molecular dynamics of benzimidazole analogues. Computational Biology and Chemistry. 2018;**76**:1-16. DOI: 10.1016/j.comp biolchem.2018.05.021

[12] Kumar K, Awasthi D, Lee S-Y, Cummings JE, Knudson SE, Slayden RA, et al. Benzimidazole-based antibacterial agents against Francisella tularensis. Bioorganic & Medicinal Chemistry. 2013;**21**:3318-3326. DOI: 10.1016/j.bmc.2013.02.059

[13] Jiménez-Juárez R, Cruz-Chávez W, de Jesús-Ramírez N, Castro-Ramírez GI, Uribe-González I, Martínez-Mejía G, et al. Synthesis and antimycobacterial activity of 2,5-disubstituted and 1,2,5-trisubstituted benzimidazoles. Frontiers in Chemistry. 2020;**8**:433. DOI: 10.3389/fchem.2020.00433

[14] Keng Yoon Y, Ashraf Ali M, Choon TS, Ismail R, Chee Wei A, Suresh Kumar R, et al. Antituberculosis: Synthesis and antimycobacterial activity of novel benzimidazole derivatives. BioMed Research International. 2013; **2013**:1-6. DOI: 10.1155/2013/926309

[15] Kaur G, Silakari O. Benzimidazole scaffold based hybrid molecules for various inflammatory targets: Synthesis and evaluation. Bioorganic Chemistry. 2018;**80**:24-35. DOI: 10.1016/j.bioorg.2018.05.014

[16] Rathore A, Sudhakar R, Ahsan MJ, Ali A, Subbarao N, Jadav SS, et al. In vivo anti-inflammatory activity and docking study of newly synthesized benzimidazole derivatives bearing oxadiazole and morpholine rings. Bioorganic Chemistry. 2017;**70**: 107-117. DOI: 10.1016/j.bioorg.2016.11.014

[17] Fang Y-M, Zhang R-R, Shen Z-H, Wu H-K, Tan C-X, Weng J-Q, et al. Synthesis, antifungal activity, and SAR study of some new 6-Perfluoropropanyl quinoline derivatives: Quinolines with antifungal activities. Journal of Heterocyclic Chemistry. 2018;**55**: 240-245. DOI: 10.1002/jhet.3031

[18] Yang Y-D, He Y-H, Ma K-Y, Li H, Zhang Z-J, Sun Y, et al. Design and discovery of novel antifungal quinoline derivatives with Acylhydrazide as a promising pharmacophore. Journal of Agricultural and Food Chemistry. 2021; **69**:8347-8357. DOI: 10.1021/acs.jafc.1c00670

[19] Elsayed Khidre R, Fathy Abdel-Wahab B, Abdel-Rehem BF. New quinoline-based compounds for analgesic and anti-inflammatory evaluation. Letters in Drug Design & Discovery. 2011;**8**:640-648. DOI: 10.2174/157018011796235194

[20] Wen X, Wang S-B, Liu D-C, Gong G-H, Quan Z-S. Synthesis and evaluation of the anti-inflammatory activity of quinoline derivatives. Medicinal Chemistry Research. 2015;**24**:2591-2603. DOI: 10.1007/s00044-015-1323-y

[21] Ibrahim DA, Abou El Ella DA, El-Motwally AM, Aly RM. Molecular design and synthesis of certain new quinoline derivatives having potential anticancer activity. European Journal of Medicinal Chemistry. 2015;**102**:115-131. DOI: 10.1016/j.ejmech.2015.07.030

[22] Afzal O, Kumar S, Haider MR, Ali MR, Kumar R, Jaggi M, et al. A review on anticancer potential of bioactive heterocycle quinoline. European Journal of Medicinal Chemistry. 2015;**97**:871-910. DOI: 10.1016/j.ejmech.2014.07.044

[23] Pham TDM, Ziora ZM, Blaskovich MAT. Quinolone antibiotics. MedChemComm. 2019;**10**:1719-1739. DOI: 10.1039/C9MD00120D

[24] Desai NC, Rajpara KM, Joshi VV. Synthesis and characterization of some new quinoline based derivatives endowed with broad spectrum antimicrobial potency. Bioorganic & Medicinal Chemistry Letters. 2012;**22**:6871-6875. DOI: 10.1016/j.bmcl.2012.09.039

[25] Medvedeva SM, Potapov AY, Gribkova IV, Katkova EV, Sulimov VB, Shikhaliev KS. Synthesis, docking, and anticoagulant activity of new factor-Xa inhibitors in a series of Pyrrolo[3,2,1-ij] Quinoline-1,2-Diones. Pharmaceutical Chemistry Journal. 2018;**51**:975-979. DOI: 10.1007/s11094-018-1726-4

[26] Novichikhina NP, Skoptsova AA, Shestakov AS, Potapov AY, Kosheleva EA, Kozaderov OA, et al. Synthesis and anticoagulant activity of

new ethylidene and spiro derivatives of Pyrrolo[3,2,1-ij]quinolin-2-ones. The Journal of Organic Chemistry. 2020;**56**: 1550-1556. DOI: 10.1134/ S1070428020090080

[27] Zemtsova MN, Zimichev AV, Trakhtenberg PL, Klimochkin YN, Leonova MV, Balakhnin SM, et al. Synthesis and antiviral activity of several quinoline derivatives. Pharmaceutical Chemistry Journal. 2011;**45**:267. DOI: 10.1007/s11094-011-0613-z

[28] Kaur R, Kumar K. Synthetic and medicinal perspective of quinolines as antiviral agents. European Journal of Medicinal Chemistry. 2021;**215**:113220. DOI: 10.1016/j.ejmech.2021.113220

[29] Vandekerckhove S, D'hooghe M. Quinoline-based antimalarial hybrid compounds. Bioorganic & Medicinal Chemistry. 2015;**23**:5098-5119. DOI: 10.1016/j.bmc.2014.12.018

[30] Chanquia SN, Larregui F, Puente V, Labriola C, Lombardo E, García LG. Synthesis and biological evaluation of new quinoline derivatives as antileishmanial and antitrypanosomal agents. Bioorganic Chemistry. 2019;**83**: 526-534. DOI: 10.1016/j. bioorg.2018.10.053

[31] Zhang H, Collins J, Nyamwihura R, Ware S, Kaiser M, Ogungbe IV. Discovery of a quinoline-based phenyl sulfone derivative as an antitrypanosomal agent. Bioorganic & Medicinal Chemistry Letters. 2018;**28**: 1647-1651. DOI: 10.1016/j. bmcl.2018.03.039

[32] Maia EHB, Assis LC, de Oliveira TA, da Silva AM, Taranto AG. Structure-based virtual screening: From classical to artificial intelligence. Frontiers in Chemistry. 2020;**8**:343. DOI: 10.3389/ fchem.2020.00343

[33] Ferreira LG, Dos Santos RN, Oliva G, Andricopulo AD. Molecular docking and structure-based drug design strategies. Molecules. 2015;**20**:13384-13421. DOI: 10.3390/molecules200713384

[34] Hranjec M, Pavlović G, Marjanović M, Kralj M, Karminski-Zamola G. Benzimidazole derivatives related to 2,3-acrylonitriles, benzimidazo [1,2-a]quinolines and fluorenes: Synthesis, antitumor evaluation in vitro and crystal structure determination. European Journal of Medicinal Chemistry. 2010;**45**:2405-2417. DOI: 10.1016/j.ejmech.2010.02.022

[35] Perin N, Uzelac L, Piantanida I, Karminski-Zamola G, Kralj M, Hranjec M. Novel biologically active nitro and amino substituted benzimidazo[1,2-a]quinolines. Bioorganic & Medicinal Chemistry. 2011;**19**:6329-6339. DOI: 10.1016/ j.bmc.2011.09.002

[36] Perin N, Alić J, Liekens S, Aerschot AV, Vervaeke P, Gadakh B, et al. Different positions of amide side chains on the benzimidazo[1,2-a] quinoline skeleton strongly influence biological activity. New Journal of Chemistry. 2018;**42**:7096-7104. DOI: 10.1039/C8NJ00416A

[37] Perin N, Martin-Kleiner I, Nhili R, Laine W, David-Cordonnier M-H, Vugrek O, et al. Biological activity and DNA binding studies of 2-substituted benzimidazo[1,2-a] quinolines bearing different amino side chains. Medicinal Chemistry Communications. 2013;**4**:1537. DOI: 10.1039/c3md00193h

[38] Tantawy A, Barghash A, Badr S, Gomaa R. Synthesis of new heterocyclic compounds containing benzimidazole moiety as inhibitors of breast cancer cell

growth. Heterocyclic Communications. 2013;**19**:125-131. DOI: 10.1515/hc-2013-0013

[39] Shi L, Wu T-T, Wang Z, Xue J-Y, Xu Y-G. Discovery of N-(2-phenyl-1H-benzo[d]imidazol-5-yl)quinolin-4-amine derivatives as novel VEGFR-2 kinase inhibitors. European Journal of Medicinal Chemistry. 2014;**84**: 698-707. DOI: 10.1016/j.ejmech.2014.07.071

[40] Brajša K, Vujasinović I, Jelić D, Trzun M, Zlatar I, Karminski-Zamola G, et al. Antitumor activity of amidino-substituted benzimidazole and benzimidazo[1,2- *a*]quinoline derivatives tested in 2D and 3D cell culture systems. Journal of Enzyme Inhibition and Medicinal Chemistry. 2016;**31**:1139-1145. DOI: 10.3109/14756366.2015.1101093

[41] Kuang W-B, Huang R-Z, Qin J-L, Lu X, Qin Q-P, Zou B-Q, et al. Design, synthesis and pharmacological evaluation of new 3-(1H-benzimidazol-2-yl)quinolin-2(1H)-one derivatives as potential antitumor agents. European Journal of Medicinal Chemistry. 2018; **157**:139-150. DOI: 10.1016/j.ejmech.2018.07.066

[42] Kuang W-B, Huang R-Z, Fang Y-L, Liang G-B, Yang C-H, Ma X-L, et al. Design, synthesis and pharmacological evaluation of novel 2-chloro-3-(1 *H* -benzo[*d*]imidazol-2-yl)quinoline derivatives as antitumor agents: *in vitro* and *in vivo* antitumor activity, cell cycle arrest and apoptotic response. RSC Advances. 2018;**8**:24376-24385. DOI: 10.1039/C8RA04640A

[43] Meščić Macan A, Perin N, Jakopec S, Mioč M, Stojković MR, Kralj M, et al. Synthesis, antiproliferative activity and DNA/RNA-binding properties of mono- and bis-(1,2,3-triazolyl)-appended benzimidazo[1,2-a]quinoline derivatives. European Journal of Medicinal Chemistry. 2020;**185**:111845. DOI: 10.1016/j.ejmech.2019.111845

[44] Pragathi YJ, Veronica D, Anitha K, Rao MVB, Raju RR. Synthesis and biological evaluation of chalcone derivatives of (1,2,4-thiadiazol-benzo[d]imidazol-2-yl)quinolin-2(1H)-one as anticancer agents. Chemical Data Collections. 2020;**30**:100556. DOI: 10.1016/j.cdc.2020.100556

[45] Gaikwad NB, Bansode S, Biradar S, Ban M, Srinivas N, Godugu C, et al. New 3-(1 *H* -benzo[*d*]imidazol-2-yl) quinolin-2(1 *H*)-one-based triazole derivatives: Design, synthesis, and biological evaluation as antiproliferative and apoptosis-inducing agents. Archiv der Pharmazie. 2021;**354**:2100074. DOI: 10.1002/ardp.202100074

[46] Sonar SS, Sadaphal SA, Pokalwar RU, Shingate BB, Shingare MS. Synthesis and antibacterial screening of new 4-((5-(difluoromethoxy)-1 *H* -benzo[d]imidazol-2-ylthio)methyl) tetrazolo[1,5 a]quinoline derivatives. Journal of Heterocyclic Chemistry. 2010; **47**:441-445. DOI: 10.1002/jhet.340

[47] Chaudhari R, Rindhe S. Synthesis and antimicrobial activities of novel n-substituted 8-(1-alkyl/alkylsulphonyl/alkoxycarbonyl-benzimidazol-2-ylmethoxy)-5-chloroquinolines. Journal of the Serbian Chemical Society. 2011;**76**:1199-1206. DOI: 10.2298/JSC100817105C

[48] Gowda J, Khader AMA, Kalluraya B, Hidayathulla S. Synthesis, characterization and antibacterial activity of benzimidazole derivatives carrying quinoline moiety. Indian Journal of Chemistry. 2011;**50B**: 1491-1495

[49] Mungra DC, Patel MP, Patel RG. Microwave-assisted synthesis of some new tetrazolo[1,5-a]quinoline-based benzimidazoles catalyzed by p-TsOH and investigation of their antimicrobial activity. Medicinal Chemistry Research. 2011;**20**:782-789. DOI: 10.1007/s00044-010-9388-0

[50] Garudachari B, Satyanarayana MN, Thippeswamy B, Shivakumar CK, Shivananda KN, Hegde G, et al. Synthesis, characterization and antimicrobial studies of some new quinoline incorporated benzimidazole derivatives. European Journal of Medicinal Chemistry. 2012;**54**:900-906. DOI: 10.1016/j.ejmech.2012.05.027

[51] de Souza IOP, Schrekker CML, Lopes W, Orru RVA, Hranjec M, Perin N, et al. Bifunctional fluorescent benzimidazo[1,2-α]quinolines for Candida spp. biofilm detection and biocidal activity. Journal of Photochemistry and Photobiology B: Biology. 2016;**163**:319-326. DOI: 10.1016/j.jphotobiol.2016.08.037

[52] Villa P, Arumugam N, Almansour AI, Suresh Kumar R, Mahalingam SM, Maruoka K, et al. Benzimidazole tethered pyrrolo[3,4-b]quinoline with broad-spectrum activity against fungal pathogens. Bioorganic & Medicinal Chemistry Letters. 2019;**29**:729-733. DOI: 10.1016/j.bmcl.2019.01.006

[53] El Faydy M, Dahaieh N, Ounine K, Lakhrissi B, Warad I, Tüzün B, et al. Synthesis, identification, antibacterial activity, ADME/T and 1BNA-docking investigations of 8-quinolinol Analogs bearing a benzimidazole moiety. Arabian Journal for Science and Engineering. 2022;**47**:497-510. DOI: 10.1007/s13369-021-05749-7

[54] El-Feky SA, Thabet HK, Ubeid MT. Synthesis, molecular modeling and anti-inflammatory screening of novel fluorinated quinoline incorporated benzimidazole derivatives using the Pfitzinger reaction. Journal of Fluorine Chemistry. 2014;**161**:87-94. DOI: 10.1016/j.jfluchem.2014.02.012

[55] Madawali IM, Das KK, Kalyane NV. Synthesis And Anti-Inflammatory Activity of Some New 2- Chloro-3-[3-(6-Nitro-1H-Benzimidazol-2-Yl)-1H-Pyrazol-5- Yl]Quinolines. IOSR Journal of Pharmacy and Biological Sciences. 2018;**13**:83-88. DOI: 10.9790/3008-1305018388

[56] Ukrainets IV, Grinevich LA, Tkach AA, Gorokhova OV, Kravchenko VN, Sim G. 4-hydroxy-2-quinolones. 191.* synthesis, tautomerism and biological activity of benzimidazol-2-ylamides of 1-r-4-hydroxy-2-oxo-1,2-dihydroquinoline-3-carboxylic acids. Chemical Heterocyclic Compound. 2011; **46**:1364-1370. DOI: 10.1007/s10593-011-0673-8

[57] Mantu D, Antoci V, Moldoveanu C, Zbancioc G, Mangalagiu II. Hybrid imidazole (benzimidazole)/pyridine (quinoline) derivatives and evaluation of their anticancer and antimycobacterial activity. Journal of Enzyme Inhibition and Medicinal Chemistry. 2016;**31**: 96-103. DOI: 10.1080/14756366.2016.1190711

[58] Baartzes N, Stringer T, Seldon R, Warner DF, Taylor D, Wittlin S, et al. Bioisosteric ferrocenyl aminoquinoline-benzimidazole hybrids: Antimicrobial evaluation and mechanistic insights. European Journal of Medicinal Chemistry. 2019;**180**:121-133. DOI: 10.1016/j.ejmech.2019.06.069

[59] Pomel S, Dubar F, Forge D, Loiseau PM, Biot C. New heterocyclic compounds: Synthesis and antitrypanosomal properties. Bioorganic

& Medicinal Chemistry. 2015;**23**: 5168-5174. DOI: 10.1016/j.bmc.2015.03.029

[60] Bharadwaj SS, Poojary B, Nandish SKM, Kengaiah J, Kirana MP, Shankar MK, et al. Efficient synthesis and in silico studies of the benzimidazole hybrid scaffold with the Quinolinyloxadiazole skeleton with potential α-glucosidase inhibitory, anticoagulant, and antiplatelet activities for type-II diabetes mellitus management and treating thrombotic disorders. ACS Omega. 2018;**3**: 12562-12574. DOI: 10.1021/acsomega.8b01476

Chapter 8

Construction of Biologically Active Five- and Six-Membered Fused Ring Pyrimidine Derivatives from 1,3-Diarylthiobarbituric Acids (DTBA)

Warjeet S. Laitonjam and Nimalini Moirangthem

Abstract

Several derivatives of fused pyrimidines were synthesized in maximum yields by using the respective condensation products, namely, 5-ethoxymethylene-1, 3-diaryl-2-thiobarbituric acids and 5-phenyl-methylene-1,3-diaryl-2-thiobarbituric acids, which can be obtained from 1,3-diarylthiobarbituric acids (DTBA). These condensation products possessing three electrophilic centres could undergo cyclo-condensation with various binucleophiles to give various fused heterocycles of pyrimidine derivatives, such as pyrazolo[3,4-*d*]pyrimidine-6-thiones, 5,7-diaryl-4-oxo-isoxazolo[5,4-*d*]pyrimidine-6-thiones, 5-oxo-pyrimido[4,5-*d*]pyrimidine-7-thiones, 2-thioxo-pyrano[2,3-d]pyrimidine-4-ones, pyrido[2,3-*d*]pyrimidines, quinazoline-4-oxo-2-thiones.

Keywords: diarylthiobarbituric acids, pyrazolopyrimidines, isoxazolopyrimidines, pyrimidopyrimidines, pyranopyrimidines, quinazolines

1. Introduction

The development of physiologically highly potent fused pyrimidines is a challenging task for synthetic organic chemists [1]. It is well known that pyrimidines either in isolated or in fused state are associated with a number of biological activities [2–6]. Moreover the pyrimidine nucleus containing thiouriedo linkage (-NH-C(S)-NH-) is pharmaceutically important as the development of medicine mainly arose from the heterocyclic compounds containing nitrogen and sulphur atoms [7–9]. Due to a wide range of biological activities exhibited by pyrimidine derivatives, these compounds occupy a unique place in the field of biological and medicinal chemistry. In view of such wide applications, several derivatives of fused pyrimidines were synthesized in maximum yields by using 1,3-diarylthiobarbituric acids, **1** (DTBA) which can be prepared in one pot reaction by treating 1,3-diarylthioureas with malonic acid in presence of acetyl chloride [10]. They are generally stable at room temperature and can be stored indefinitely without apparent

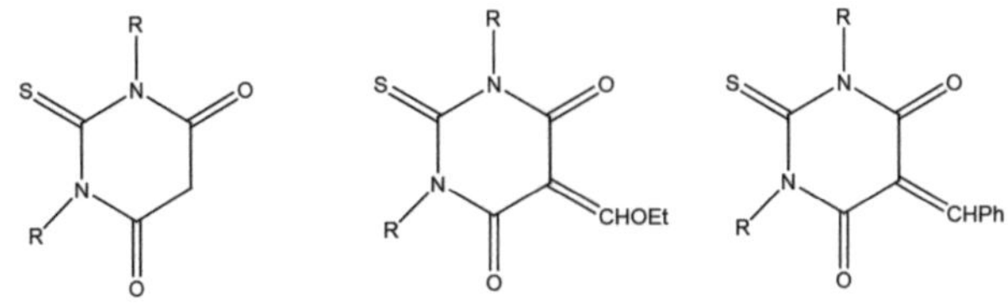

1 **2** **3**

Figure 1.
Thiobarbituric acid derivatives.

decomposition. Having an active methylene group, they can furnish several condensation products for easy cyclization to give various fused heterocyclic compounds of interest. When DTBA was reacted with ethyl orthoformate and benzaldehyde to the condensation products, namely, 5-ethoxymethylene-1,3-diaryl-2-thiobarbituric acids (**2**) and 5-phenyl-methylene-1,3-diaryl-2-thiobarbituric acids (**3**), respectively were formed (**Figure 1**) [11]. As **2** and **3** possess three electrophilic centres, they could undergo cyclocondensation with various binucleophiles to give various fused heterocycles of pyrimidine derivatives [12].

2. Pyrazolo-pyrimidines

The methods reported for the synthesis of pyrazolo-pyrimidines involved a number of steps and yields were poor [13–18]. A convenient route for the synthesis of pyrazolo-pyrimidines was described [11, 19]. The condensation products, **2** and **3** have been utilized as three carbon fragments for the synthesis of 4-oxo-pyrazolo[3,4-*d*]pyriimidines, **4** and **6**, respectively. When **2** was treated with hydrazine hydrate in the presence of ethanol and acetic acid, the corresponding pyrazolo[3,4-*d*]pyrimidin-6-thione (**4**) was yielded in 65–82% overall yields (**Figure 2**). Similarly, when **2b** was treated with phenyl hydrazine in ethanol and acetic acid, 2-phenyl-5,7-bis(2′-methylphenyl)-4-oxo-pyrazolo[3,4-*d*]pyrimidin-6-thione (**5b**) was produced in 55% yield. However, when **3** was refluxed with hydrazine hydrate in ethanol and acetic acid, the corresponding (3*H*)-3-phenyl-5,7-diaryl-4-oxo-pyrazolo[3,4-*d*]pyrimidin-6-thiones (**6**) were obtained in 60–75% overall yields (**Figure 2**). It was observed that **6** could be oxidized to give the respective 3-oxo-pyrazolo-[3,4-*d*]pyrimidin-6-thiones (7) [11, 12, 19].

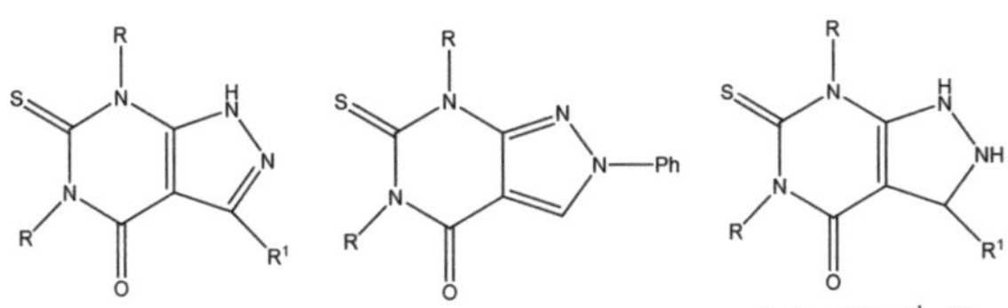

4a, R = C_6H_5, R^1 = H;
4b, R = 4-$CH_3C_6H_4$, R^1 = H;
4c, R = 4-ClC_6H_4, R^1 = H;
7a, R = C_6H_5, R^1 = Ph;
7b, R = 4-$CH_3C_6H_4$, R^1 = Ph;
7c, R = 4-ClC_6H_4, R^1 = Ph

5b, R = 4-$CH_3C_6H_4$, R^1 = H

6a, R = C_6H_5, R^1 = Ph;
6b, R = 4-$CH_3C_6H_4$, R^1 = Ph;
6c, R = 4-ClC_6H_4, R^1 = Ph

Figure 2.
Pyrazolo-pyrimidines.

It is well known that the fused heterocycles possessing the pyrazolo[3,4-*d*]pyrimidine nucleus serve as a class of compounds which exhibited a remarkable variety of biological activities [20–23]. The pyrazolo[3,4-*d*]pyrimidine compounds have been found to show similar characteristics to purines, since they are isomeric structural purine analogues, which have resulted in several potent antagonists in biological systems [24, 25].

3. Isoxazolo-pyrimidines

Most of the methods for the preparation of the isoxazolo[5,4-*d*]pyrimidine system are initiated from the construction of an isoxazole ring first and followed by pyrimidine ring closure [26, 27]. DTBAs were also used as precursors for the synthesis of various isoxazolo[5,4-*d*]pyrimidines [28]. When **2** and **3** were treated with hydroxylamine hydrochloride in dehydrated alcohol and acetic acid gave 5,7-diaryl-4-oxo-isoxazolo[5,4-*d*]pyrimidin-6-thiones **(8)** in 65–85% overall yields and 3,5,7-triaryl-2,3-dihydro-4-oxo-isoxazolo[5,4-*d*]pyrimidin-6-thones (**9**) in 60–70% overall yields, respectively (**Figure 3**) [28].

It was reported that the isoxazolo-pyrimidine derivatives exhibited antifungal, antibacterial and many other important biological properties [29–31]. The biological activities in terms of the antifungal and antibacterial properties of the compounds (**8** and **9**) were determined by the standard disc diffusion method. The bacteria (*Escherichia coli* and *Bacillus subtilis*) and fungi (*Penicillium* species, *Aspergillus* sp., *Fusarium* sp. and *Trichoderma* sp) were grown in nutrient agar and potato dextrose agar (PDA) plates, respectively. Compound **9c** was found to have the highest activity against coli form organisms whereas other compounds were moderated and have less activity. The studied compounds were found to be inactive against some species of fungi, except **8b** was found to have antibiotic property in *penicillium* species.

4. Furo-pyrimidines

It was reported [32] that furo[2,3-*d*]pyrimidines **10** were prepared by condensing 1,3-diarylthiobarbituric acids (**1**) with benzoin in presence of *p*-toluene sulphonic acid. In another reaction, derivatives of furo[2,3-*d*]pyrimidines **11** were prepared by direct condensation of the corresponding 1,3-diarythiobarbituric acids, **1** with chloroacetone in presence of triethylamine (**Figure 4**).

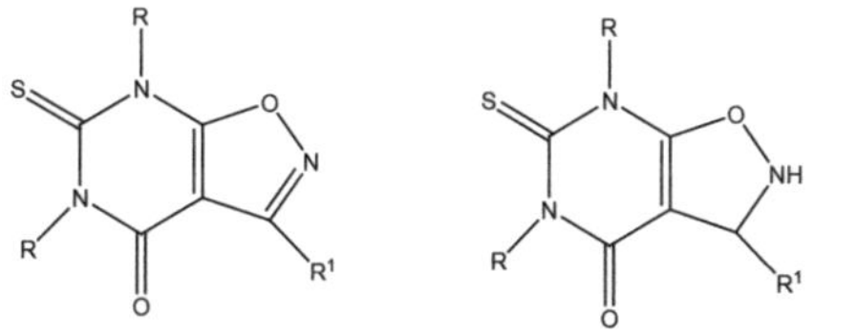

8a, R = C_6H_5, R^1 = H;
8b, R = 4-$CH_3C_6H_4$, R^1 = H;
8c, R = 4-ClC_6H_4, R^1 = H

9a, R = C_6H_5, R^1 = Ph;
9b, R = 4-$CH_3C_6H_4$, R^1 = Ph;
9c, R = 4-ClC_6H_4, R^1 = Ph

Figure 3.
Isoxazolo-pyrimidines.

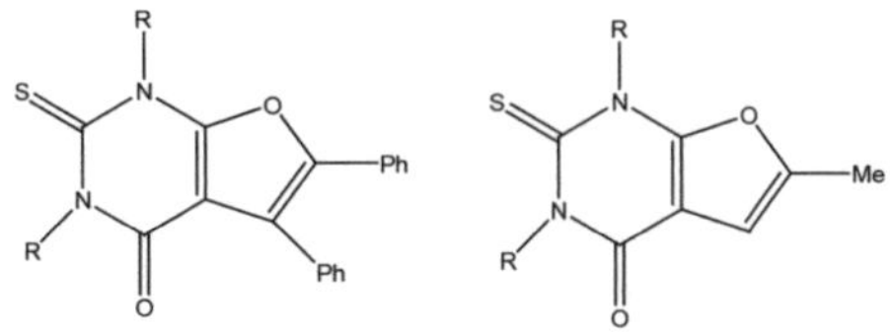

10a, R = C_6H_5;
10b, R = 4-$CH_3C_6H_4$;
10c, R = 4-ClC_6H_4

11a, R = C_6H_5;
11b, R = 4-$CH_3C_6H_4$;
11c, R = 4-ClC_6H_4

Figure 4.
Furo-pyrimidines.

5. Thieno-pyrimidines

The synthesis of thieno[2,3-*d*] pyrimidines **13** which are found to be associated with varied biological and pharmacological activities, on condensation of **12** with mesityl thioglycolate was reported (**Figure 5**) [33].

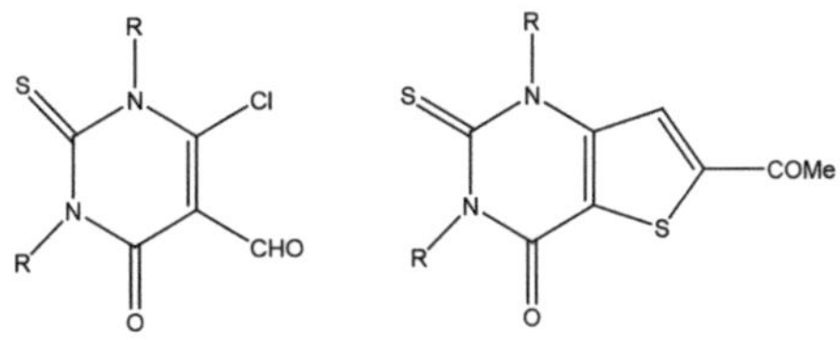

12a, R = C_6H_5;
12b, R = 4-$CH_3C_6H_4$;
12c, R = 4-ClC_6H_4

13a, R = C_6H_5;
13b, R = 4-$CH_3C_6H_4$;
13c, R = 4-ClC_6H_4

Figure 5.
Thieno-pyrimidines.

6. Pyrimido-pyrimidines

In view of the various physiological properties, the pyrimido[4,5-*d*]pyrimidines were generally prepared from derivatives of enamines [34, 35]. Various substituted 2-amino-6,8-diaryl-5-oxo-pyrimido[4,5-*d*]pyrimidin-7-thiones (**14**) in 62–75% overall yields can be synthesized by reacting the respective compounds **2** with guanidine nitrate in the presence of sodium methoxide and methanol (**Figure 6**) [36]. Similarly, the condensation products **3** were reacted with guanidine nitrate in sodium methoxide and methanol affording the corresponding substituted 3,4-dihydro-5-oxo-pyrimido[4,5-*d*] pyrimidine-7-thiones, **15** in good yields (**Figure 6**) [36].

The derivatives of pyrimido-pyrimidines are quite effective in the inhibition of cancer cell sickness [37], exhibition of diuretic activities and anti-inflammatory activities [38]. The antifungal activity for the compounds **14** and **15** was screened against *Aspergillus niger* at different concentrations by agar growth paper disc method using Czapeck's nutrient medium and DMF was used as control. The average percentage inhibition after 96 hr. was determined and the results were compared with those obtained using commercial fungicide Carbendazim. Most of the compounds were found to be fairly active at the concentrations of 1000 ppm comparable with standard fungicide. It is noteworthy that introduction of pyrimido moiety in the pyrimidine nucleus enhances the fungitoxicity to some extent and furthermore,

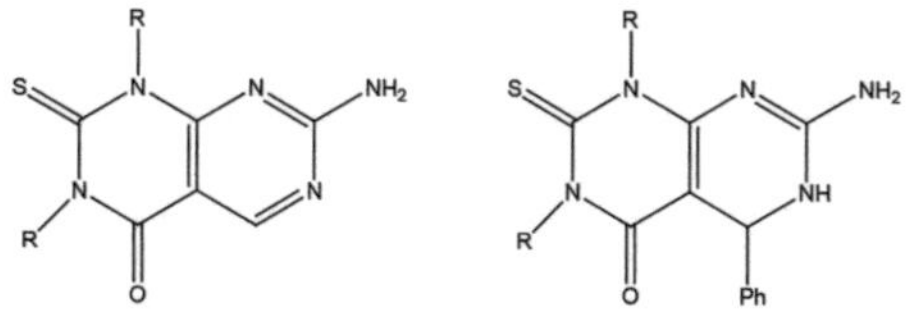

14a, R = C_6H_5;
14b, R = 4-$CH_3C_6H_4$;
14c, R = 4-ClC_6H_4

15a, R = C_6H_5;
15b, R = 4-$CH_3C_6H_4$;
15c, R = 4-ClC_6H_4

Figure 6.
Pyrimido-pyrimidines.

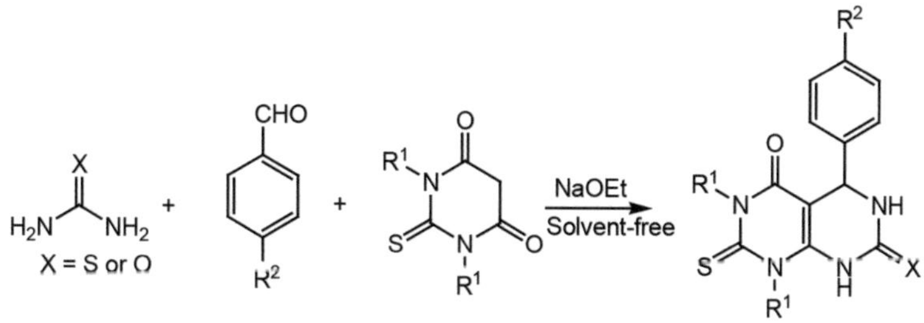

Figure 7.
Synthesis of thioxo pyrimido[4,5-d]pyrimidinone derivatives.

chloro-substituted heterocycles (**14c** and **15c**) were more active than methyl-substituted compounds (**14b** and **15b**).

Many of the earlier methods reported for the syntheses of pyrimido[4,5-*d*]pyrimidine derivatives require toxic chemicals, long reaction times, afford low to moderate yields and require many steps. A simple one-pot method for the synthesis of thioxo pyrimido[4,5-*d*]pyrimidinone derivatives by treatment of *N,N*-diethylthiobarbituric acids, benzaldehydes and thiourea/urea with NaOEt as the catalyst under solvent-free conditions was reported (**Figure 1**). Several 5-aryl-2,7-dithioxo-pyrimido[4,5-d]pyrimidine-4-ones; 4-aryl-7-thioxo-pyrimido[4,5-d]pyrimidine-2,5-diones were synthesized using an eco-friendly and efficient, multi-component reaction (MCR) under solvent free conditions. The reactions proceed *via* Biginelli type condensation of aromatic aldehyde, thiobarbituric acid and urea or thiourea in presence of catalytic amount of NaOEt. The synthesized compounds were screened for anti-inflammatory activity. It was observed that the compounds containing fluoro-substituents gave good yields and showed high anti-inflammatory activities. There are no reports available on the formation of Biginelli products using NaOEt as a catalyst under solvent-free conditions (**Figure 7**).

7. Pyrano-pyrimidines

Pyrano[2,3-*d*]pyrimidine derivatives could be synthesized *via* a multicomponent domini Knoevenagel/hetero Diels-Alder reaction of 1,3-dimethyl barbituric acid with an aromatic aldehyde and ethyl vinyl ether or 2,3-dihydrofuran in presence of 1 mol % of indium (III) chloride. The reaction also proceeds in aqueous media without using any catalyst, but the yield is comparatively less (65–75%). Preparation of naturally occurring complex molecules containing a uracil ring possesses significant synthetic challenges. The development of clinically useful anticancer (5-fluorouracil)

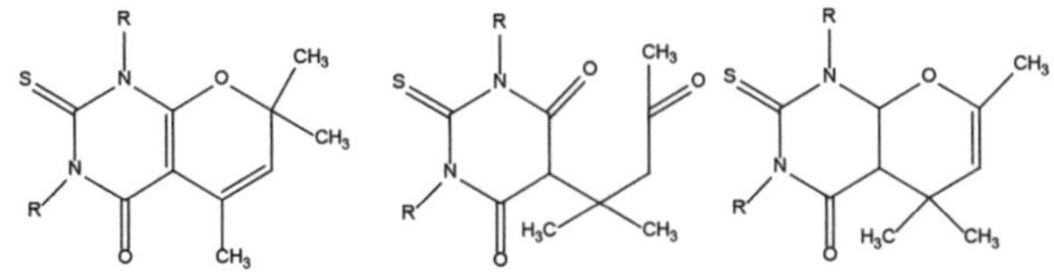

16 17 18

Figure 8.
Pyrano-pyrimidines.

and antiviral drugs (AZT, DDC, DDI, BVDV) has renewed interest in the synthetic manipulation of uracils.

Ahluwalia *et al.* [39] reported the synthesis of pyrano[2,3-*d*] pyrimidines **16** by reacting thiobarbturic acids, **1** with mesityl oxide in the presence of pyridine (**Figure 8**). The products were identical with the products obtained by the reaction of appropriate thiobarbituric acids with acetone in presence of triethylamine. The reaction of thiobarbituric acids with mesityl oxide in the absence of any base gave the corresponding open chain compounds **17** which on heating with glacial acetic acids and phosphorus pentoxide gave the corresponding cyclic compounds **18** [39].

Although a variety of routes for the synthesis of these compounds have been appeared in the literature, the majority of them involve a number of steps, drastic conditions, long reaction time and low yields [40–44]. Moreover, very few methods are reported for the synthesis of 2-thioxo-pyrano[2,3-d]pyrimidine-4-ones, as most of the methods reported are of pyrano[2,3-*d*]pyrimidines. In search of an efficient method and in continuation to our studies on fused pyrimidine derivatives, we report the full details of the work and studies related to the synthesis of 7-amino-1,3-diaryl-5-phenyl-2-thioxo-pyrano[2,3-d]pyrimidin-4(1*H*)-ones (**19**) from DTBA by reacting with various arylidenes, such as, phenylmethylenemalononitrile, ethyl phenylmethylene-cyanoacetate and phenylmethylenecyanoacetamide in presence of sodium methoxide and methanol [45]. Thus, the reaction of DTBA with arylidenes and sodium methoxide in presence of methanol by refluxing for 6 hr. afforded the compounds **19** in 70–82% overall yields (**Figure 9**).

Compounds containing dihydro-5H-pyrano[2,3-*d*]pyrimidines moiety have interesting biological properties [46]. In addition, compounds having a chalcone

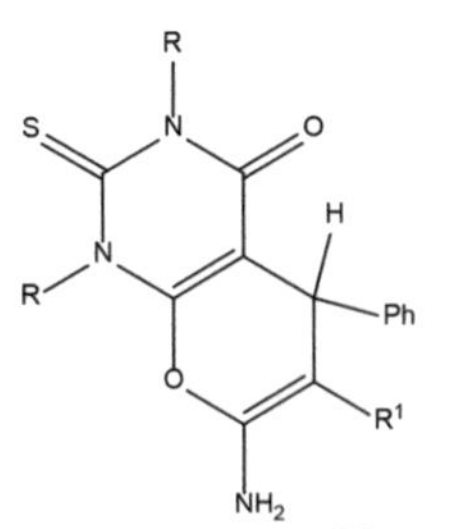

19a, R = C_6H_5, R_1 = CN;
19b, R = 4-ClC_6H_4, R_1 = CN;
19c, R = 4-$CH_3C_6H_4$, R_1 = CN;
19d, R = 2-$CH_3C_6H_4$, R_1 = CN;
19e, R = 4-ClC_6H_4, R_1 = OH;
19f, R = 4-$CH_3C_6H_4$, R_1 = OH

Figure 9.
2-Thioxo-pyrano[2,3-d]pyrimidine-4-ones.

unit attached to the pyranopyrimidine ring are efficient herbicides [47]. The antifungal screening of the synthesized compounds (**19**) were done *in vitro* at 100 and 300 μg/ml solution against the two types of fungi, namely, *Fusarium oxysporum* and *Helminthosporium oryzae* using acetone as solvent. The zones of inhibition were measured in mm. The result showed that most of the compounds exhibit moderate to high activity against both fungi. Incorporation of chlorine increased the activity of high order [48]. The synthesized compounds, **19** were also screened against gram negative bacteria *Escherichia coli* (MTCC 739) and gram positive bacterium *Bacillus subtilis* (MTCC 121). The agar cup-plate method was used and nutrient agar was culture medium for the antibacterial activity. The solvent acetone served as a control. The zone of inhibition frame was measured in mm. The result of antibacterial activity showed that compounds **19b** and **19e** showed the most active compound against *B. subtilis* only, while **19a** and **19c** do not show any zone of inhibition against these bacteria [48].

8. Pyrido-pyrimidines

Very few substituted pyrido[2,3-*d*]pyrimidines were synthesized from 6-aminouracils [49, 50]. The synthesis of substituted 3-cyano-6,8-diaryl-2,5-dioxo-pyrido[2,3-*d*]pyrimidin-7-thiones (**20**) from the corresponding condensation product (**2**) derived from 1,3-diarythiobarbituric acids were also reported [12]. Thus, the reaction of **2** with cyanoacetamide in presence of sodium isopropoxide and isopropanol afforded the corresponding pyrido[2,3-*d*]pyrimidines (**20**) in 62–74% overall yields (**Figure 10**).

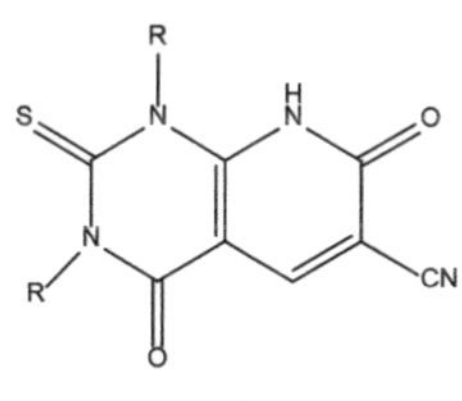

20a, R = C_6H_5;
20b, R = 4-$CH_3C_6H_4$;
20c, R = 4-ClC_6H_4

Figure 10.
Pyrido-pyrimidines.

9. Quinazolines

Quinazolines are interesting targets for new method development due to their importance in a broad range of therapeutic areas [51, 52]. Quinazoline derivatives, which possess a wide range of biological activities contain the 4-oxo-2-thioxo-1,2,3,4-tetrahydropyrimidine structural moiety in their heterocyclic rings [53–58].

The synthetic methods available for the preparation of quinazolines involved the amidation of 2-aminobenzoic acid or its derivatives, i.e. 2-aminobenzonitrile, 2-aminobenzoate, and 2-arylnitrilium salts, followed by oxidative ring closure [59–63]. Other synthetic pathways include the cyclization of anthranilamides with aldehydes [64], and with ketones or acid chlorides under acidic or basic conditions [65–67]. These methods involved multistep processes, poor yields, toxic reagents and

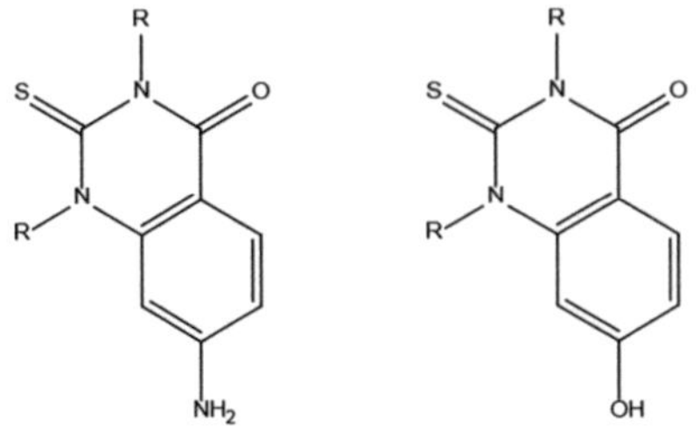

21a, R = C_6H_5; 21b, R = 4-$CH_3C_6H_4$; 21c, R = 4-ClC_6H_4

22a, R = C_6H_5; 22b, R = 4-$CH_3C_6H_4$; 22c, R = 4-ClC_6H_4

Figure 11.
Quinazolines.

time-consuming experimental procedures. Moreover, very few methods are reported for the synthesis of 2-thioxoquinazoline-4-ones, as most of the methods reported are of quinazoline-2,4(1*H*,3*H*)-diones, from various sources [61, 62, 68–75].

Recently, Saeed *et al.* [76] reported the base catalysed intramolecular nucleophilic cyclization of substituted thioureas in the presence of DMF to afford the 2-thioxoquinazoline-4-ones. The solid-phase synthesis for the preparation of2-thioxoquinazoline-4-one had been reported [77–80]. Treatment of 5-ethoxymethylene-1,3-diaryl-2-thiobarbituric acids (**2**) with malononitrile in presence of NH_4OAc and acetic acid with $ZnCl_2$ as catalyst in refluxing condition gave the corresponding 2-thioxoquinazolin-4-ones, **21** in 78–85% overall yields [81] (**Figure 11**). However, the reaction of **2** with ethyl cyanoacetate in presence of ammonium acetate and acetic acid with $ZnCl_2$ as a catalyst gave 7-hydroxy-2,3-dihydro-2-thioxo-1,3-diarylquinazolin-4(*1H*)-ones, **22** in 76–87% overall yields [81].

The synthesized compounds **21** and **22** were screened *in vitro* for their antimicrobial activities [82, 83]. Cytotoxicity studies were performed for the compounds on human lung cancer cell line A549 using an MTT assay. The A549 cells were grown at 37°C, 5% CO_2 and 100% relative humidity.

10. Benzo[5,6]chromeno[2,3-d]pyrimidines

Novel 2-thioxo-benzo[5,6]chromeno[2,3-d]pyrimidin-4-one derivatives were synthesized in aqueous media using cetylpyridinium chloride (CPC) as micellar catalyst in three-component one-pot reaction involving thiobarbituric acids, aromatic aldehydes and β-napthol [84]. The synthesized compounds were found to show antioxidant and cytotoxic activities (**Figure 12**).

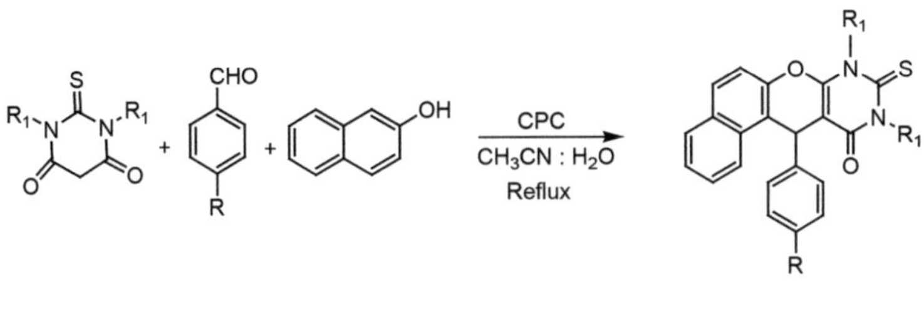

Figure 12.
Synthesis of 2-thioxo-benzo[5,6]chromeno[2,3-d]pyrimidin-4-one derivatives.

11. Pyrazolopyrano-pyrimidinones

Tricyclic fused pyrazolopyranopyrimidines were synthesized by one-pot, four-component reaction of ethyl acetoacetate, hydrazine hydrate, aromatic aldehydes and barbituric acid in good to excellent yields (88–95%) [85]. Another method for the synthesis of pyrazolopyranopyrimidines was employed using DABCO as catalyst in water [86].

A new series of triheterocyclic compounds containing pyrazole, pyran, and pyrimidinone rings was synthesized via a one-pot condensation of ethylacetoacetate, hydrazine hydrate, barbituric acid, and aromatic aldehydes in the presence of catalytic amounts of titanium dioxide nanowires [87]. Various functional groups were well tolerated under the optimized reaction conditions. A highly efficient, green protocol, one-pot four-component reaction involving thiobarbituric acid, hydrazine hydrate, ethyl acetoacetate and aromatic aldehydes for the synthesis of 7-thioxo-pyrazolopyrano-pyrimidinone derivatives has been accomplished using SDS (sodium dodecyl sulphate) as a catalyst (**Figure 13**) [88]. The procedure offers the advantages of green solvent, easy work-up avoiding the chromatographic separation and use of inexpensive, biodegradable, reusable catalyst. These novel 7-thioxo-pyrazolopyrano-pyrimidinone derivatives were screened for antimicrobial, and antioxidant activities. It was found that 3-methyl-4-(2,4-dichlorophenyl)-6,8-diethyl-7-thioxo4,6,7,8-tetrahydropyrazolo[4′,3′:5,6]pyrano-[2,3-d]pyrimidin-5(1H)-one has shown high antifungal and anti-bacterial activities against the tested fungi and bacteria, which may be due to the presence of chlorine atoms [88]. All the prepared pyrazolopyranopyrimidines were tested as anti-inflammatory agents and some of them revealed moderate to potent anti-inflammatory activity [89].

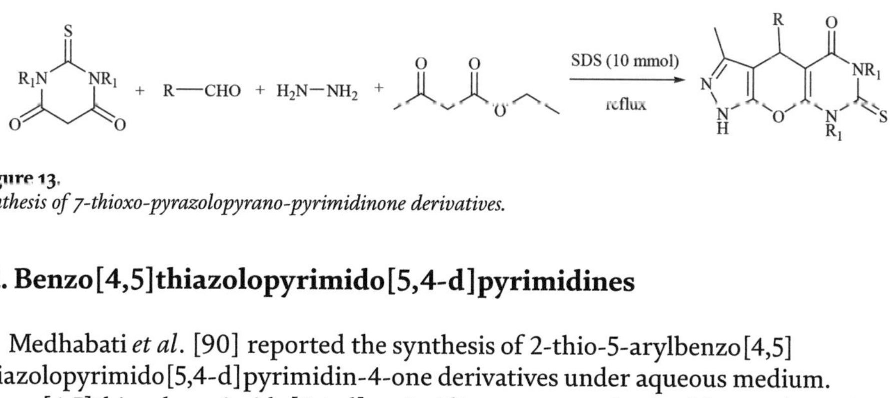

Figure 13.
Synthesis of 7-thioxo-pyrazolopyrano-pyrimidinone derivatives.

12. Benzo[4,5]thiazolopyrimido[5,4-d]pyrimidines

Medhabati *et al.* [90] reported the synthesis of 2-thio-5-arylbenzo[4,5]thiazolopyrimido[5,4-d]pyrimidin-4-one derivatives under aqueous medium. Benzo[4,5]thiazolopyrimido[5,4-d]pyrimidines were synthesized by condensation

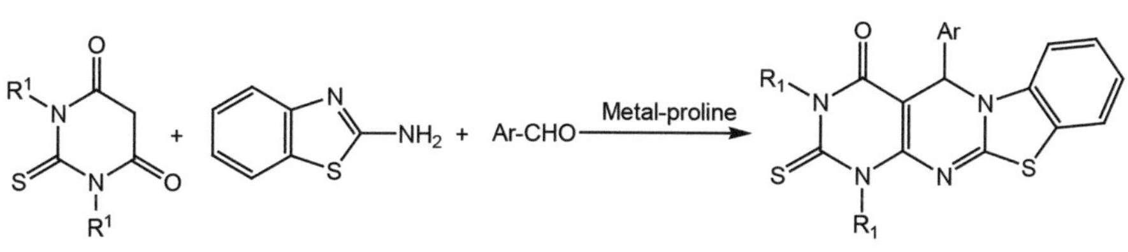

Figure 14.
Synthesis of 2-thio-5-arylbenzo[4,5]thiazolopyrimido[5,4-d]pyrimidin-4-ones.

of thiobarbituric acids, 2-aminobenzothiazole and aldehydes using metal-proline as catalyst (**Figure 14**).

13. Pyrimido[5,4-e[1,3,4]thiadiazolo[3,2-a]pyrimidines

An efficient one-pot method for the synthesis of 2-methyl-5,7,9-triphenyl-8-thioxo-8,9-dihydro-5H-pyrimido[5,4-e[1,3,4]thiadiazolo[3,2-a] pyrimidin-6(7H)-one and its derivatives was reported. The present protocol is also extendable to a wide variety of substrates. The advantages of this protocol are the use of easily available catalyst, short reaction time, easy work-up, ease of product isolation, and high yield (**Figure 15**).

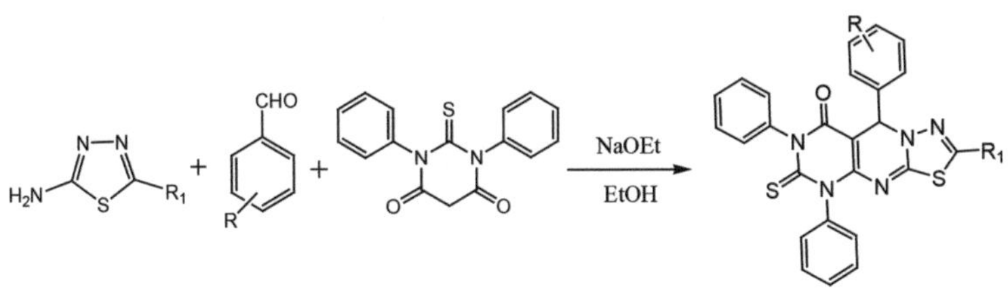

Figure 15.
Synthesis of Pyrimido[5,4-e[1,3,4]thiadiazolo[3,2-a]pyrimidines.

14. Conclusion

Many fused pyrimidine derivatives are belonged to the most important heterocyclic systems, and there are numerous synthetic preparative methods for these compounds. However, in some cases, difficult access to key intermediates, or to their precursors, was a serious limitation for the above syntheses. Various fused pyrimidines were synthesized in maximum yields by using the respective condensation products, namely, 5-ethoxymethylene-1,3-diaryl-2-thiobarbituric acids and 5-phenyl-methylene-1,3-diaryl-2-thiobarbituric acids, which can be obtained from 1,3-diarylthiobarbituric acids (DTBA). These condensation products possessing three electrophilic centres could undergo cyclocondensation with various binucleophiles to give various fused heterocycles of pyrimidine derivatives, such as, pyrazolo[3,4-*d*] pyrimidine-6-thiones, 5,7-diaryl-4-oxo-isoxazolo[5,4-*d*]pyrimidine-6-thiones, 5-oxo-pyrimido[4,5-*d*]pyrimidine-7-thiones, 2-thioxo-pyrano[2,3-d]pyrimidine-4-ones, pyrido[2,3-*d*]pyrimidines, quinazoline-4-oxo-2-thiones, etc.

Author details

Warjeet S. Laitonjam[1*] and Nimalini Moirangthem[2]

1 Department of Chemistry, Manipur University, Imphal, Manipur, India

2 ARS Dharma College, University of Delhi, New Delhi, India

*Address all correspondence to: warjeet@yahoo.com

References

[1] Katritzky AR, Rees CW. Comprehensive Heterocyclic Chemistry. New York: Pergamon Press; 1984. p. 173

[2] Moriguchi I, Kaneniwa N. Spectroscopic studies on molecular interactions. IV. Charge-transfer properties and antibacterial activity of sulfonamides. Chemical Pharmaceutical Bulletin. 1969;**17**(12):2554-2557

[3] Arakawa K, Miyasaka T, Ochi H. Heterocyclic compounds. II. Acetyl transfer reactions of 3-acetoxy-1-acetyl-5-methylpyrazole and related compounds. Chemical and Pharmaceutical Bulletin. 1974;**22**:214-223

[4] Ahluwalia VK, Das U. A facile entry to pyrimidine-derivatives - synthesis of 5-(1,3-dithiolan dithian-2-ylidene) and pyrano[2,3-d]pyrimidines. Industrial Journal of Chemistry. 1996;**35B**:673-676

[5] Zeid IF, Abdel-Rahman AAH, Abdel-Megeid AES, El-Etrawy ASH. Synthesis of new thiolated acyclonucleosides with potential anti HBV activity. Nucleosides, Nucleotides. 1995;**18**:95-111

[6] Sengupta AK, Agarwal KC, Mushtaq M. Synthesis and anticonvulsant activity of some substituted Thiosemicarbazides and triazoles. Industrial Journal of Chemistry. 1979;**17B**:184-185

[7] Singh H, Yadav LDS. Synthesis of some 5-Aryl-2-heteroaryl/heteroarylamino-1, 3, 4-oxadiazoles as potential fungicides. Agricultural Biological Chemistry. 1976;**40**:759-764

[8] Rashad AE, Heikal OA, El-Nezhawy AOH, Abdel-Megeid FME. Synthesis and isomerization of thienotriazolopyrimidine and thienotetrazolopyrimidine derivatives with potential anti-inflammatory activity. Heteroatom Chemistry. 2005;**16**:226-234

[9] Shukla HK, Desai NC, Astik RR, Thaker KA. Studies on some thiosemicarbazones and 1, 3, 4-thiadiazoles as potential antitubercular and antibacterial agents. Journal of Industrial Chemical Society. 1984;**61**:168-171

[10] Das IND, Dutt S. Colour in relation to chemical constitution of the organic and inorganic salts of Jsonitroso-diphenyl-thio-barbituric acid and its higher homologues and analogues. Proceedings of the Indian Academy Science. 1938;**8A**:145-159

[11] Laitonjam WS, Nongmeikapam AD. Synthesis of substituted 4-oxopyrazole[3,4-d]pyrimidine-6-thiones from 1,3-diaryl-5-ethoxymethylene-2-thiobarbituric acids. Indian Journal of Chemistry. 1994;**33B**:1091-1092

[12] Laitonjam WS, Thokchom HS, Nongmeikapam AD. Synthesis of fused pyrazolo-, isoxazolo-, pyrimido- and pyridopyrimidines. Canadian Journal of Chemistry. 2005;**83**:1056-1062

[13] Kohra S, Tomonaga Y, Hosomi A. Synthesis of pyrimidine derivatives by the reaction of ketene dithioacetals with amides. Journal of Heterocycling Chemistry. 1988;**25**:959-968

[14] Swelam SA, Abd El-Salam OS, Zaki MEA. Synthesis of some pyrazolo[3,4-d]pyrimidines and their fused triazole and tetrazole derivatives. Journal of the Serbian Chemical Society. 1999;**64**:655-662

[15] Mect SM, Werner D. A concise synthesis of substituted 4-alkylaminopyrazolo[3,4-d] pyrimidines. Journal of the Chemical Society, Perkin Transactions. 1973;**1**:1903-1906

[16] Papesch V, Dodson RM. Isomeric Pyrazolo[4,3-d]pyrimidinediones. The Journal of Organic Chemistry. 1965;**30**:199-203

[17] Maki Y, Izuta K, Suzuki M. A new synthesis of pyrazolo[3,4-d]pyrimidines. Journal of Chemical Society. 1971;**1971**:1442-1443

[18] Yoneda F, Nagamatsu T, Nagamura T. Synthesis of pyrazolo[3,4-d]pyrimidines by intramolecular cycloaddition of azahexatrienes. Journal of the Chemical Society, Perkin Transactions. 1977;**1**:765-767

[19] Laitonjam WS, Nongmeikapam AD. A facile synthesis of 2(3H)-3,5,7-triphenyl-2(3H)-3-phenyl-5,7-bis(substituted phenyl)-4-oxo-pyrazolo[3,4-d]pyrimidine-6-thiones. Indian Journal of Heterocycling Chemistry. 1995;**5**:139-140

[20] Robins RKP, Purine Antagonists I. Synthesis of some 4,6-substituted pyrazolo [3,4-d] pyrimidines[1]. Journal of the American Chemical Society. 1956;**78**:784-790

[21] White FR. 4-Aminopyrazolo9(3,4-d) pyrimidine and three derivatives. Cancer Chemotherapy Reports. 1959;**3**:26-36

[22] Jun Scott JL, Foye LV. Clinical evaluation of 4-propylamino-1-H-pyrazolo (3,4-d) pyrimidine. Cancer Chemotherapy Reports. 1962;**20**:73-80

[23] Robins RK. Antitumor activity and structural relationships of Purine derivatives and related compounds against neoplasms in experimental animals. Journal of Medicinal Chemistry. 1964;7:186-199

[24] Kobayashi S. The synthesis and xanthine oxidase inhibitory activity of pyrazolo[3, 4-d]pyrimidines. Chemical Pharmaceutical Bulletin Japan. 1973;**21**:941-951

[25] Marr JJ, Berens RL, Nelson DJ, Lafon SW. Antileishmanial effect of allopurinol and allopurinol ribonucleoside on intracellular forms of *Leishmania donovani*. Biochemical Pharmacology. 1980;**29**:2397-2398

[26] Taylor EC, Garcia EE. The synthesis of 4-Aminoisoxazolo[5,4-d]pyrimidines[1]. The Journal of Organic Chemistry. 1964;**29**:2116-2120

[27] Rajagopalan P, Talaty CN. Synthesis of 4-hydroxy-, 4-chloro-, 4-amino- and 4-substituted aminoisoxazolo [5.4-d] pyrimidines. Tetrahedron. 1967;**23**:3541-3543

[28] Laitonjam WS, Nongmeikapam AD. Synthesis of substituted 4-oxo and substituted 2,3-dihydro-4-oxoisoxazolo[5,4-d]pyrimidin-6-thiones. Indian Journal of Chemistry. 1996;**35B**:478-479

[29] Kaim L, Grimaud L, Patil P. Three-component strategy toward 5-membered heterocycles from isocyanide dibromides. Organic Letters. 2011;**13**:1261-1263

[30] Tu S-J, Zhang XH, Han Z-G, Cao XD, Wu S-S, Yan S, et al. Synthesis of isoxazolo[5,4-b]pyridines by microwave-assisted multi-component reactions in water. Journal of Combinatorial Chemistry. 2009;**11**:428-432

[31] Bellary JM, Badiger VV. Synthesis of some Isoxazolo[4,5-e]-s-triazolo[4,3-c] pyrimidines. Indian Journal of Chemistry. 1981;**20B**:654

[32] Kaur M. Ph.D. Thesis, Delhi University, 1989

[33] Ahluwalia VK, Dahiya A, Bala M. Novel synthesis of thieno[2,3-c]pyrazoles and thieno[2,3-d]pyrimidines. Indian Journal of Chemistry. 1996;**35B**:715

[34] Taminaga Y, Okuda H, Matsuda Y. Synthesis of Pyrimido[4,5-d] pyrimidines. Heterocycles. 1979;**12**:503-504

[35] Jiang B, Shi F, Tu S-J. Microwave-assisted multicomponent reactions in the heterocyclic chemistry. Current Organic Chemistry. 2010;**14**:357-378

[36] Laitonjam WS, Nongmeikapam AD. Synthesis of pyrimido[4,5-d]pyrimidines as antifungal agents. Indian Journal of Heterocycling Chemistry. 1998;7:193-196

[37] Kruithof A, Ruijter E, Orru RVA. Microwave-assisted multicomponent synthesis of heterocycles. Current Organic Chemistry. 2011;**15**:204-236

[38] Hirota K, Huang J, Sajiki H, Maki Y. Pyrimidines. Part 57. A Versatile Synthesis of Pyrimido[4,5-d]pyrimidine-2,4,5-trione Derivatives. Heterocycles. 1986;**24**:2293

[39] Ahluwalia VK, Sharma HR, Tyagi R. A novel one step synthesis of pyrano (2,3-d) pyrimidines. Tetrahedron. 1986;**42**:4045-4048

[40] Ahmed MG, Romman UKR, Ahmed SM, Akhter K, Halim ME, Salauddin M. A study on the synthesis of 5,7-diaryl-1,2,3,4-tetrahydro-2,4-dioxo-5H-pyrano[2,3-d]pyrimidines. Bangladesh Journal of Science Industrial Research. 2006;**41**(3-4):119

[41] Andotra CS, Khajuria J, Kaur S. Synthesis of 1,2,3,4-tetrahydro-2,4-dioxo-7-(2,4-dialkoxyphenyl)-5-substituted phenyl-5H-pyrano-[2,3-d]-pyrimidines and 3-(2,4-dialkoxyphenyl)-5-substituted phenylisoxazoles. Journal of Industrial Chemical Society. 2006;**83**:509-512

[42] Yu J, Wang H. Green synthesis of pyrano[2,3-d]pyrimidine derivatives in ionic liquids. Synthetic Communications. 2005;**35**:3133-3140

[43] Devi I, Bhuyan PJ. A novel microwave-assisted one-pot synthesis of Pyrano[2,3-d]pyrimidines and pyrido[2,3-d]pyrimidines via a three component reaction in solvent-free condition. Synlett. 2004:283-286

[44] Prajapati D, Gohain M. An efficient synthesis of novel pyrano[2,3-d]- and furopyrano[2,3-d]pyrimidines via indium-catalyzed multi-component domino reaction. Beilstein Journal of Organic Chemistry. 2006;**2**:1-5

[45] Laitonjam WS, Moirangthem N. A facile synthesis of 7-amino-1,3-diaryl-5-phenyl-2- thioxo-pyrano[2,3-d] pyrimidine-4(5H)-one. American Chemical Science Journal. 2011;**1**:58-70

[46] Yun M, Changtao Q, Limin W, Min Y. Convenient selective synthesis of pyrano[2,3-d]pyrimidines. Russian Chemical Bulletin. 2000;**57**:2223-2226

[47] Antonio H, Roberto MA, Pedro R, John A. An easy synthesis of pyranopyrimidines. Monatshefte fur Chem. 2006;**137**:1421-1430

[48] Moirangthem N. Ph. D. Thesis, Manipur University, Manipur, 2010

[49] Broom AD, Shin JL, Anderson GL. Pyrido[2,3-d]pyrimidines. IV. Synthetic studies leading to various oxopyrido[2,3-d]pyrimidines. The Journal of Organic Chemistry. 1976;**41**:1095-1099

[50] Bhuyan P, Boruah RC, Sandhu JS. Studies on uracils. 10. A facile one-pot synthesis of pyrido[2,3-d]- and pyrazolo[3,4-d]pyrimidines. The Journal of Organic Chemistry. 1990;**55**:568-571

[51] Connolly DJ, Cusack D, O'Sullivian TP, Guiry PJ. Synthesis of quinazolinones and quinazolines. Tetrahedron. 2005;**61**:10153-10202

[52] Mhaske SB, Argade NP. The chemistry of recently isolated naturally occurring quinazolinone alkaloids. Tetrahedron. 2006;**62**:9787-9826

[53] Xia Y, Yang ZY, Hour MJ, Kuo SC, Xia P, Bastow KF, et al. Part 204: Synthesis and biological evaluation of substituted 2-aryl quinazolinones. Bioorganic & Medicinal Chemistry Letters. 2001;**11**:1193-1196

[54] De Clercq E. New developments in anti-HIV chemotherapy. Current Medicinal Chemistry. 2001;**8**:1543-1572

[55] Corbett JW. A review of recent HIV-1 non-nucleoside reverse transcriptase inhibitor research. Current Medicinal Chemistry. 2002;**1**:119-140

[56] Le Mahieu RA, Carson M, Welton AF, Baruth HW, Yaremko B. Synthesis and antiallergy activity of 5-oxo-5H-thiazolo[2,3-b]quinazolinecarboxylic acids. Journal of Medicinal Chemistry. 1983;**26**:107-110

[57] Vandenberk J, Kennis L. V. der Heertum, A. V. U.S. Patent 4, 522, 945. 1985

[58] Sohda T, Makino H, Boba A. Eur. Patent E.P. 0567107. 2001

[59] Jiarong L, Xian C, Daxin S, Shuling M, Qing L, Qi Z, et al. A new and facile synthesis of Quinazoline-2,4(1H,3H)-diones. Organic Letters. 2009;**11**:1193-1196

[60] Li Z, Huang H, Sun H, Jiang H, Liu H. Microwave-assisted efficient and convenient synthesis of 2,4(1H,3H)-Quinazolinediones and 2-Thioxoquinazolines. Journal of Combinatorial Chemistry. 2008;**10**:484-486

[61] Kornet MJ. Journal of Combinatorial Chemistry. 2008;**10**:1533

[62] Couture A, Cornet H, Grandclaudon P. An expeditious synthesis of 2-aryl- and 2-Alkylquinazolin-4(3H)-ones. Synthesis. 1991:1009-1010

[63] Kotsuki H, Sakai H, Morimoto H, Suenaga H. A new quinazoline synthesis. Synlett. 1999:1993-1995

[64] Abdel-Jalil RL, Volter W, Saeed M. A novel method for the synthesis of 4(3H)-quinazolinones. Tetrahedron Letters. 2004;**45**:3475-3476

[65] Feldman JR, Wagner EC. Some reactions of methylene-bis-amines as ammono-aldehydes. The Journal of Organic Chemistry. 1942;**7**:31-42

[66] Yale HL. Substituted 2,3-dihydro-4(1H)quinazolinones, 2. Journal of Heterocycling Chemistry. 1977;**14**:1357-1359

[67] Mhaske SB, Argade NP. Regioselective quinazolinone-directed ortho lithiation of Quinazolinoylquinoline: Practical synthesis of naturally occurring human DNA topoisomerase I poison Luotonin a and Luotonins B and E. The Journal of Organic Chemistry. 2004;**69**:4563-4566

[68] Petrov JS, Andreev GN. Synthesis of 2,4(1h,3h)-quinazolinedione and 3-substituted 2,4(1h,3h)-quinazolinediones. Organic Preparations and Procedures International. 2005;**37**:560-565

[69] Willis MC, Snell RH, Fletcher AJ, Woodward RL. Tandem palladium-catalyzed urea arylation–intramolecular Ester Amidation: Regioselective synthesis of 3-alkylated 2,4-Quinazolinediones. Organic Letters. 2006;**8**:5089-5091

[70] Taylor EC, Shvo Y. Heterocyclic synthesis from o-aminonitriles. XXX. Synthesis of some diazasteroids. The Journal of Organic Chemistry. 1968;**33**:1719-1727

[71] Papadopoulos EP. Reactions of o-aminonitriles with isocyanates. 2. A facile synthesis of imidazo[1,2-c]quinazoline-2,5-(3H,6H)dione. Journal of Heterocycling Chemistry. 1981;**18**:515-518

[72] Papadopoulos EP. Reactions of o-aminonitriles with isocyanates. 1. A two-step synthesis of 2,6-dihydroimidazo[1,2-c]quinazolin-5-(3H)one. Journal of Heterocycling Chemistry. 1980;**17**:1553-1558

[73] Mizuno T, Okamoto T, Ito T, Miyata T. Synthesis of 2,4-dihydroxyquinazolines using carbon dioxide in the presence of DBU under mild conditions. Tetrahedron Letters. 2000;**41**:1051-1053

[74] Mizuno T, Ishino Y. Highly efficient synthesis of 1H-quinazoline-2,4-diones using carbon dioxide in the presence of catalytic amount of DBU. Tetrahedron. 2002;**58**:3155-3158

[75] Mizuno T, Iwai T, Ishino Y. The simple solvent-free synthesis of 1H-quinazoline-2,4-diones using supercritical carbon dioxide and catalytic amount of base. Tetrahedron Letters. 2004;**45**:7073-7075

[76] Saeed A, Shaheen U, Bolte M. Synthesis, characterization and crystal structure of some novel 1-Aryl-2-Thioxo-2,3-dihydro-1H-Quinazolin-4-ones. Journal of Chinese Chemical Society. 2010;**57**:82-88

[77] Makino S, Suzuki N, Nakanishi E, Tsuji T. Efficient solid-phase synthesis of quinazoline-2-thioxo-4-ones with SynPhase™ lanterns. Tetrahedron Letters. 2000;**41**:8333-8337

[78] Makino S, Nakanishi E, Tsuji T. Solid-phase synthesis of 1,3-disubstituted 2-thioxoquinazoline-4-ones using S_NAr reaction. Tetrahedron Letters. 2001;**42**:1749-1752

[79] Makino S, Nakanishi E, Tsuji T. Solid-phase synthesis, combinatorial chemistry, Heterocyle, SNAr, 2,1,3-Benzothiadiazin-4-one 2-oxide. Bulletin of the Korean Chemical Society. 2003;**24**:389-392

[80] Ivachtchenko AV, Kovalenko SM, Drushlyak OG. Synthesis of substituted 4-Oxo-2-thioxo-1,2,3,4-tetrahydroquinazolines and 4-Oxo-3,4-dihydroquinazoline-2-thioles. Journal of Combinatorial Chemistry. 2003;**5**:775-788

[81] Laitonjam WS, Moirangthem N. A new and facile synthetic approach to substituted 2-thioxoquinazolin-4-ones by the annulation of a pyrimidine derivative. Journal of Organic Chemistry. 2010;**6**:1056-1060

[82] Hewitt W, Vincent S. Theory and Application of Microbiological Assay. New York: Academic Press, Inc.; 1989

[83] Cremer A. Antibiotic Sensitivity and Assay Tests. 4th ed. London: Butterworths; 1980. p. 521

[84] Dini A, Medhabati DT, Reena H, Warjeet SL. Synthesis and screening for antioxidant and cytotoxic activities of novel 2-thioxobenzo[f]chromeno[2,3-d]pyrimidin-4-ones derived by

cetylpyridinium chloride catalyzed multicomponent reactions in aqueous micellar media. Industrial Journal of Chemistry. 2021;**60B**:1243-1257

[85] Li XT, Zhao AD, Mo LP, Zhang ZH. Meglumine catalyzed expeditious four-component domino protocol for synthesis of pyrazolopyranopyrimidines in aqueous medium. RSC Advances. 2014;**4**:51580-51588

[86] Heravi MM, Mousavizadeh F, Ghobadi N, Tajbakhsh M. A green and convenient protocol for the synthesis of novel pyrazolopyranopyrimidines via a one-pot, four-component reaction in water. Tetahedron Letters. 2014;**55**:1226-1228

[87] Dastkhoon S, Tavakoli Z, Khodabakhshi S, Baghernejad M, Abbasabadi MK. Nanocatalytic one-pot, four-component synthesis of some new triheterocyclic compounds consisting of pyrazole, pyran, and pyrimidinone rings. New Journal of Chemistry. 2015;**39**:7268-7271

[88] Dini A, Shyamkesho MS, Warjeet SL. Synthesis, antimicrobial and antioxidant activities of novel 7-thioxo4,6,7,8-tetrahydro-pyrazolo[4′,3′,5,6] pyrano[2,3-d]pyrimidin-5(1H)-ones derived by SDS catalyzed multicomponent reactions in aqueous micellar media. Natural Products Journal. 2018;**8**:1-11

[89] Zaki MEA, Soliman HA, Hiekal OA, Rashad AE. Pyrazolopyranopyrimidines as a Class of Anti-Inflammatory Agents. Tübingen: Verlag der Zeitschrift für Naturforschung; 2005

[90] Medhabati DT, Ranjana DT, Warjeet SL. Synthesis of biologically active 2-thio-5-arylbenzo[4,5] thiazolopyrimido [5,4-d]pyrimidin-4-one derivatives catalyzed by metal proline in water. Industrial Journal of Chemistry. 2021;**60B**:1230-1242

Chapter 9

Synthetic Strategies and Biological Activities of 1,5-Disubstituted Pyrazoles and 2,5-Disubstituted Thiazoles

Erika Lozano, Melissa M. Lewis-Bakker and Lakshmi P. Kotra

Abstract

Pyrazoles and thiazoles belong to 5-membered aromatic heterocycles called azoles. In addition to a nitrogen, pyrazoles contain an additional nitrogen in a 1,2 linkage and thiazoles contain a sulfur atom in a 1,3 linkage. These compounds are useful pharmacophores that offer a broad range of therapeutic applications. Pyrazoles can be synthesized by (i) the condensation of 1,3 dipolar compounds and alkenes/alkynes, (b) cyclocondensation of hydrazines and dicarbonyl reagents, and (c) multi component reactions. Access to thiazoles is typically via (a) the condensation of α-haloketones with nucleophilic thioamides containing the N-C-S fragment, (b) a reaction between α-aminonitriles and various reactants containing an X-C-S fragment, and (c) a reaction of acylaminocarbonyls and phosphorus pentasulfide. This chapter will focus on the strategies and our perspectives on the synthesis of pyrazoles and thiazoles including derivatives at the 1,5 positions and 2, 4, 5 positions respectively, reported during 2015–2022. Additionally, their therapeutic and biological evaluations will be discussed.

Keywords: Pyrazoles, 1,5-subsititutions, thiazoles, 2,4,5-subsititutions, therapeutic, biological evaluations

1. Introduction

Heterocycles are cyclic compounds containing atoms other than carbons and hydrogens and may be synthesized or occur naturally in plants, animals and human beings. In particular, nitrogen-containing heterocycles are considered very important building blocks in the chemical and pharmaceutical industries (among others) since they can be used as catalysts, protecting groups, chiral auxiliaries and chemical intermediates.

Azoles are 5-membered nitrogen-containing aromatic heterocycles, to which thiazoles and pyrazoles belong. Pyrazoles contain two nitrogens in a 1,2 linkage while thiazoles contain nitrogen and sulfur in a 1,3 linkage. Pyrazoles and thiazoles are useful pharmacophores with a potential for various therapeutic applications.

This chapter will focus on the following as it relates to pyrazoles and thiazoles reported during 2015–2022: (i) strategies and perspectives on the synthesis of 1,5-disubstituted pyrazoles and 2,4,5-trisubstituted thiazoles derivatives and (ii) their therapeutic and biological evaluations.

2. Pyrazoles

2.1 Structure

The pyrazole structure exists as two tautomers due to the movement of the C-N double bond (**Figure 1**). When both nitrogens are unsubstituted, each has a similar reactivity. However, when any carbon or nitrogen atom is substituted, the substitution(s) influences the reactivity of the other positions of the ring. This in turn may lead to several regioisomers of the substitituted pyrazole.

2.2 Biological applications of Pyrazoles

Pyrazoles form core structures of many pharmaceutical products They have therapeutic properties such as antibacterial, anticancer, anti-inflammatory, anti-oxidant, anti-tuberculosis and anti-depressive agents [1, 2]. Examples of drugs containing the pyrazole moiety are Fomepizole for alcohol poisoning, Sildenafil or Viagra® for erectile dysfunction and Rimonabant for obesity treatment (**Figure 2**) [1].

2.3 Conventional synthesis of 1, 5-Pyrazoles

Pyrazoles are conventionally formed by (i) cycloadditions between 1,3 dipolar compounds and alkenes/alkynes and (ii) cyclocondensations of hydrazines and dicarbonyls (first reported in 1883 using dicarbonyls) as shown in **Figure 3** [2]. Additionally, multi-component or one-pot reactions utilizing a combination of consisting of at least three of the afore-mentioned compounds can also generate various substituted pyrazoles. Disadvantages of these methods include the use of unstable

Figure 1.
Tautomeric structures of an unsubstituted pyrazole.

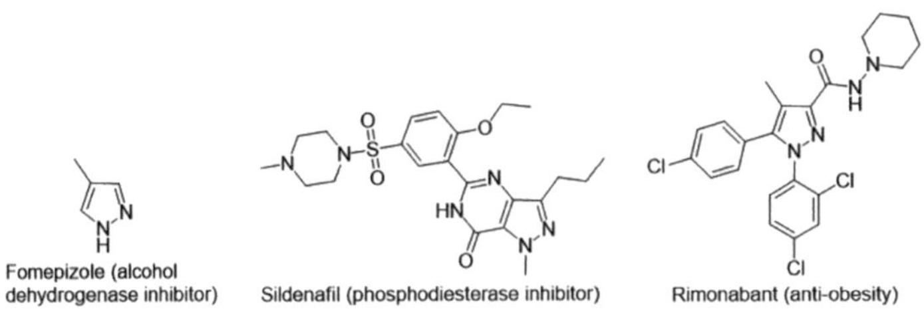

Figure 2 .
Examples of drugs containing the pyrazole moiety.

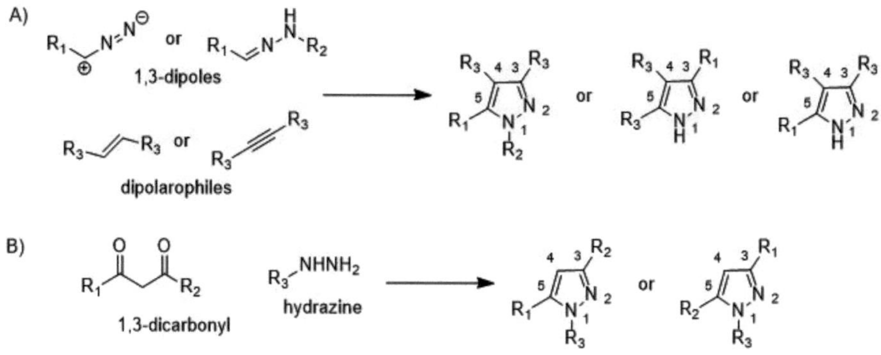

Figure 3.
Conventional reactions to form pyrazoles. A) Cycloadditions B) Cyclocondensations.

reagents such as the diazo compounds in cycloadditions and product obtained as a mixture of regioisomers when substituted starting materials are used.

2.4 Recent synthetic routes towards 1, 5-Pyrazoles

2.4.1 Ruthenium-catalyzed Cyclocondensations of Propargyl alcohols and Hydrazines

Metal catalysts have proved to be quite efficient in organic chemistry as means of carrying out various and sometimes difficult transformations. In the synthesis of pyrazoles, the bifunctional ruthenium cyclopentadienone complex (Ru Catalyst) catalyzes the reaction between secondary propargyl alcohols and nucleophiles such as hydrazine giving 1,5-disubstituted pyrazoles along with byproducts [3]. This cascade conversion likely proceeds via cycloisomerization, Michael addition, cyclocondensation and dehydrogenation/oxidation) steps [3]. The reaction proceeds via microwave irradiation in toluene and when the nucleophile is phenyl hydrazine (**2**), both the 1,5 and 1,3-disubstituted regioisomers (**a** and **b**) are formed along with the 3-phenylbutene byproduct **c** (**Figure 4**) [3]. The trends in the yields indicates that when the R group of the propargyl alcohol **1** is bulky or has unsaturated side chains, the reaction favors the 1,5-disubstituted pyrazoles leading to higher yield of this regioisomer.

From a mechanistic perspective as shown in **Figure 5**, Kaufmann *et al* [3] reported that the propargyl alcohol is activated by the Ru catalyst to form the by the chelated π-complex **I**, which is then converted to the alkynyl species **II**. A 1,2-hydrogen shift forms the alkenyl complex **III** which then undergoes a regioselective nucleophilic attack by phenyl hydrazine to give complex **IV**. Cyclocondensation of complex **IV** gives the alkyl complex **V** which subsequently undergoes a β-hydride elimination to give the 1,5-pyrazole. The active catalyst is regenerated by the elimination of hydrogen from the hydride complex **VI**. The 1,3- regioisomer is formed from the hydrazone of intermediate **III** while the byproduct is formed when the internal carbon of the alkynyl complex **II** undergoes a nucleophilic attack, when non-bulky propargyl alcohols are used.

2.4.2 Ruthenium-catalyzed Cyclocondensation of Diazonium salts and Cyclopropanols

Another Ruthenium-catalyzed reaction to synthesize pyrazoles was reported by Cardinale *et al* [4]. The photocatalysis of arenediazoniums (**10** and **12**) and

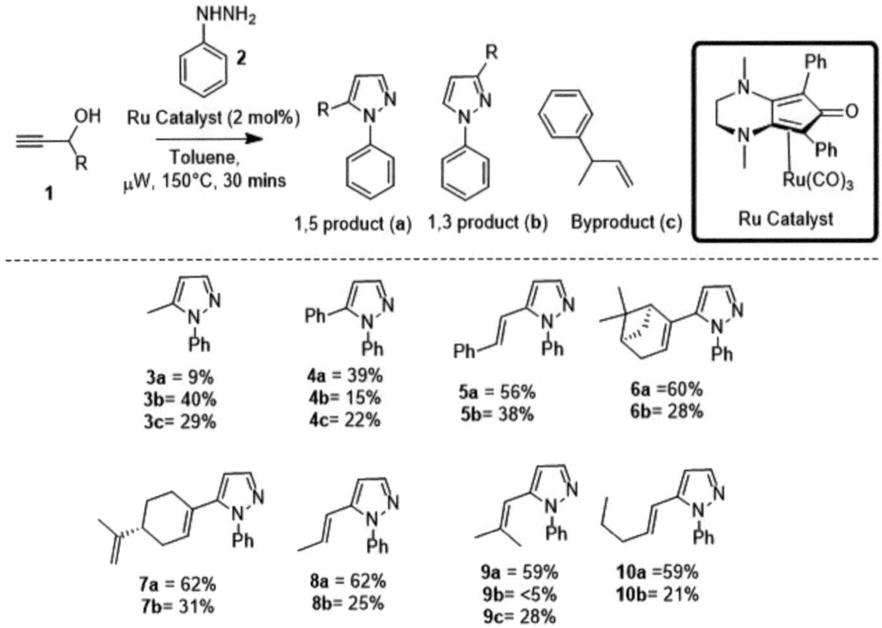

Figure 4.
Regioisomeric pyrazoles produced by the ruthenium catalyzed cascade conversions between secondary propargyl alcohols and phenyl hydrazine [3].

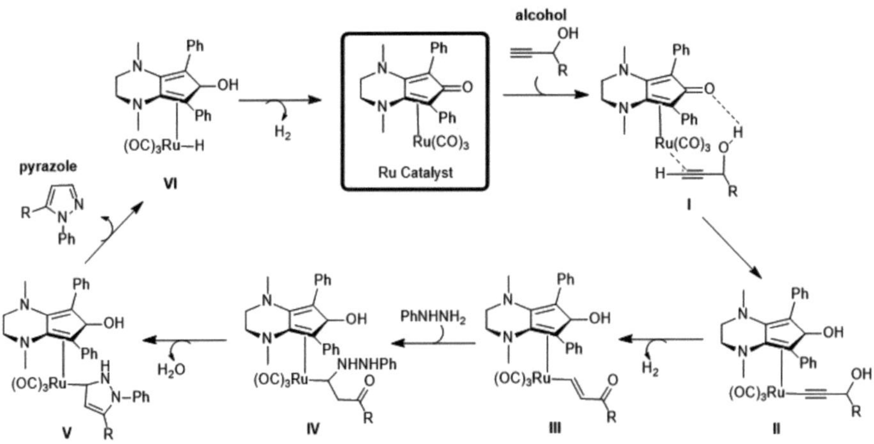

Figure 5.
Proposed mechanisms for the ruthenium catalyzed cascade conversions [3].

arylcyclopropanols (**12 and 13**) using $[Ru(bpy)3]^{2+}$ under blue light irradiation in acetonitrile yielded 1, 5 -disubstituted pyrazoles **14–21** (**Figure 6**). This photocatalysis provides several advantages such mild reaction conditions (20 mins at room temperature), compatibility with functional groups such as I, SF_5, SO_2NH_2, N_3, CN and high regioselectivity and yields.

The proposed mechanism of the photocatalysis is seen in **Figure** 7 [4]. The reaction is proposed to be initiated when the $[Ru^{2+}]^*$ is the oxidatively quenched by the arenediazonium salt **10** (or **12**) to give $[Ru^{3+}]$ which oxidizes the cyclopropanol

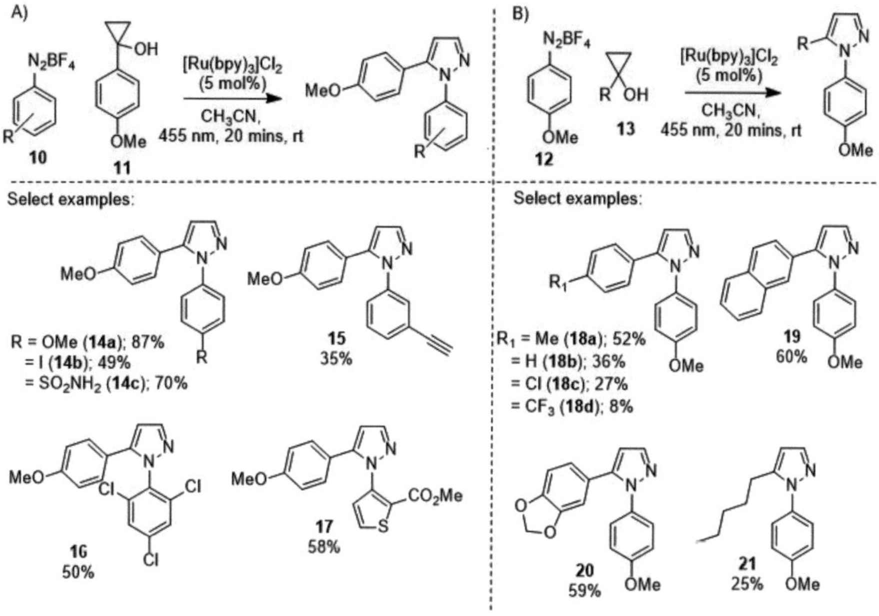

Figure 6.
Ruthenium-catalyzed synthesis of pyrazoles under blue light. Synthetic scope of (a) diazonium salts and (B) cyclopropanols [4].

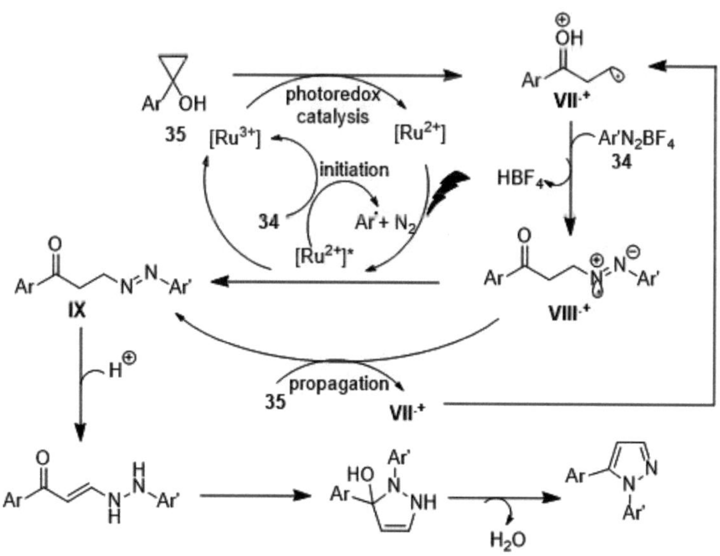

Figure 7.
Proposed mechanism for the ruthenium photocatalysis of pyrazoles [4].

11 (or **13**) to give **VII•+**. The radical cation **VIII•+** is formed when **VII•+** undergoes a fast ring-opening and trapping with the arenediazonium salt **10** (or **12**). The newly formed **VIII•+** can either (i) quench the excited catalyst thereby closing the photocatalytic cycle or (ii) become involved in a radical chain process by oxidizing another molecule of **11** (or **13**). As a result of the dominance of the radical chain process, the arenediazonium **10** (or **12**) is used in slight excess to activate the catalyst and produce the oxidant [Ru(bpy)3^{3+}] salt. The intermediate **IX** then undergoes cyclization followed by loss of water to produce the pyrazole.

2.4.3 Copper-catalyzed reaction of α, β-Cyanoesters and Hydrazines

In another recent report, Cu(I) catalyst, $Cu(PPh_3)_2NO_3$ was used to synthesize 1,5-disubstituted pyrazoles from hydrazines **22** and α,β-unsaturated cyanoester **23** under ultrasound irradiation (**Figure 8**) [5]. This efficient route uses short reaction times, moderate temperatures, facile work-ups, high regioselectivity and yields [5].

From a mechanistic perspective, the authors proposed that copper coordinates easily to π-bonds and heteroatoms. This would lead to the formation of the C-N bond via an oxidative addition followed by reductive elimination, similar to that observed in Pd(0) cross coupling reactions [5].

2.4.4 Reaction of Ferrocenes and Hydrazines

Ferrocene is an organometallic compound that has excellent redox properties and may be incorporated as part of a drug's structural moiety, leading to additional therapeutic applications. Ferrocenyl compounds have shown potential as antimalarial, anticancer, cytotoxic and DNA-cleaving agents [6]. Ferrocenyl pyrazoles have also been shown to have potential therapeutic applications as well.

Recently, ferrocenyl pyrazoles **30** and **32** were prepared from ferrocenes **29** and **31**, respectively, and hydrazines in a simple two-step procedure to regioselectively give the desired 1,5-pyrazoles in good yields (**Figure 9A** and **9B**) [7].

The ferrocenyl pyrazoles **30** and **32** were then evaluated in four antimicrobial assays involving two strains each of fungi (*Aspergillus niger*, *Trichophyton rubrum*)

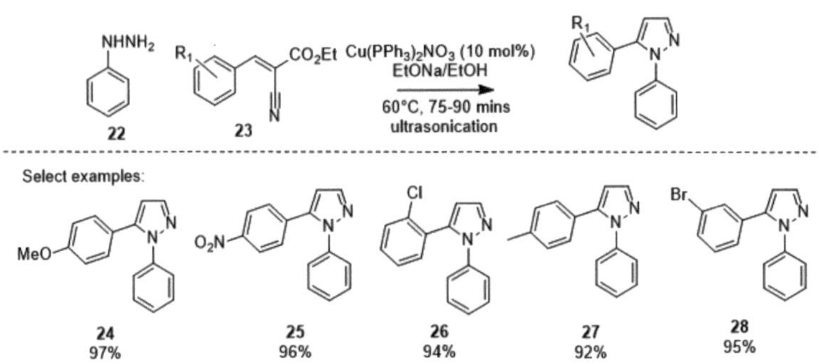

Figure 8.
Synthesis of the 1,5 pyrazoles using ultrasonication and a Cu(I) catalyst [5].

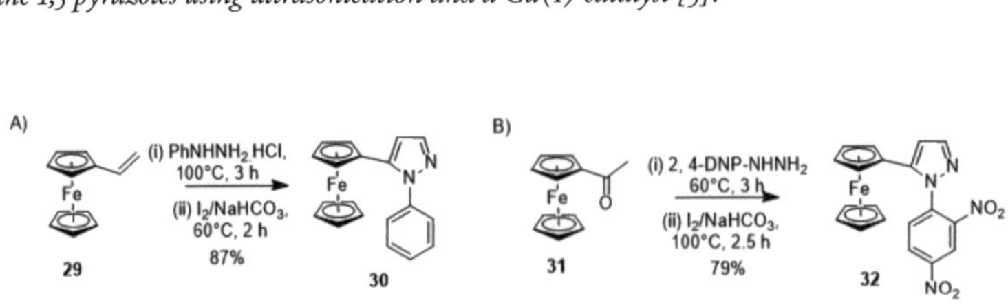

Figure 9.
Synthesis of ferrocenyl pyrazoles [6].

and bacteria (*Staphylococcus aureus*, *Klebsiella pneumoniae*). Standard antimicrobial agents fluconazole and neomycin were used in the fungal and bacterial studies, respectively. In these assays, the zone of inhibition (**Tables 1** and **2**) was measured and the corresponding minimum inhibitory concentrations (MICs) were calculated. The MIC values ranged from 85 to 95 μg/mL indicating potential as antimicrobial agents.

DNA photo-cleavage activity of the ferrocenyl pyrazoles was also evaluated using super coiled plasmid DNA by gel electrophoresis [6]. When the faster super-coiled DNA (form I) was treated with the pyrazoles, it converted to the slower-moving nicked DNA (form II), in comparison to the untreated normal DNA. These results confirmed the DNA cleavage activity of both ferrocenyl pyrazoles.

2.4.5 C5-electrophilic fluorination of Pyrazoles

The 5-fluoropyrazole core is an important structural moiety in the agrochemical and pharmaceutical industries. Synthesis of 5-fluoropyrazoles typically occurs by reaction of (i) 1,3-dicarbonyl containing CF_3/CF_2 (fluorinated synthon) with hydrazines followed by HF-elimination [8]; (ii) copper-catalyzed click reaction of fluorosydnones with alkynes [9]; (iii) Selectfluor and pyrazoles containing carboxylic acids followed by decarboxylation [10] and (iv) KF with pyrazoles [11, 12]. The electrophilic fluorination of pyrazoles to give 5-fluoropyrazoles was recently described using N-fluorobenzenesulfonimide (NFSI) [13]. As shown in **Figure 10** 1-substituted pyrazoles were subjected to C5-deprotonation by lithium base, which was then fluorinated using NFSI to yield 1-substituted-5-fluoropyrazoles [13].

		Bacterial strains							
		S. aureus				*Klebsiella pneumoniae*			
Conc. (ppm)		100	150	200	250	100	150	200	250
Zone of Inhibition (mm)	**30**	12 ± 0.5	16 ± 0.7	17 ± 0.6	18 ± 0.6	11 ± 0.5	13 ± 0.5	18 ± 0.5	20 ± 0.7
	32	12 ± 0.5	15 ± 0.7	17 ± 0.6	21 ± 0.6	10 ± 0.5	14 ± 0.5	16 ± 0.5	19 ± 0.7
Standards		21 mm (Fluconazole)				23 mm (Fluconazole)			

Table 1.
Antibacterial activities of **30** *and* **32** *[6].*

		Fungal strains							
		A. niger				*Trichophyton rubrum*			
Conc. (ppm)		100	150	200	250	100	150	200	250
Zone of Inhibition (mm)	**30**	10 ± 0.5	12 ± 0.5	15 ± 0.5	18 ± 0.5	9 ± 0.7	12 ± 0.4	14 ± 0.5	18 ± 0.6
	32	10 ± 0.5	13 ± 0.5	16 ± 0.5	18 ± 0.5	9 ± 0.7	12 ± 0.4	15 ± 0.5	17 ± 0.6
Standards		24 mm (Fluconazole)				9 mm (Fluconazole)			

Table 2.
Antifungal activities of **30** *and* **32** *[6].*

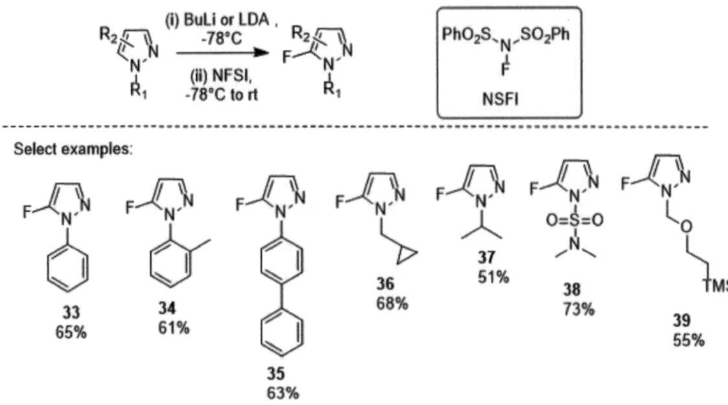

Figure 10.
Synthesis of 1-substituted-5-fluoropyrazoles [13].

3. Thiazoles

3.1 Structure

Thiazoles are five-membered aromatic heterocycles belonging to the azole group. While azoles are generally characterized by the presence of a nitrogen atom, the thiazole ring features the N in a 1,3-linkage with sulfur. The sulfur atom bears a lone pair of electrons which are delocalized throughout the ring, while C2 bears an acidic proton allowing for a range of reactions to occur in this position (**Figure 11**) [14].

3.2 Biological applications of Thiazoles

Thiazoles are noted for their utility in medicinal chemistry as an active and often potent pharmacophore [15] and are investigated for their therapeutical potential. Thiazoles are found in biologically relevant compounds such as the natural product Thiamin or Vitamin B1 and the first effective antibiotic drug series, Penicillin (**Figure 12**).

Additionally, there has been a growing interest in the synthesis of thiazole-containing compounds for applications as photosensitizers, sensors, catalysts, pigments, and more [16].

Certain biological applications of thiazoles were first identified in 1887 by Hantzsch and Weber. Since this discovery, thiazoles are a notable moiety present in several modern compounds of interest for their antitubercular [17], antidiabetic [18], antimalarial [19], antibacterial [19–31], antiviral [26, 32–34], antifungal [22, 24–27, 29, 30, 33, 34], antioxidant [35], anti-inflammatory [30, 35], anti-cancer [36, 37], and anti-proliferative [27, 38–40], activities.

Figure 13 represents a few examples of approved drugs that contain a thiazole ring. Epothilones are a class of anti-neoplastic agents first identified in 1995,

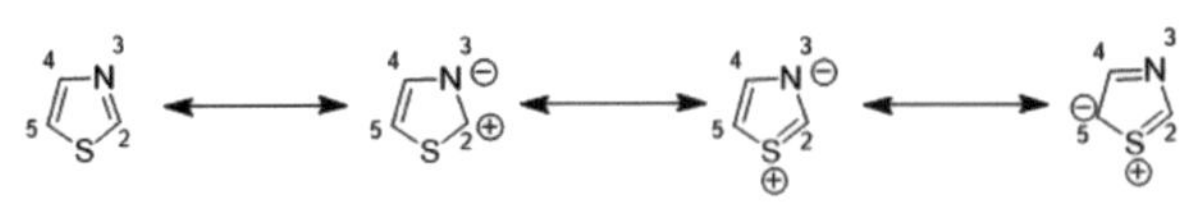

Figure 11.
Resonance of unsubstituted 1,3-thiazole.

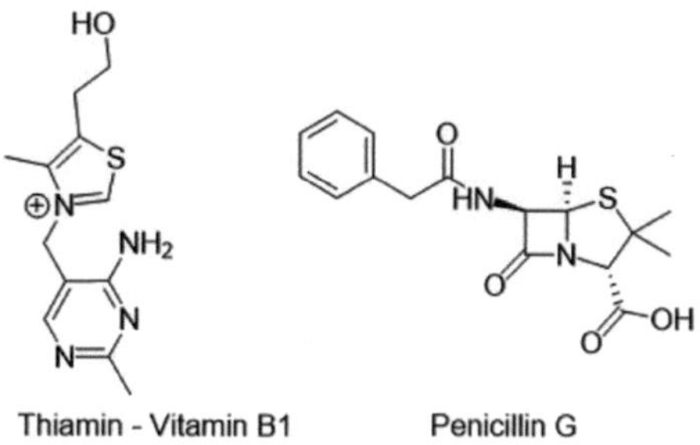

Figure 12.
Biologically relevant thiazole derivatives.

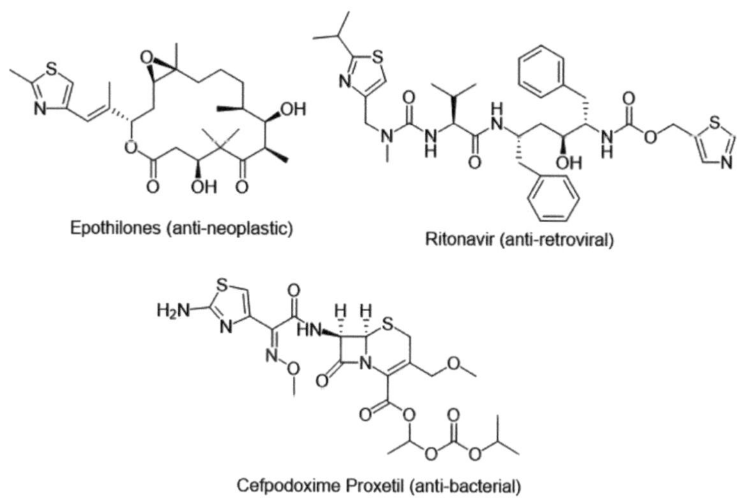

Figure 13.
Approved drugs bearing thiazole moieties.

which act by stabilizing microtubules and inducing apoptosis [41, 42]. Ritonavir was discovered in 1998 as a potent and effective HIV protease inhibitor [43] and today, is used in combination with other antiretroviral agents in the treatment of HIV infection. More recently, it has been used in combination therapy with lopinavir to treat severe Covid-19 [44, 45]. The prodrug Cefpodoxime Proxetil was identified in 1993 as a safe and effective broad-spectrum antibiotic [46–48]. The thiazole moiety is unquestionably a vital functionality in the structure of many drugs.

3.3 Conventional synthesis of 2, 4, 5-Thiazoles

The synthesis of thiazoles has been broadly described in literature over the years. The main methods used to prepare thiazoles are the Hantzsch, Cook-Heilbron and Gabriel's syntheses as shown in **Figure 14** [16].

Hantzsch's synthesis of thiazole rings was first published in 1887 by Hantzsch and Weber [49]. Hantzsch's synthesis involves the condensation and aromatization of α-haloketones with nucleophilic thioamides containing the N-C-S fragment, where cyclization yields the thiazole moiety [16].

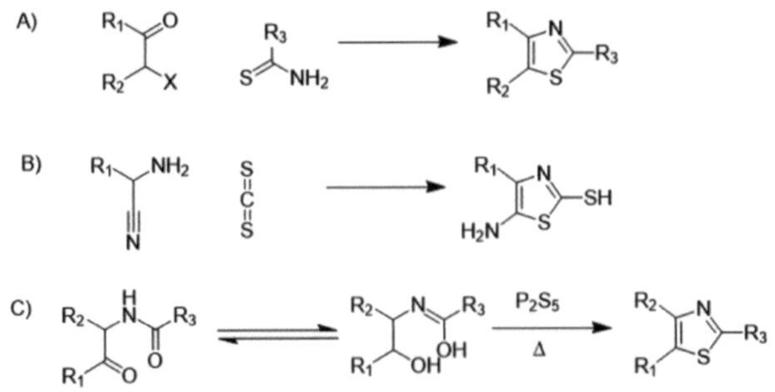

Figure 14.
General synthetic scheme of 2,4,5-subsituted thiazoles by a) Hantzsch's synthesis B) Cook-Heilbron's synthesis and C) Gabriel's synthesis [16].

The Cook-Heilbron reaction involves the interaction of aminonitrile derivatives with esters or various reactants containing a X-C-S fragment (such as dithioacids, carbon disulfide) under mild conditions to yield aminothiazoles [16, 50–53]. This synthesis generally introduces different moieties to the 2-position of 4- or 5-aminothiazoles.

Gabriel's synthesis was introduced in 1910 and describes the reaction between acylaminocarbonyls and phosphoruspentasulfide (such as Lawesson's reagent) to yield a thiazole ring with phenyl and alkyl substitutions in the 2- and 5-positions [16, 54].

3.4 Recent synthetic routes towards 2,4,5-substituted thiazoles

3.4.1 Hantzsch's synthesis from Thioamide andα-Halocarbonyls

Figure 15 describes the synthesis of MSI-1 (3-(5-isopropyl-4-(4-methylpyridin-3-yl) thiazol-2-yl) benzamide), a natural product monomer which acts as a SREBP-1 inhibitor in the treatment for lung squamous cell carcinoma (LUSC). It also enhances the sensitivity of these cancer cells to antitumor agents. Compound **40** was brominated to obtain **41** and in a parallel reaction, the thioamide derivative **43** was generated from **42**. Subsequently, MSI-1 was obtained from the cyclocondensation reaction between α -halocarbonyl **41** and thioamide derivative **43** [55].

The reaction mechanism may proceed as shown in **Figure 16** as described by Hantzsch A and Weber [49], where bromine of **45** acts as a leaving group of allowing the coupling of sulfur of **46** to the α-position of the carbonyl. Following a series of several proton transfers, cyclization occurs via the nucleophilic attack of nitrogen to

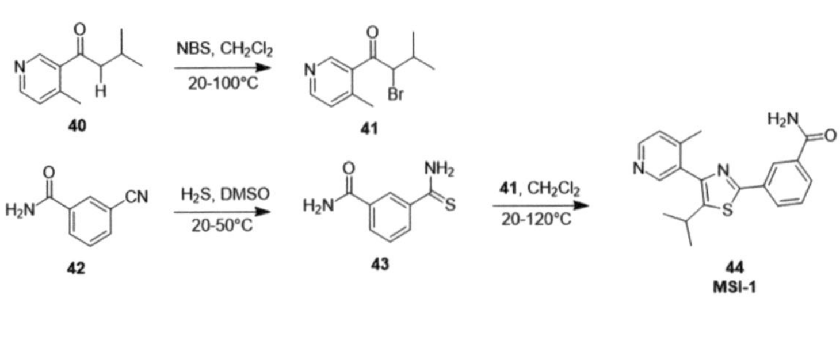

Figure 15.
Hantszch's reaction between α-halocarbonyl **41** *and thioamide* **43** *[55].*

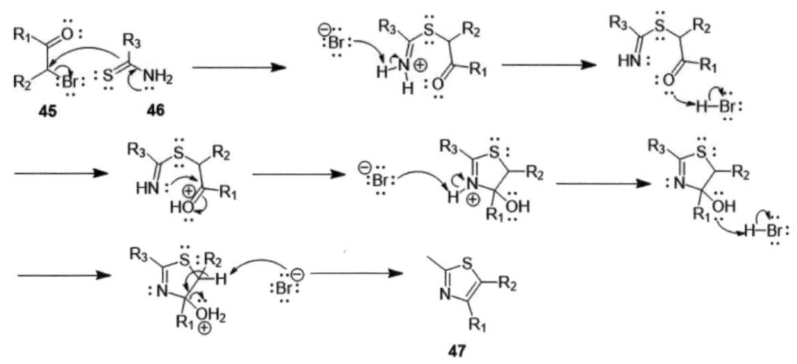

Figure 16.
Plausible mechanism of Hantzsch's reaction [49].

the electron-deficient carbonyl carbon and the desired product **47** is obtained after elimination of water [55].

Molecular docking studies show that MSI-1 enters the hydrophobic pocket of SREBP-1, binding through π-π conjugation. The complex is stabilized by further π-π interactions with three amino acid residues PHE271, TYR335, and PHE349. MSI-1 was shown to impedes the activation of SREBP-1 by inhibiting downstream genes of SREBP-1 associated to lipid metabolism in a dose dependent manner. Additionally, MSI-1 inhibits the Warburg Effect of cancerous and malignant cells and the Epithelial-Mesenchymal Transition process (indicative of chemo-resistance) in LUSC cell line NCI-H226. This effect was demonstrated by a decrease in glucose uptake and lactate production, as well as a reduction in ATP production and LDH activation [55].

3.4.2 Hantzsch's synthesis from Thiourea andα-Halocarbonyls

Wang *et al* recently reported the synthesis of thiazole-naphthalene hybrids and their antiproliferative activities as tubulin polymerization inhibitors from thiourea and α-halocarbonyls (**Figure 17**) [56]. The condensation of 1-methoxynaphtalene **48** with phenylacetic acid **49** in presence of trifluoroacetic anhydride (TFAA) in trifluoroacetic acid (TFA) yielded deoxybenzoin **50**. Compound **50** was then treated with pyridinium tribromide ($PyBr_3$) in CH_2Cl_2 to brominate the α-position of the carbonyl. Finally, under reflux in ethanol, the cyclocondensation reaction of **51** and thiourea **52** produced desired compound **53**. Three thiazole-naphthalene derivatives were prepared with this protocol. Moreover, fourteen compounds were generated by introducing substitutions to the amine in the presence of acid anhydride [56].

Compound **53** showed potent antiproliferative activity on the human breast cancer (MCF-7) and human lung adenocarcinoma (A549) cell lines when compared to standard treatments (cisplatin, 5-fluorouracil, tamoxifen, and CA-4). They determined that the 4-ethoxyphenyl substitution was more favorable than 4-methoxyphenyl. Additionally, replacement with 2-bromo-3,4,5-trimethoxyphenyl reduced the antiproliferative activity [56]. Compound **53** exhibited low toxicity in human normal cell line (IC_{50} = 16.37 ± 4.61 μM). Its *in vitro* tubulin polymerization inhibitory activity of was investigated and revealed that the compound acts as a tubulin destabilizing agent with an IC_{50} of 3.3 μM (compared to reference colchicine IC_{50} of 9.1 μM). Furthermore, it was demonstrated that compound **53** leads to cell cycle arrest at the metaphase in a dose-dependent manner, and an Annexin V-FITC/PI assay showed

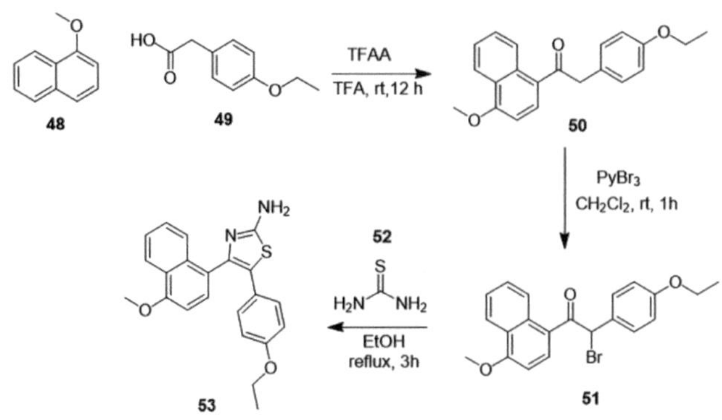

Figure 17.
Hantszch reaction between α-halocarbonyl **51** *and thiourea* **52** *[56].*

that it can effectively induce apoptosis in MCF-7 cells. Molecular docking studies revealed that **53** can bind with tubulin by adopting an "L-shaped" conformation. The naphthalene moiety can accommodate in the hydrophobic pocket of the protein, and it binds to the colchicine site of tubulin [56].

3.4.3 Holzapfel: Meyers: Nicolaou modification of Hantzsch's synthesis

The Holzapfel-Meyers-Nicolaou modification is based on the Hantzsch reaction between thioamide and an α-halocarbonyl, however, it involves the generation of a hydroxythiazoline intermediate under basic conditions. This intermediate is then dehydrated in the presence of trifluoroacetic anhydride (TFAA) and pyridine, followed by the addition of triethylamine (TEA) to yield the desired thiazole [57].

Figure 18 shows a recent example of the Holzapfel-Meyers-Nicolau modification reported in the synthesis of 5-acylamino-1,3-thiazoles from α-chloroglycinates and thioamide derivatives [40]. ESI-MS monitoring revealed the presence of the hydroxy intermediate **X**. Dehydration of **X** results in the desired 5-acylaminothiazole derivative **56** in a high yield (87%). Phenylic substitutions were introduced successfully to the 2 and 4-positions using this approach [58].

3.4.4 Cook-Heilbron's synthesis

Avadhani *et al* developed an efficient one-pot reaction for the preparation of 4-amino 2-aryl-5-substituted thiazoles. **Figure 19** presents the reaction of cyanamide **57** with dithioester **58a** or **58b** with excess NaH, followed by the addition of halocarbonyl derivative **59** to yield potent antiproliferative agents 4-amino-2-phenyl and 2-(thiophne-3-yl)- 5-(2,3,4-trimethoxybenzoyl)-thiazoles **60a** and **60b** in high yields

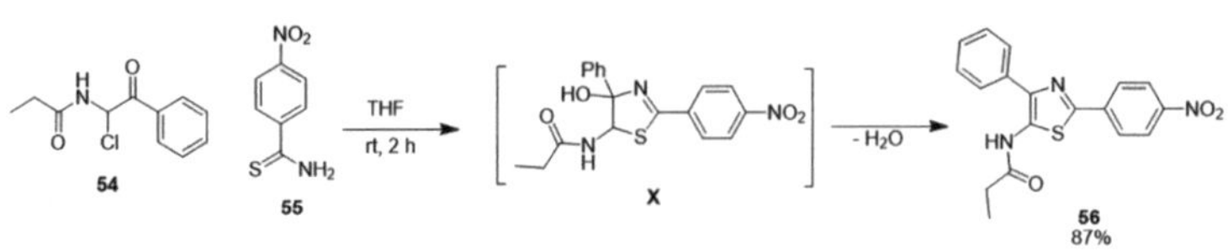

Figure 18.
Holzapfel-Meyers-Nicolau reaction between α-chloroglycinate **54** *and thioamide* **55** *[58].*

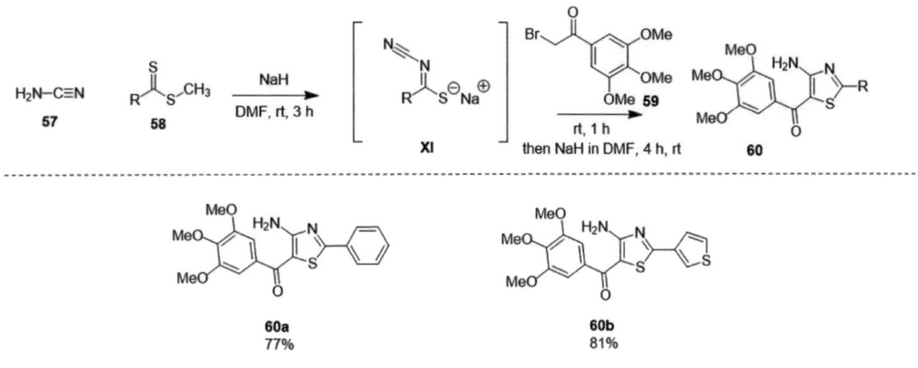

Figure 19.
Cook-Heilbron's reaction between cyanamide 57 *and dithioate derivative* 58 *[59].*

(77% and 81% respectively). The Thorpe-Ziegler-type reaction proceeds via base-mediated intermolecular cyclization of N-cyanothioimidate intermediate **XI** [59].

The 2-methylthio and arylamino substitutions can be introduced at the 2-position using this method, while the 5-position supports ester, nitrile, and carbonyl substituents. The latter can be utilized to introduce reactive functionalities, expanding the possible range of substitutions achieved by this protocol. The novel methodology was further validated by the synthesis of 26 4-amino-2-(het)aryl/alkyl 5-substituted thiazole derivatives [59]. It should be noted that this example introduces the amino moiety to the 4-position. The 5-aminothiazoles can be synthesized from the reaction with 2-aminoacetonitrile instead of cyanamide.

3.4.5 Lactic acid-mediated one-pot reaction

Figure 20 shows the one-pot Hantzsch synthesis of the 2-aminothiazole **63** using lactic acid as a green solvent and catalyst [60]. The ketone **61** was brominated *in-situ* followed by heterocyclization with thiourea **52** to produce the substituted aminothiazole **63**. Lactic acid was selected as a solvent and catalyst due to its ability to solubilize all the reactants, its increased product yield and shorter reaction times in comparison to acetic acid. Maximum yields of 2-aminothiazoles (up to 96%) were obtained when the temperature was increased from room temperature (45% in 1.2 h) to 90–100°C [60]. Lower yields were reported yields when strong deactivating groups were introduced (such as -NO_2), likely due to the reduced *in-situ* formation of the α-brominated

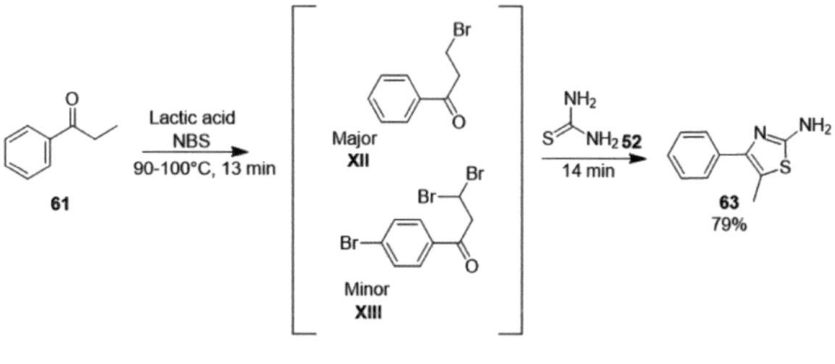

Figure 20.
Lactic acid-mediated one-pot synthesis of 2-aminothiazoles [60].

ketone. Overall, this work developed an effective, rapid, and sustainable one-pot synthesis of 2-aminothiazoles in excellent yields [60].

3.4.6 Aluminum oxide/PVA thin film-catalyzed reaction

The catalytic activity of hybrid PVA/Al_2O_3 nanofilms in the synthesis of thiazole derivatives was recently reported by Riyadh *et al* [61]. The reaction between the thiosemicarbazone 2-benzylidenehydrazine **64** and the α-haloester ethyl-2-chloro-3-oxobutanoate **65** in the presence of the PVA/Al_2O_3 nanocomposite as a basic catalyst is demonstrated in **Figure 21**. Under thermal conditions, the optimal loading catalyst was 10 wt% and the desired compounds were obtained after 180 min [61].

Figure 22 describes a plausible reaction mechanism where the Al_2O_3 particles act as a base in the deprotonation of the thiol group from carbothioamide tautomeric intermediate **XVIII**. The thiolate anion **XIX** then attacks α-halocarbonyl **65**,

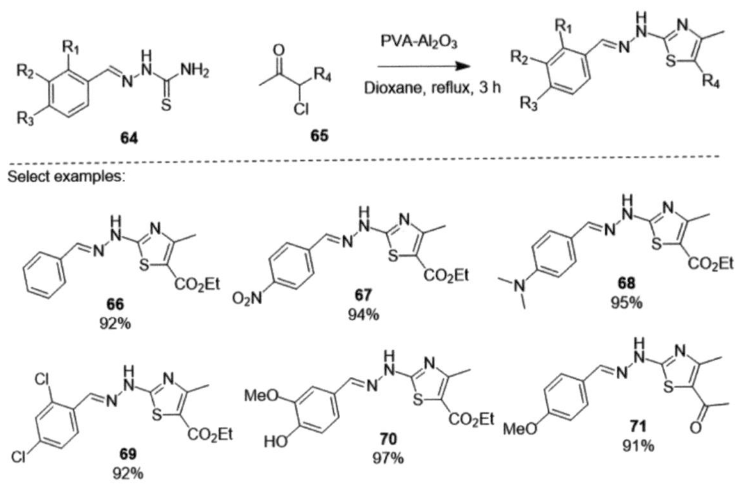

Figure 21.
Polyvinyl alcohol/aluminum oxide (PVA/Al_2O_3) thin film nanocomposite as a catalyst in the synthesis of 2,4,5-thiazoles [61].

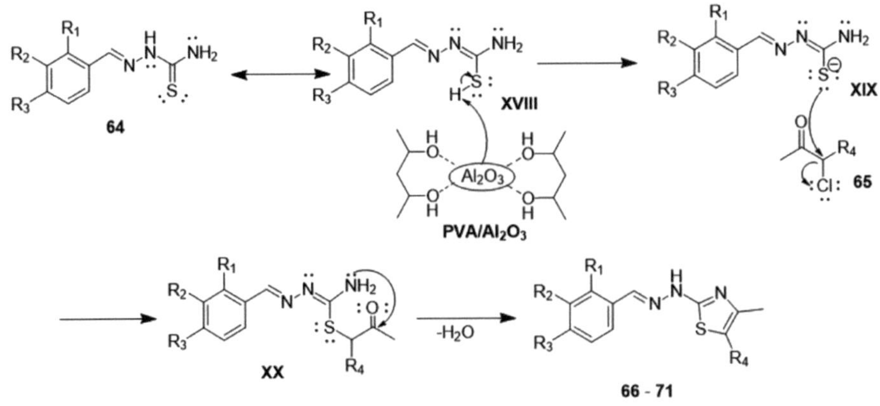

Figure 22.
Proposed mechanism of PVA/Al2O3-catalyzed reaction [61].

displacing -Cl and generating the intermediate **XX**. The cyclocondensation of **XX** yielded the desired thiazole derivatives. The catalyst was recycled three times and recovered in excellent yields (90%) [61].

4. Conclusion

This chapter reviewed novel synthetic methods involving the presence of catalysts, solvent-free conditions, and green chemistry approaches reported during 2015–2022, for the preparation of pyrazoles and thiazoles. Additionally, their therapeutic and biological evaluations were discussed. These structural moieties will continue to play an important role in therapeutic and other applications, therefore further development, versatility and diversification of their syntheses will continue to be of great interest to chemists worldwide.

Conflict of interest

The authors declare no conflict of interest.

Author details

Erika Lozano[1,2], Melissa M. Lewis-Bakker[1,2] and Lakshmi P. Kotra[1,2*]

1 Leslie Dan Faculty of Pharmacy, University of Toronto, Toronto, Canada

2 Krembil Research Institute, University Health Network, Toronto, Canada

*Address all correspondence to: p.kotra@utoronto.ca

References

[1] Ansari A, Ali A, Shamsuzzaman AM. Review: Biologically active pyrazole derivatives. New England Journal of Chemistry. 2017;**41**:16-41. DOI: 10.1039/C6NJ03181A

[2] Karrouchi K, Radi S, Ramli Y, Taoufik J, Mabkhot YN, Al-aizari FA, et al. Synthesis and pharmacological activities of Pyrazole derivatives: A review. Molecules. 2018;**23**:134. DOI: 10.3390/molecules23010134

[3] Kaufmann J, Jäckel E, Haak E. Ruthenium-catalyzed formation of pyrazoles or 3-hydroxynitriles from propargyl alcohols and hydrazines. ARKIVOC. 2019;**iv**:91-101. DOI: 10.24820/ark.5550190.p010.893

[4] Cardinale L, Neumeier M, Majek M, Jacobi von Wangelin A. Aryl Pyrazoles from photocatalytic cycloadditions of Arenediazonium. Organic Letters. 2020;**22**:7219-7224. DOI: 10.1021/acs.orglett.0c02514

[5] Pise AS, Bururngale AS, Devkate SS. Ultrasound assisted synthesis of 1,5-Disubstituted Pyrazole using Cu(I) catalyst. Asian Journal of Chemistry. 2020;**32**:575-579. DOI: 10.14233/ajchem.2020.22432

[6] Patra M, Gasser G. The medicinal chemistry of ferrocene and its derivatives. Nature Reviews Chemistry. 2017;**1**:0066. DOI: 10.1038/s41570-017-0066

[7] Kumar M, Sharma S, Singh Tuli H, Khare R, Parkash V. A novel synthesis, characterization and biological studies of Ferrocenyl substituted Pyrazoles. Asian Journal of Chemistry. 2019;**31**:2729-2732. DOI: 10.14233/ajchem.2019.22161

[8] Ohtsuka Y, Uraguchi D, Yamamoto K, Tokuhisa K, Yamakawa T. Syntheses of 2-(trifluoromethyl)-1,3-dicarbonyl compounds through direct trifluoromethylation with CF3I and their application to fluorinated pyrazoles syntheses. Tetrahedron. 2012;**68**:2636-2649. DOI: 10.1016/j.tet.2012.01.075

[9] Liu H, Audisio D, Plougastel L, Decuypere E, Buisson DA, Koniev O, et al. Ultrafast click chemistry with Fluorosydnones. Angewandte Chemie, International Edition. 2016;**55**:12073-12077. DOI: 10.1002/anie.201606495

[10] Yuan X, Yao JF, Tang ZY. Decarboxylative fluorination of electron-rich Heteroaromatic carboxylic acids with Selectfluor. Organic Letters. 2017;**19**:1410-1413. DOI: 10.1021/acs.orglett.7b00335

[11] Neeff A, Pazenok S. Method for the production of 5-fluoro-1,3-dialkyl-1h-pyrazol-4-carbonyl fluorides. 2009

[12] Pazenok S, Lui N, Gary S. Process for the preparation of 5-Fluoro-1-alkyl-3-fluoroalkyl-1H-pyrazole-4-carbonyl chlorides and fluorides. 2011

[13] Levchenko V, Dmytriv YV, Tymtsunik AV, Liubchak K, Rudnichenko A, Melnyk AV, et al. Preparation of 5-Fluoropyrazoles from PyrazolesandN-Fluorobenzenesulfonimide (NFSI). The Journal of Organic Chemistry. 2018;**83**:3265-3274. DOI: 10.1021/acs.joc.8b00199

[14] Arshad MF, Alam A, Alshammari AA, Alhazza MB, Alzimam IM, Alam MA, et al. Thiazole: A versatile standalone moiety contributing to the development of various drugs and biologically active agents. Molecules. 2022;**27**:3994. DOI: 10.3390/molecules27133994

[15] Ali SH, Sayed AR. Review of the synthesis and biological activity of thiazoles. Synthetic Communications. 2021;**51**:670-700. DOI: 10.1080/00397911.2020.1854787

[16] Ibrahim SA, Rizk HF. Synthesis and biological evaluation of thiazole derivatives. In: Kuznetsov A, editor. Azoles. Rijeka, London, UK: IntechOpen; 2020. p. Ch. 2. DOI: 10.5772/intechopen.93037

[17] Omar MA, Masaret GS, Abbas EMH, Abdel-Aziz MM, Harras MF, Farghaly TA. Novel anti-tubercular and antibacterial based benzosuberone-thiazole moieties: Synthesis, molecular docking analysis, DNA gyrase supercoiling and ATPase activity. Bioorganic Chemistry. 2020;**104**:104316. DOI: 10.1016/j.bioorg.2020.104316

[18] Li Z, Xu X, Hou J, Wang S, Jiang H, Zhang L. Structure-based optimization of free fatty acid receptor 1 agonists bearing thiazole scaffold. Bioorganic Chemistry. 2018;**77**:429-435. DOI: 10.1016/j.bioorg.2018.01.039

[19] Vekariya RH, Patel KD, Vekariya MK, Prajapati NP, Rajani DP, Rajani SD, et al. Microwave-assisted green synthesis of new imidazo [2,1-b] thiazole derivatives and their antimicrobial, antimalarial, and antitubercular activities. Research on Chemical Intermediates. 2017;**43**:6207-6231. DOI: 10.1007/s11164-017-2985-5

[20] Eryılmaz S, Türk Çelikoğlu E, İdil Ö, İnkaya E, Kozak Z, Mısır E, et al. Derivatives of pyridine and thiazole hybrid: Synthesis, DFT, biological evaluation via antimicrobial and DNA cleavage activity. Bioorganic Chemistry. 2020;**95**:103476. DOI: 10.1016/j.bioorg.2019.103476

[21] Lu Q, Yu Q, Zhu Y, Weng J, Yuan J, Hu D, et al. Novel stilbene analogues containing thiazole moiety: Synthesis, biological evaluation and docking study. Journal of Molecular Structure. 2019;**1180**:780-786. DOI: 10.1016/j.molstruc.2018.12.068

[22] Eissa SI, Farrag AM, Abbas SY, El Shehry MF, Ragab A, Fayed EA, et al. Novel structural hybrids of quinoline and thiazole moieties: Synthesis and evaluation of antibacterial and antifungal activities with molecular modeling studies. Bioorganic Chemistry. 2021;**110**:104803. DOI: 10.1016/j.bioorg.2021.104803

[23] Reddy GM, Garcia JR, Reddy VH, de Andrade AM, Camilo A, Pontes Ribeiro RA, et al. Synthesis, antimicrobial activity and advances in structure-activity relationships (SARs) of novel tri-substituted thiazole derivatives. European Journal of Medicinal Chemistry. 2016;**123**:508-513. DOI: 10.1016/j.ejmech.2016.07.062

[24] Nastasă C, Tiperciuc B, Duma M, Benedec D, Oniga O. New hydrazones bearing thiazole scaffold: Synthesis, characterization, antimicrobial, and antioxidant investigation. Molecules. 2015;**20**:17338. DOI: 10.3390/molecules200917325

[25] Rajagopal K, Dhandayutham S, Nandhagopal M, Narayanasamy M, Elzagheid I, Rhyman ML, et al. Thiazole derivatives: Synthesis, characterization, biological and DFT studies. Journal of Molecular Structure. 2022;**1255**:132374. DOI: 10.1016/j.molstruc.2022.132374

[26] Farghaly TA, Althagafi I, Ibrahim MH, Al-Qurashi NT, Farooq U. Synthesis under microwaves irradiation, structure elucidation, docking study for inhibiting COVID-19 and DFT calculations of novel azoles incorporated indole moiety. Journal of Molecular Structure. 2021;**1244**:131263. DOI: 10.1016/j.molstruc.2021.131263

[27] Bera P, Aher A, Brandao P, Manna SK, Mondal G, Jana A, et al. Induced apoptosis against U937 cancer cells by fe(II), co(III) and ni(II) complexes with a pyrazine-thiazole ligand: Synthesis, structure and biological evaluation. Polyhedron. 2020;**182**:114503. DOI: 10.1016/j.poly.2020.114503

[28] Althagafi I, El-Metwaly N, Farghaly TA. New series of thiazole derivatives: Synthesis, structural elucidation, antimicrobial activity, molecular modeling and MOE docking. Molecules. 2019;**24**:1741. DOI: 10.3390/molecules24091741

[29] Blaja SP, Lungu LV, Kuchkova KI, Ciocarlan AG, Barba AN, Vornicu N, et al. Norlabdane compounds containing thiosemicarbazone or 1,3-thiazole fragments: Synthesis and antimicrobial activity. Chemistry of Natural Compounds. 2021;**57**:101-110. DOI: 10.1007/s10600-021-03292-3

[30] Nadaf AA, Barretto DA, Najare MS, Mantur S, Garbhagudi M, Gaonkar S, et al. Design, synthesis and biological evaluation of novel N,4-diphenylthiazol-2-amine derivatives. Medicinal Chemistry Research. 2020;**29**:442-458. DOI: 10.1007/s00044-019-02495-2

[31] Makam P, Kannan T. 2-aminothiazole derivatives as antimycobacterial agents: Synthesis, characterization, in vitro and in silico studies. European Journal of Medicinal Chemistry. 2014;**87**:643-656. DOI: 10.1016/j.ejmech.2014.09.086

[32] Sokolova AS, Yarovaya OI, Bormotov NI, Shishkina LN, Salakhutdinov NF. Synthesis and antiviral activity of camphor-based 1,3-thiazolidin-4-one and thiazole derivatives as orthopoxvirus-reproduction inhibitors. Medicinal Chemistry Communications. 2018;**9**:1746-1753. DOI: 10.1039/C8MD00347E

[33] Farghaly TA, Alsaedi AMR, Alenazi NA, Harras MF. Anti-viral activity of thiazole derivatives: An updated patent review. Expert Opinion on Therapeutic Patents. 2022;**32**:791-815. DOI: 10.1080/13543776.2022.2067477

[34] Stankova I, Chuchkov K, Chayrov R, Mukova L, Galabov A, Marinkova D, et al. Adamantane derivatives containing thiazole moiety: Synthesis, antiviral and antibacterial activity. International Journal of Peptide Research and Therapeutics. 2020;**26**:1781-1787. DOI: 10.1007/s10989-019-09983-4

[35] Ramagiri RK, Vedula RR, Thupurani MK. A facile one-step multi-component approach toward the synthesis of 3-(2-amino-4-thiazolyl) coumarins by using trimethylsilyl isothiocyanate and their antioxidant and anti-inflammatory activity. Phosphorus, Sulfur, and Silicon and the Related Elements. 2015;**190**:1393-1397. DOI: 10.1080/10426507.2014.990016

[36] Ramalingam A, Sarvanan J. Synthesis, docking and anti-cancerous activity of some novel thiazole derivatives of biological interest. IJPI. 2020;**10**:594-603. DOI: 10.5530/ijpi.2020.4.104

[37] Coman F, Mbaveng AT, Leonte D, Bencze LC, Vlase L, Imre S, et al. Heterocycles 44. Synthesis, characterization and anticancer activity of new thiazole ortho-hydroxychalcones. Medicinal Chemistry Research. 2018;**27**:1396-1407. DOI: 10.1007/s00044-018-2156-2

[38] Łączkowski KZ, Anusiak J, Świtalska M, Dzitko K, Cytarska J, Baranowska-Łączkowska A, et al. Synthesis, molecular docking, ctDNA

interaction, DFT calculation and evaluation of antiproliferative and anti-toxoplasma gondii activities of 2,4-diaminotriazine-thiazole derivatives. Medicinal Chemistry Research. 2018;**27**:1131-1148. DOI: 10.1007/s00044-018-2136-6

[39] Guerrero-Pepinosa NY, Cardona-Trujillo MC, Garzón-Castaño SC, Veloza LA, Sepúlveda-Arias JC. Antiproliferative activity of thiazole and oxazole derivatives: A systematic review of in vitro and in vivo studies. Biomedicine & Pharmacotherapy. 2021;**138**:111495. DOI: 10.1016/j.biopha.2021.111495

[40] Spanò V, Attanzio A, Cascioferro S, Carbone A, Montalbano A, Barraja P, et al. Synthesis and antitumor activity of new thiazole nortopsentin analogs. Marine Drugs. 2016;**14**:226. DOI: 10.3390/md14120226

[41] Forli S. Epothilones: From discovery to clinical trials. Current Topics in Medicinal Chemistry. 2014; **14**:2312-2321. DOI: 10.2174/1568026614666141130095855

[42] Bollag DM. Epothilones: Novel microtubule-stabilising agents. Expert Opinion on Investigational Drugs. 1997;**6**:867-873. DOI: 10.1517/13543784.6.7.867

[43] Kempf DJ, Sham HL, Marsh KC, Flentge CA, Betebenner D, Green BE, et al. Discovery of ritonavir, a potent inhibitor of HIV protease with high oral bioavailability and clinical efficacy. Journal of Medicinal Chemistry. 1998;**41**:602-617. DOI: 10.1021/jm970636+

[44] Hung Y, Lee J, Chiu C, Lee C, Tsai P, Hsu I, et al. Oral nirmatrelvir/ritonavir therapy for COVID-19: The dawn in the dark? Antibiotics. 2022;**11**:220. DOI: 10.3390/antibiotics11020220

[45] Cao B, Wang Y, Wen D, Liu W, Wang J, Fan G, et al. A trial of Lopinavir–ritonavir in adults hospitalized with severe covid-19. New England Journal of Medicine. 2020;**382**:1787-1799. DOI: 10.1056/NEJMoa2001282

[46] Chugh K, Agrawal S. Cefpodoxime: Pharmacokinetics and therapeutic uses. The Indian Journal of Pediatrics. 2003;**70**:227-231. DOI: 10.1007/BF02725589

[47] Chocas EC, Paap CM, Godley PJ. Cefpodoxime proxetil: A new, broad-spectrum, oral cephalosporin. The Annals of Pharmacotherapy. 1993;**27**:1369-1377. DOI: 10.1177/106002809302701111

[48] Borelli C, Schaller M, Niewerth M, Nocker K, Baasner B, Berg D, et al. Modes of action of the new arylguanidine abafungin beyond interference with ergosterol biosynthesis and in vitro activity against medically important fungi. Chemotherapy. 2008;**54**:245-259. DOI: 10.1159/000142334

[49] Hantzsch A, Weber JH. Ueber verbindungen des thiazols (pyridins der thiophenreihe). Ber. Dtsch. Chem. Ges. 1887;**20**:3118-3132. DOI: 10.1002/cber.188702002200

[50] Cook AH, Heilbron I, Levy AL. Studies in the azole series. Part II. The interaction of α-amino-nitriles and carbon disulphide. Journal of the Chemical Society. 1947;**2**:1598-1609. DOI: 10.1039/JR9470001598

[51] Cook AH, Heilbron I, Levy AL. Studies in the azole series. Part I. a novel route to 5-aminothiazoles. Journal of the Chemical Society. 1947;**1**:1594-1598. DOI: 10.1039/JR9470001594

[52] Cook AH, Heilbron I, Stern ES. 412. Studies in the azole series. Part X. some

5-amino-2-mercapto-4-alkylthiazoles and 2: 4-dithio-5-alkylhydantions. Journal of the Chemical Society. 1948;**1948**:2031-2033. DOI: 10.1039/JR9480002031

[53] Capp CW, Cook AH, Downer JD, Heilbron I. 268. Studies in the azole series. Part VIII. The interaction of α-amino-nitriles and carbethoxyisothiocyanate. Journal of the Chemical Society. 1948:1340-1345. DOI: 10.1039/JR9480001340

[54] Vernin G. General synthetic methods for thiazole and thiazolium salts. In: Metzger JV, editor. Chemistry of Heterocyclic Compounds. New Jersey, USA: Wiley; 1979. pp. 165-335. DOI: 10.1002/9780470187081.ch3

[55] Ma D, Liu X, Jia H, Zhang Y, Jiang Q, Sun H, et al. A novel small-molecule inhibitor of SREBP-1 based on natural product monomers upregulates the sensitivity of lung squamous cell carcinoma cells to antitumor drugs. Frontiers in Pharmacology. 2022;**13**:895744. DOI: 10.3389/fphar.2022.895744

[56] Wang G, Liu W, Fan M, He M, Li Y, Peng Z. Design, synthesis and biological evaluation of novel thiazole-naphthalene derivatives as potential anticancer agents and tubulin polymerisation inhibitors. Journal of Enzyme Inhibition and Medicinal Chemistry. 2021;**36**:1693-1701. DOI: 10.1080/14756366.2021.1958213

[57] Merritt EA, Bagley MC. Holzapfel-meyers-nicolaou modification of the hantzsch thiazole synthesis. Synthesis. 2007;**2007**:3535-3541. DOI: 10.1055/s-2007-990851

[58] Tomassetti M, Lupidi G, Piermattei P, Rossi FV, Lillini S, Bianchini G, et al. Catalyst-free synthesis of polysubstituted 5-acylamino-1,3-thiazoles via hantzsch cyclization of α-chloroglycinates. Molecules. 2019;**24**:3846. DOI: 10.3390/molecules24213846

[59] Avadhani A, Iniyavan P, Kumar Y, Ila H. Single-pot preparation of 4-amino-2-(het)aryl-5-substituted thiazoles employing functionalized dithioesters as thiocarbonyl precursors. The Journal of Organic Chemistry. 2021;**86**:8508-8515. DOI: 10.1021/acs.joc.1c00616

[60] Bodireddy MR, Khaja Mohinuddin PM, Gundala TR, Gangi Reddy NC. Lactic acid-mediated tandem one-pot synthesis of 2-aminothiazole derivatives: A rapid, scalable, and sustainable process. Cogent Chemistry. 2016;**2**:1154237. DOI: 10.1080/23312009.2016.1154237

[61] Riyadh SM, Khalil KD, Bashal AH. Structural properties and catalytic activity of binary poly (vinyl alcohol)/Al2O3 nanocomposite film for synthesis of thiazoles. Catalysts. 2020;**10**:100. DOI: 10.3390/catal10010100

Chapter 10

Greener Approach towards the Synthesis of Nitrogen Based Heterocycles

Monica Dinodia

Abstract

The preferable application of green chemistry in research is to utilize environment benign, mild, non toxic, reproducible catalyst and efficient solvents in synthesis of molecules. Use of green chemistry techniques had enabled in dramatically reducing chemical waste and reaction times as has recently been reported in several organic syntheses reactions. Greener routes are required in the synthesis of N-heterocycles, due to the remarkable importance of these compounds in medicinal chemistry. This chapter is dedicated to the synthesis of N containing heterocyclic compounds using eco-friendly solvent like water and bio-derived solvents (glycerol, ethyl lactate, and gluconic acid aqueous solution). Water and bio-based solvents for the synthesis of aromatic nitrogen heterocycles was chosen due to the negligible toxicity associated with these solvents. Apart from being eco-friendly, water also has the potential to become a universally acceptable solvent due to its abundance and low cost. Work on microwave synthesis is also reported as it is an eco-friendly and faster process for the synthesis of these N-based heterocyclic compounds. Due to its rapid action to produce products with greater purity and yield, it is now being used worldwide.

Keywords: heterocycles, green chemistry, synthesis, eco-friendly, sustainable

1. Introduction

Green chemistry, also known as sustainable chemistry, is an area of chemistry and chemical engineering which focus on the design of products and processes that minimize or eliminate the use and generation of hazardous substances [1]. In 1998, Paul Anastas and John C. Warner reported a set of 12 green chemistry principles [2]. Use of green chemistry can help us to create alternatives to hazardous sub stances. Chemical processes can be designed which reduce waste and prevent diminishing of natural resources. Processes can be designed that use lesser amounts of energy. Heterocyclic compounds had a special place among pharmaceutically important natural products and synthetic molecules. Among the heterocycles, nitrogen based molecules are abundant in nature and is of utmost importance to life because their structural subunits exists in natural products like hormones, vitamins, antibiotics, alkaloids, herbicides, Nucleic acid (DNA and RNA) etc [3]. The construction of complicated cyclic target molecules having minimum environment impact is a challenging task for academicians, scientists and industry people. Greener methods are required for the synthesis of N-heterocycles as these compounds are medicinally important [4, 5]. Environment friendly protocols have been explored all over the globe for heterocyclic synthesis to improve energy consumption, atom economy and reaction yields [6, 7]. In the past few decades, numerous research papers have reported the use of water as green solvent [8, 9]. Researchers are also making use of polyethylene glycol for its low price and low acute toxicity [10, 11]. More recently, research teams have demonstrated that the use of bio-based solvents is also a solvent of choice [12]. All these results prove that the concepts of green chemistry have made remarkable progress. The conventional methods for the synthesis of N-heterocyclic skeleton require the use of expensive starting materials and high temperature. Chemists play an important role in the construction of a sustainable future through the application of greener chemical processes. As so, the development of new synthetic methods using more efficient energy sources and less hazardous solvents as well as renewable and eco-friendly catalysts to attain the N-heterocyclic core can provide significant environmental and economic advantages.

2. Use of green solvents

2.1 Reactions in water

A novel and direct method for the synthesis of pharmacologically promising spiro [indoline-3,7′-pyrrolo[1,2-*c*]imidazole]-6′-carbonitrile derivatives has been developed via a three-component one pot reaction reaction. The reaction was cleaner and proceeded in good yields using isatin, malononitrile, and hydantoin or thiohydantoin in the presence of Et3N/water (**Figure 1**) [13].

A regioselective synthesis of polysubstituted pyrroles in good yields from α-azido chalcones and 1,3-dicarbonyl compounds using indium trichloride in water as an efficient catalyst is discussed (**Figure 2**) [14]. Under conventional heating $InCl_3$ in water has been found to complete the reaction in 30 min. Microwave irradiation on the other side reduced the reaction time from 30 min to 10 min.

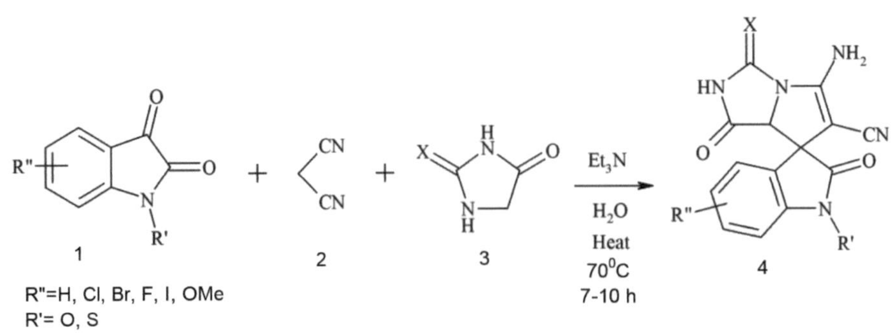

Figure 1.
Synthesis of spiro[indoline-3,7′-pyrrolo[1,2-c]imidazole]-6′-carbonitrile derivatives.

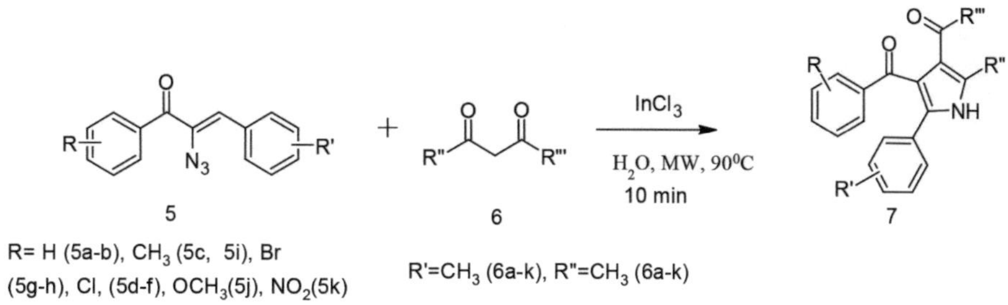

Figure 2.
Synthetic route for polysubstituted pyrroles.

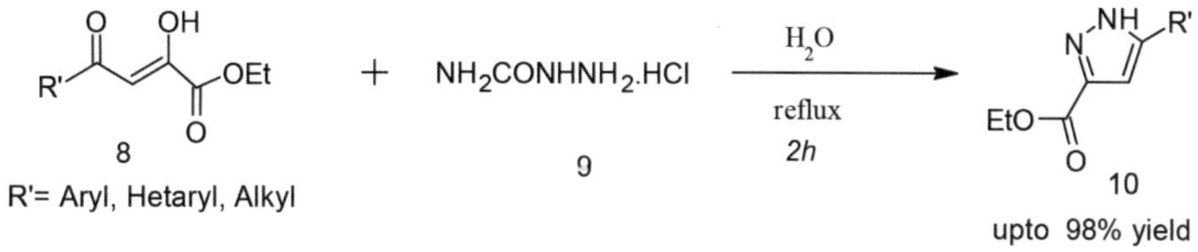

Figure 3.
Synthesis of 3,5-disubstituted pyrazoles.

Highly efficient, green and simple method for the construction of pyrazole-3-carboxylates and 3,5-disubstituted pyrazoles by cyclization of 4-aryl(hetaryl, alkyl)-2,4-diketoesters and 1,3-diketones with semicarbazide hydrochloride using water has been developed (**Figure 3**) [15]. This synthesis did not require toxic hydrazine and product purification, eliminating the use of toxic liquid chemicals.

Synthesis of diarylthiazoles and diarylimidazoles utilizing a reaction between α-tosyloxyketones with a variety of thioamide/amidine in water without any additives is reported. This methodology demonstrated several advantages such as being simple, efficient and high yielding, also it a greener protocol (**Figure 4** and **Table 1**) [16].

An environmentally friendly, green, practical, attractive and effective method to construct 4*H*-pyrido[1,2-*a*]pyrimidin-4-one has been developed (**Figure 5**) [17]. The reaction using water as the solvent in absence of catalyst under the mild conditions makes these transformations very efficient (**Table 2**).

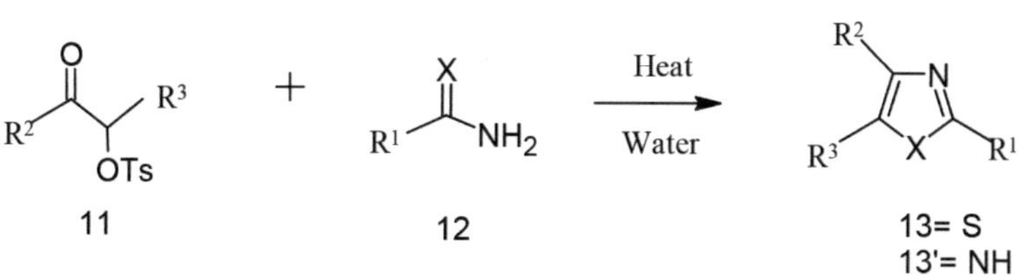

Figure 4.
Synthesis of diarylthiazoles and diarylimidazoles.

Compound	R^2	R^3	R^1	Yield (%)	Time (h)
13a	C_6H_5	H	C_6H_5	87	5.0
13b	C_6H_5	H	4-ClC_6H_4	90	6.0
13c	C_6H_5	H	NH_2	92	1.0
13d	C_6H_5	H	CH_3	82	5.0
13e	C_6H_5	H	4-O$CH_3C_6H_4$	80	4.0
13f	C_6H_5	H	3-Indolyl	70	4.0
13g	C_6H_5	H	4-Pyridyl	65	5.0
13h	C_6H_5	H	1-Methyl-3-indolyl	85	4.0
13i	CH_3	COCH_3	C_6H_5	85	5.0
13j	CH_3	COCH_3	4-ClC_6H_4	90	3.0
13k	CH_3	COCH_3	4-O$CH_3C_6H_4$	90	3.0
13l	4-O$CH_3C_6H_4$	H	C_6H_5	88	3.0
13m	4-O$CH_3C_6H_4$	H	4-ClC_6H_4	93	3.0
13n	4-O$CH_3C_6H_4$	H	NH_2	80	4.0
13o	4-ClC_6H_4	H	C_6H_5	85	3.0
13p	4-ClC_6H_4	H	4-ClC_6H_4	90	4.0
13q	4-ClC_6H_4	H	3-Indolyl	87	4.0
13r	1-PhSO_2-3-indolyl	H	C_6H_5	75	6.0
13s	1-PhSO_2-3-indolyl	H	1-Methyl-3-indolyl	86	5.0
13'a	C_6H_5	H	C_6H_5	70	3.0
13'b	C_6H_5	H	4-ClC_6H_4	65	3.5
13'c	C_6H_5	H	4-Pyridyl	65	2.5
13'd	4-ClC_6H_4	H	C_6H_5	70	3.0
13'e	4-ClC_6H_4	H	4-ClC_6H_4	63	3.0
13'f	4-ClC_6H_4	H	4-Pyridyl	55	2.0
13'g	4-O$CH_3C_6H_4$	H	4-ClC_6H_4	55	3.0
13'h	4-O$CH_3C_6H_4$	H	4-Pyridyl	60	2.5
13'i	4-O$CH_3C_6H_4$	H	C_6H_5	50	3.5

Table 1.
*Synthesis of diarylthiazoles [13] and diarylimidazoles **13'** [16].*

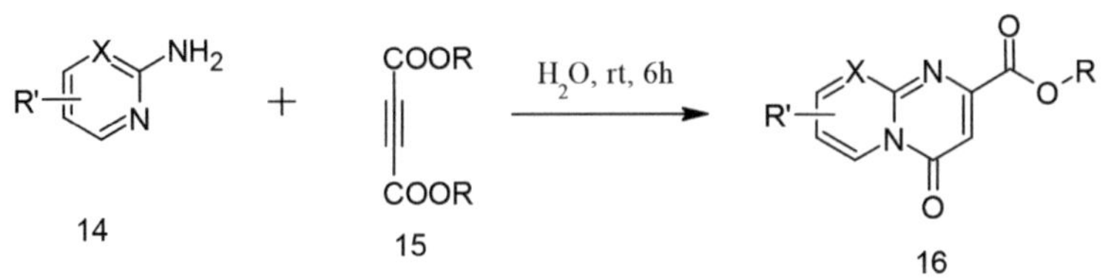

Figure 5.
*Construction of 4*H*-pyrido[1,2-a]pyrimidin-4-one.*

Entry* [17]	R^1/X		R	Product	Yield (%)
14a	H/C	**15a**	Me	**16aa**	82
14b	3-Me/C	**15a**	Me	**16ba**	60
14c	4-Me/C	**15a**	Me	**16ca**	74
14d	5-Me/C	**15a**	Me	**16da**	83
14e	6-Me/C	**15a**	Me	—	—
14f	5-F/C	**15a**	Me	**16fa**	45
14g	4-Cl/C	**15a**	Me	**16ga**	71
14h	5-Cl/C	**15a**	Me	**16ha**	39
14i	5-Br/C	**15a**	Me	**16ia**	47
14j	6-Br/C	**15a**	Me	—	—
14k	3-BnO/C	**15a**	Me	**16ka**	40
14l	4-COOEt/C	**15a**	Me	**16la**	72
14a	H/C	**15′b**	Et	**16ab**	82
14b	3-Me/C	**15′b**	Et	**16bb**	35
14c	4-Me/C	**15b**	Et	**16cb**	86
14d	5-Me/C	**15b**	Et	**16db**	97
14e	6-Me/C	**15b**	Et	—	—
14f	5-F/C	**15b**	Et	**16fb**	52
14g	4-Cl/C	**15b**	Et	**16gb**	53
14i	5-Br/C	**15b**	Et	**16ib**	50
14j	6-Br/C	**15b**	Et	—	—
14k	3-BnO/C	**15b**	Et	**16kb**	15
14l	4-COOEt/C	**15b**	Et	**16lb**	56
14m	H/N	**15a**	Me	—	—

Table 2.
* *Reaction conditions:* ***14a*** *(0.32 mmol),* ***15a*** *(0.38 mmol), catalyst-free, in 30 mL water for 6 h under air.*

2.2 Reactions in bio-based solvents

Bio based green solvent Ethyl lactate was used, for the 1,3-dipolar cycloaddition reaction to generates a series of medicinally important spiro[benzo[*f*]pyrrolo[2,1-*a*]isoindole-5,3′-indoline]-2′,6,11-trione derivatives in excellent yields at room temperature (**Figure 6**) [18]. The product was obtaind in high yield (approximately 90%) with the 1,3-dipolar cycloaddition reaction of substituted isatin and proline with napthaquinone.

The first investigation of the use of eucalyptol as a new solvent for organic synthesis was reported by a group [19]. Heterocycles having oxygen, sulfur and nitrogen were chosen as targets or as starting materials for widely used palladium-catalysed cross-coupling reactions, *like* Suzuki-Miyaura and Sonogashira-Hagihara reactions. Eucalyptol turned out to be a viable sustainable solvent and was shown to be an interesting alternative to conventional solvents for the one-pot synthesis of 2,3-diarylimidazol[1,2-a]pyridines (**Figure 7**).

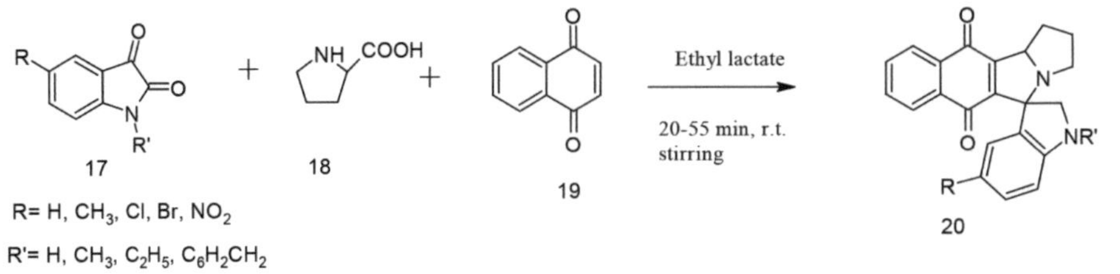

Figure 6.
Synthesis of spiro[benzo[f]pyrrolo[2,1-a]isoindole-5,3′-indoline]-2′,6,11-trione derivatives.

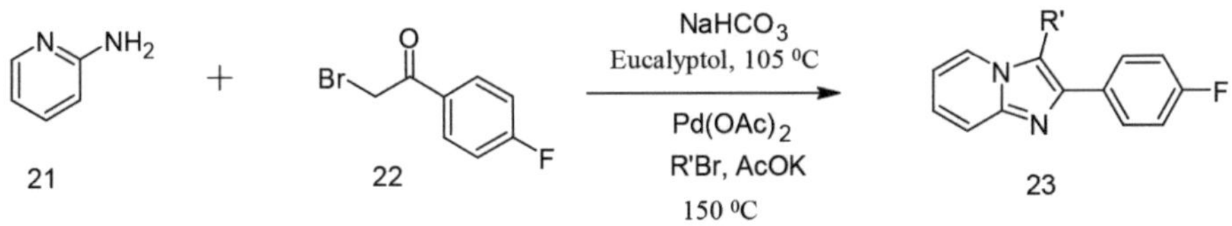

Figure 7.
One-pot synthesis of 2,3-diarylimidazol[1,2-a]pyridines.

A simple and chemoselective method of synthesizing T-shaped oxazolonaphtho [1′,2′:4,5]imidazo[1,2-*a*]pyridines (**Figure 8**) in a one-pot selective fashion was developed in good yields using lactic acid as an alternative solvent to acetic acid [20]. The reaction exhibited advantages like bio-based origin, ease of isolating the product and superior synthetic efficiency. The synthetic strategy adopted was highly compatible with various functionalities.

A Group of researchers published a paper on an efficient and environmentally benign synthetic protocol for the synthesis of pyrrole derivatives (**Figure 9**) using gluconic acid aqueous medium as an eco-friendly bio-based catalytic solvent system. The synthesis was done by the four-component coupling of amines, aldehydes, 1,3-dicarbonyl compounds, and nitromethane [21]. Gluconic acid aqueous solution could be recycled and reused several times without significant loss of its efficiency. This reaction showed excellent functional group tolerance, short reaction time, and high yield of products.

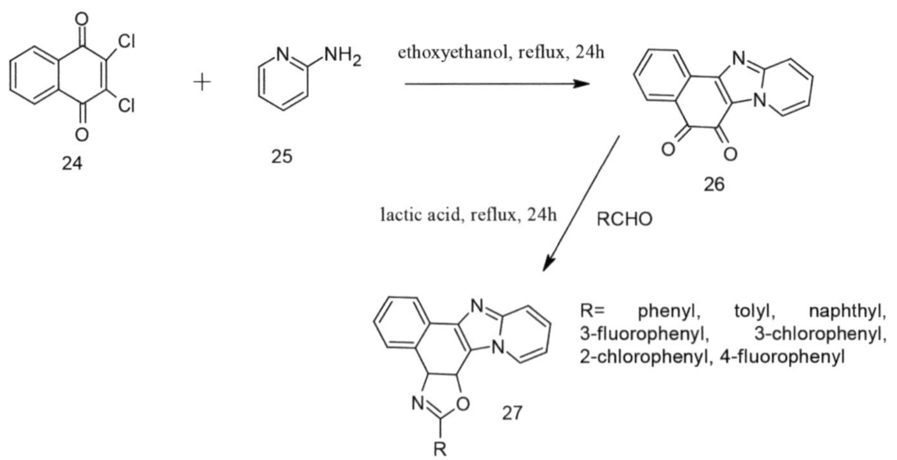

Figure 8.
Synthesis of T-shaped oxazolonaphtho[1′,2′:4,5]imidazo[1,2-a]pyridines.

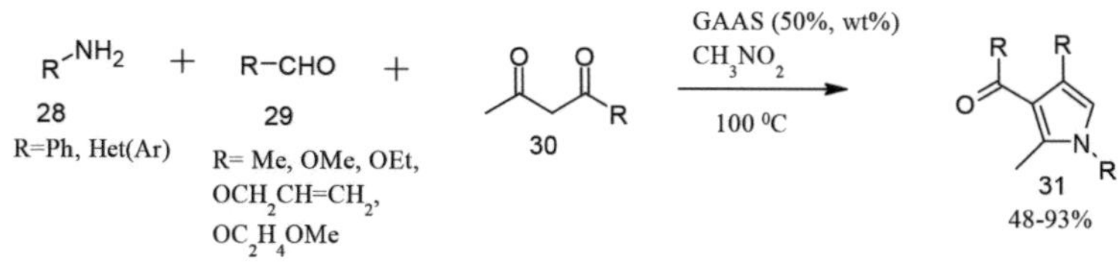

Figure 9.
Environmentally benign synthetic protocol for the synthesis of structurally diverse pyrrole derivatives.

3. Microwave irradiation

One of the potential green chemistry method using microwaves (MW) have emerged during the recent years [22, 23]. A review outlines the use of MW technique for the synthesis of N-containing heterocycles [24]. Novel 2-aryl-3,4-dihydro-2*H*-thieno[3,2-*b*]indole derivatives (**Figure 10**) has been synthesised regioselectively in good yields from the reaction of 5-aryldihydro-3(2*H*)-thiophenones and arylhydrazine hydrochloride under microwaves [25]. The group also did a comparison of the efficacies of the thermal and microwave reactions, by carrying out the reaction under microwave for the synthesis of **34a** (Ar = C_6H_5, R = H) at the reflux temperature of 80°C. At 80°C, the reaction using microwaves was completed in 5 min, that is ten times faster than that under thermal condition (50 min) using the same solvent.

A rapid solvent-less synthesis of 5-hydroxy-benzo[*g*]indole scaffolds (**Figure 11**) is accomplished from Lewis acid-catalyzed one-pot reaction of naphthoquinone, ω-morpholinoacetophenone, and urea under microwave irradiation [26]. To investigate the role of Lewis acid, the researchers also carried microwave-mediated reaction using Lewis acids like $TiCl_4$, $AlCl_3$, $ZnCl_2$, $SmCl_2$, and $InCl_3$ under similar reaction conditions. They found that in comparison to $BF_3{\cdot}OEt_2$ catalyzed reaction, all the other Lewis acids gave poor yields. Also the reaction failed to proceed in the absence of the Lewis acid (**Figure 12**).

Researchers have developed a novel, convenient, environmentally friendly one-pot synthesis of imidazo[1,2-a]pyridines (**Figure 13**) using 2-aminopyridines and in-situ generated phenacyl bromides under microwave irradiation in polyethylene glycol (PEG-400) and water (1:2) [27]. This protocol is a better alternative to the existing method as it involves use of in-situ-generated α-bromoacetophenones, utilization of

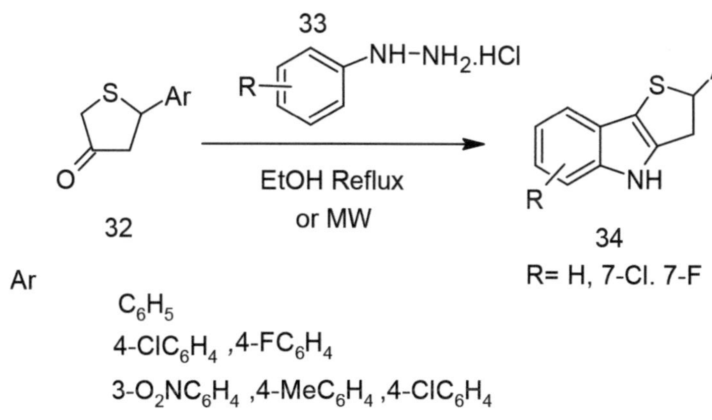

Figure 10.
*Construction of novel 2-aryl-3,4-dihydro-2*H*-thieno[3,2-*b*]indole derivatives.*

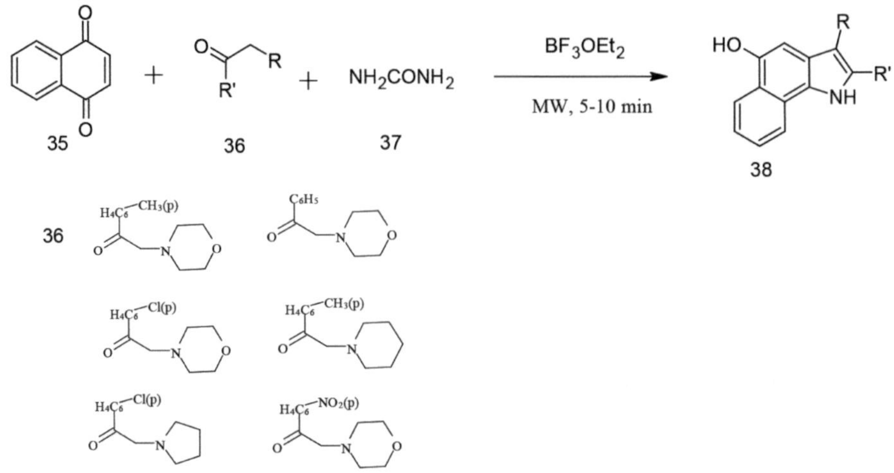

Figure 11.
Solvent-less synthesis of 5-hydroxy-benzo[g]indole scaffolds.

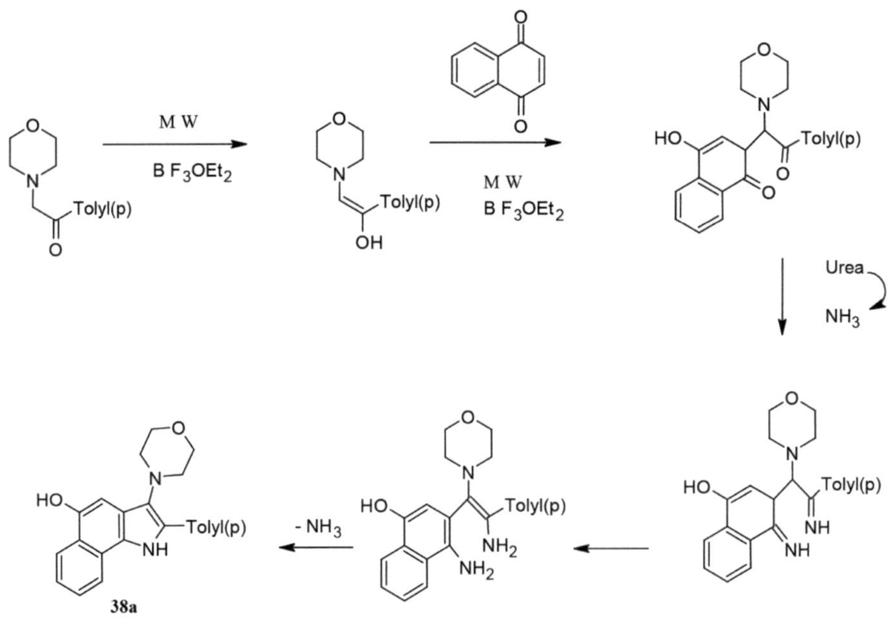

Figure 12.
Proposed mechanism for the synthesis of benzo[g]indole derivative (38a).

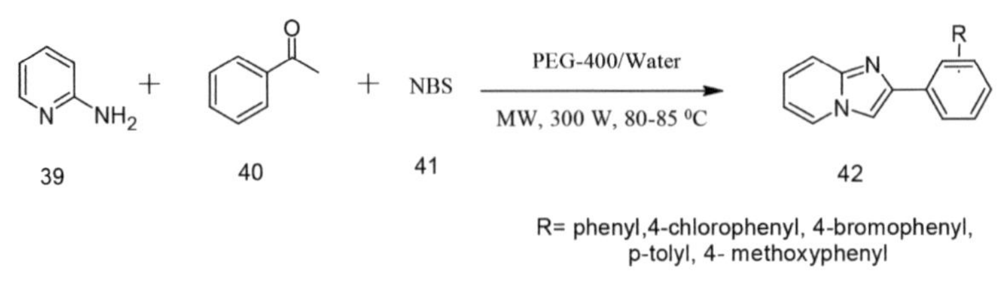

Figure 13.
One-pot synthesis of imidazo[1,2-a]pyridines.

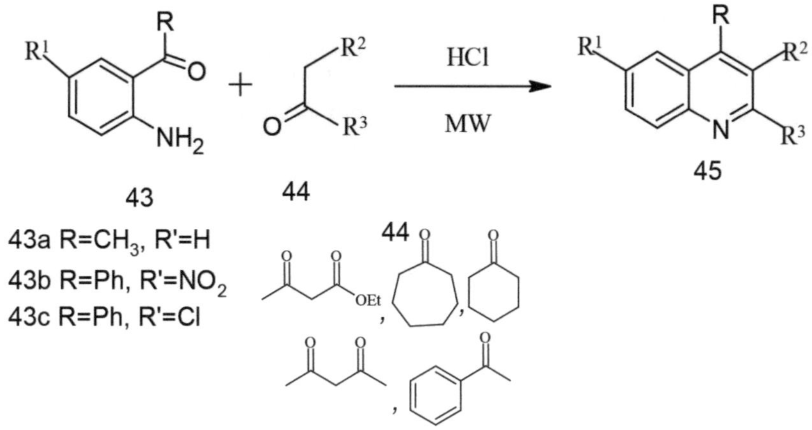

Figure 14.
Synthesis of a series of substituted quinolines.

lachrymatric α-haloketones and volatile toxic organic solvents is avoided. There is reduction in the reaction time also to obtain excellent yield.

A series of substituted quinolines (**Figure 14**) was developed via the Friedländer reaction employing microwave irradiation (MW), in the presence of a catalytic amount of hydrochloric acid [28]. The products were tested in vitro against the parasites causative of malaria, leishmaniasis, sleeping sickness and Chagas' disease (TDR, WHO). Some of these compounds exhibited activity against *Plasmodium falciparum* and others resulted moderately active against *Trypanosoma cruzi*. The MW syntheses were carried out in a domestic oven adapted for the use of a reflux condenser, with constant power of (400 W). All the reactions were completed between 1.5 and 12 min.

In search for green reactions leading to the Formation of N-Heterocycles, an excellent review was identified entitled "More Sustainable Approaches for the Synthesis of N-Based Heterocycles", published in 2009 [29]. A recent review on multicomponent green synthesis of N-containing heterocycles using mixed oxides as heterogeneous catalysts is also reported [30]. A group of workers had recently published review on "Eco-friendly and sustainable synthetic approaches to biologically significant fused N-heterocycles" [31]. Other groups had also presented well-documented work on the synthesis of heterocyclic compounds using greener methods [32–34].

4. Conclusion

In this chapter, I had listed the reactions using green methodologies for the synthesis of nitrogen heterocyclic compounds. These molecules often have interesting biological activities and are structures often generated in medicinal chemistry. Recently, numerous publications concerning the synthesis of heterocycles under solvent-free, use of bio-based solvent, reactions carried on solid support, microwave irradiation condition have appeared. Work on the use of green reactions leading to the formation of nitrogen heterocycles mainly indole, pyrrole, imidazole, pyrimidine, pyridine derivatives is reported in the present chapter using safe solvents, namely water, various bio-based systems is discussed. Green synthetic reactions using microwaves for the one pot synthesis of N-heterocyclic compounds is of immense utility in terms of yield, being faster, cheaper, eco-friendly than conventional reactions. I hope that gathering information will enable the scientific community to implement heterocyclic syntheses that respect our environment a little better.

Conflict of interest

The authors declare no conflict of interest.

Author details

Monica Dinodia
Department of Chemistry, Hansraj College, University of Delhi, Delhi, India

*Address all correspondence to: monicadinodia@yahoo.co.in

References

[1] Green Chemistry. United States Environmental Protection Agency. 2006. Available from: https://www.epa.gov/greenchemistry/basics-green-chemistry#definition [Accessed: August 12, 2022]

[2] Anastas PT, Warner JC. Green Chemistry: Theory and Practice. Oxford: Oxford University Press; 1998. p. 135

[3] Joule JA. Natural products containing nitrogen heterocycles-some highlights 1990–2015. In: Advances in Heterocyclic Chemistry. Elsevier; 2016. pp. 81-106

[4] Taylor RD, MacCoss M, Lawson ADG. Rings in drugs: Miniperspective. Journal of Medical Chemistry. 2014;**57**(14):5845-5859

[5] Vitaku E, Smith DT, Njardarson JT. Analysis of the structural diversity, substitution patterns, and frequency of nitrogen heterocycles among U.S. FDA approved pharmaceuticals: Miniperspective. Journal of Medicinal Chemistry. 2014;**57**(24):10257-10274

[6] Anastas P, Eghbali N. Green chemistry: Principles and practice. Chemical Society Review. 2010;**39**(1): 301-312

[7] Gawande MB, Bonifácio VDB, Luque R, Branco PS, Varma RS. Benign by design: Catalyst-free in-water, on-water green chemical methodologies in organic synthesis. Chemical Society Reviews. 2013;**42**(12):5522

[8] Rahmati A, Pashmforoush N. Synthesis of various heterocyclic compounds via multi-component reactions in water. Journal of the Iranian Chemical Society. 2015;**12**(6):993-1036

[9] Panda S, Jain S. "In Water" syntheses of heterocyclic compounds. Mini-Reviews in Organic Chemistry. 2011; **8**(4):455-464

[10] Pires M, Purificação S, Santos A, Marques M. The role of PEG on Pd- and Cu-catalyzed cross-coupling reactions. Synthesis. 2017;**49**(11):2337-2350

[11] Vafaeezadeh M, Hashemi MM. Polyethylene glycol (PEG) as a green solvent for carbon–carbon bond formation reactions. Journal of Molecular Liquids. 2015;**207**:73-79

[12] Paul S, Pradhan K, Das AR. Ethyl lactate as a green solvent: A promising bio-compatible media for organic synthesis. Current Green Chemistry. 2022;**3**(1):111-118

[13] Karamthulla S, Pal S, Khan MN, Choudhury LH. Synthesis of novel spiro [indoline-3,7′-pyrrolo[1,2-c]imidazole]-6′-carbonitrile derivatives in water using a regioselective sequential three component reaction. RSC Advances. 2013;**3**(36):15576

[14] Suresh R, Muthusubramanian S, Nagaraj M, Manickam G. Indium trichloride catalyzed regioselective synthesis of substituted pyrroles in water. Tetrahedron Letters. 2013;**54**(14): 1779-1784

[15] Marković V, Joksović MD. "On water" synthesis of N-unsubstituted pyrazoles: Semicarbazide hydrochloride as an alternative to hydrazine for preparation of pyrazole-3-carboxylate derivatives and 3,5-disubstituted pyrazoles. Green Chemistry. 2015;**17**(2):842-847

[16] Kumar D, Kumar NM, Patel G, Gupta S, Varma RS. A facile and eco-friendly synthesis of diarylthiazoles and diarylimidazoles in water. Tetrahedron Letters. 2011;**52**(16):1983-1986

[17] Yan H, Ma Y, Sun Y, Ma C, Wang Y, Ren X, et al. Catalyst-free synthesis of alkyl 4-oxo-4H-pyrido[1,2-a] pyrimidine-2-carboxylate derivatives on water. Tetrahedron. 2014;**70**(17): 2761-2765

[18] Dandia A, Jain AK, Laxkar AK. Ethyl lactate as a promising bio based green solvent for the synthesis of spiro-oxindole derivatives via 1,3-dipolar cycloaddition reaction. Tetrahedron Letters. 2013;**54**(30):3929-3932

[19] Campos JF, Scherrmann MC, Berteina-Raboin S. Eucalyptol: A new solvent for the synthesis of heterocycles containing oxygen, sulfur and nitrogen. Green Chemistry. 2019;**21**(6):1531-1539

[20] Balijapalli U, Thiyagarajan MD, Manickam S, Sathiyanarayanan KI. Synthesis of T-shaped Oxazolonaphthoimidazo[1,2- *a*] pyridines using lactic acid as bio-based green solvent: An insight into photophysical studies. ChemistrySelect. 2016;**1**(11):2900-2908

[21] Khandelwal S, Rajawat A, Kumar M. An efficient and environmentally benign one-pot three- component domino protocol for the synthesis of structurally diverse spiroquinazolines. Current Catalysis. 2015;**4**(3):214-223

[22] Daştan A, Kulkarni A, Török B. Environmentally benign synthesis of heterocyclic compounds by combined microwave-assisted heterogeneous catalytic approaches. Green Chemistry. 2012;**14**(1):17-37

[23] Hoz A de la, Diaz-Ortiz A, Moreno A, Sanchez-Migallon A, Prieto P, Carrillo JR, et al. Microwave-assisted reactions in heterocyclic compounds with applications in medicinal and supramolecular chemistry. Combinatorial Chemistry & High Throughput Screening. 2021;**10** (10):877-902

[24] Majumder A, Gupta R, Jain A. Microwave-assisted synthesis of nitrogen-containing heterocycles. Green Chemistry Letters and Reviews. 2013; **6**(2):151-182

[25] Karthikeyan SV, Perumal S, Shetty KA, Yogeeswari P, Sriram D. A microwave-assisted facile regioselective Fischer indole synthesis and antitubercular evaluation of novel 2-aryl-3,4-dihydro-2H-thieno[3,2-b] indoles. Bioorganic & Medicinal Chemistry Letters. 2009;**19**(11): 3006-3009

[26] Borthakur M, Gogoi S, Gogoi J, Boruah RC. Lewis acid catalyzed rapid synthesis of 5-hydroxy-benzo[g]indole scaffolds by a modified Nenitzescu reaction. Tetrahedron Letters. 2010; **51**(39):5160-5163

[27] Wagare DS, Farooqui M, Keche TD, Durrani A. Efficient and green microwave-assisted one-pot synthesis of azaindolizines in PEG-400 and water. Synthetic Communications. 2016; **46**(21):1741-1746

[28] Muscia GC, Bollini M, Carnevale JP, Bruno AM, Asís SE. Microwave-assisted Friedländer synthesis of quinolines derivatives as potential antiparasitic agents. Tetrahedron Letters. 2006; **47**(50):8811-8815

[29] Candeias NR, Branco LC, Gois PMP, Afonso CAM, Trindade AF. More sustainable approaches for the synthesis of N-based heterocycles. Chemical Reviews. 2009;**109**(6):2703-2802

[30] Bhaskaruni SVHS, Maddila S, Gangu KK, Jonnalagadda SB. A review on multi-component green synthesis of N-containing heterocycles using mixed

oxides as heterogeneous catalysts. Arabian Journal of Chemistry. 2020; **13**(1):1142-1178

[31] Gupta SS, Kumari S, Kumar I, Sharma U. Eco-friendly and sustainable synthetic approaches to biologically significant fused N-heterocycles. Chemistry of Heterocyclic Compounds. 2020;**56**(4):433-444

[32] Frecentesec F, Saccone I, Caliendo G, Corvino A, Fiorino F, Magli E, et al. Microwave assisted organic synthesis of heterocycles in aqueous media: Recent advances in medicinal chemistry. Medicinal Chemistry. 2016;**12**(8):720-732

[33] Batista VF, Pinto DCGA, Silva AMS. Synthesis of quinolines: A green perspective. ACS Sustainable Chemistry & Engineering. 2016;**4**(8):4064-4078

[34] Kaur N. Review on the synthesis of six-membered N,N-heterocycles by microwave irradiation. Synthetic Communications. 2015;**45**(10):1145-1182

Chapter 11

Recent Methods for Synthesis of Coumarin Derivatives and Their New Applications

Deepika Sharma, Vasudevan Dhayalan,
Chitrarasu Manikandan and Rambabu Dandela

Abstract

Coumarin (2*H*-1-benzopyran-2-one) and its heterocyclic derivatives are widely used as lactone scaffolds used by innovative methods for the preparation of heterocyclic molecules. Nowadays, significant biological activities, as well as properties of unique nature coumarin derivatives, have played an important role in the development of novel drugs. This chapter entitles numerous methods of one-pot construction of coumarin derivatives, together with well-known name reactions and other type reactions as well, in the presence of various metal-based homogenous and heterogeneous catalyst system. Coumarin is one of the very important heterocycles and its analogs like natural product and pharmaceutically active drug molecules are extracted/isolated from a plants, animals, and microbes. Coumarin precursors have a wide range of biological activities Hence, the synthesis of coumarins and their heterocyclic analogs have become among the most interesting compound over the last many years in the growth of improved synthetic methodologies to form different types of functional groups that are present in coumarins derivatives. The synthesis of coumarins enabled by current approaches and their most recent bio-applications are discussed in this book chapter. Corresponding complex heterocycles-based coumarin analogs are produced from substituted alkyne substrates and other starting materials as well.

Keywords: coumarin, drugs, catalysis, name reactions, arylation, heterocycles, radical reaction

1. Introduction

Coumarin heterocycle is an award gifted from nature, coumarins get the name from "Coumarou" which is the vulgarity term of the plant that belongs to the Fabaceae family named tonka bean. The natural product of coumarin and its scaffolds were isolated and purified by Vogel in 1820 [1], and it was prepared by Perkin in 1868 [2]. The lactones-based ubiquitous heterocyclic coumarins have been known as fragrance products in perfumes, because of their nature of sweet smell. Naturally appearing coumarins and their analogs are known in about 700 chemical structures in more than 100 plant families [3] and remarkably, the number of core structures of coumarin derivatives still increases. Coumarins (2*H*-1-benzopyran-2-one) are one the most important heterocyclic compounds and their scaffolds are an essential class of lactone family with a fused α-pyrone ring attached to a benzene moiety. The centre core of coumarins has shown great interest over the years because of their significant biological importance and applications [4–14]. This type of fused oxygen heterocycles is associated with a wide range of biological properties like antitubercular, anti-inflammatory, anticancer, anticonvulsant, and neuroprotective, such as antiviral, antifungal, antibacterial, antihyperglycemic, anticoagulant, antihypertensive, antiadipogenic, anti-HIV, antibacterial, antimicrobial and antioxidant, etc. effects.

Moreover, coumarins analogs are attracting the vital attention of chemists due to their broad range of materials applications like photosensitizers, fluorescent, optical brighteners [15, 16], and laser dyes [17, 18], and additives [19] in food, cosmetics, and pharmaceuticals, etc.

In this particular lactone based coumarin family of drug molecules such as warfarin [20], acenocoumarol [21], and phenprocoumon [22] are the most prominent ones, which are currently used in a many different nations (**Figure 1**). Due to their unique nature and greater half-life, notably, warfarin is used more often compare to acenocoumarol. These results highlight a significant new development in the biological evaluation of coumarins and their derivatives, as well as medicinal chemistry [23–30]. As a current result, the

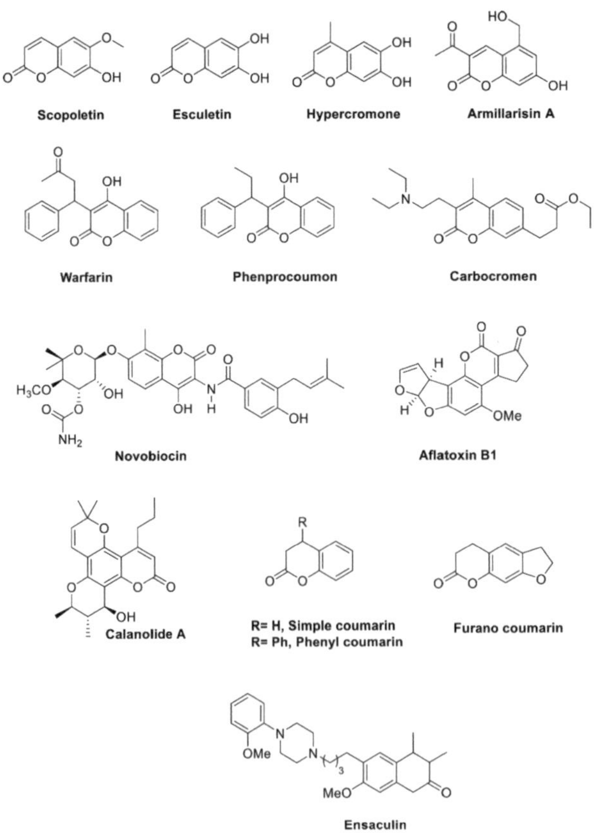

Figure 1.
Representative examples of lactone-based bioactive coumarin-based derivatives.

coumarin heterocyclic ring system is widely used mostly in pharmaceutical industry to build a various functional groups present in the drug molecules. Significant research has been shown to isolate and purify naturally present biological active coumarins from a range of plants, animals, and microbes, and to artificially design and synthesize functionalized coumarin molecules from academic and industry as well with unique heterocyclic structures and characteristics [31–36]. Given the importance of coumarin parent compounds and their derivatives in medicinal chemistry, we have gathered diverse methods for the preparation of coumarin and its scaffolds from alkyne type aryl propiolate as suitable starting materials *via* transition metal and, non-metal mediated catalysis and photo-chemical and radical based transformations. In this chapter, we mostly tried to cover the traditional name reactions mediated coumarin synthesis and described new synthetic methods to access coumarin heterocycles and also showed recent exciting applications which are reported in recent years. As a result, we think that this book chapter will be useful for not just the students but for researchers and scientists as well, who are trying to figure out new ways to build complicated coumarins, therapeutic compounds based on coumarins, and polycyclic coumarin derivatives.

2. Name reaction enabled coumarin synthesis

In the present literature, several name reaction mediated methods are reported for the synthesis of coumarin derivatives used by following well-established reaction protocols such as Perkin reaction, Pechmann reaction, Claisen rearrangement, Knoevenagel reaction, Kostanecki-Robinson coupling reaction, Reformatsky Reaction, Wittig reaction, Michael addition, Heck-lactonization reaction, and Baylis-Hillman reaction in the presence of various metal-free or metal-based homogenous and heterogeneous catalyst systems. We have demonstrated suitable reaction and mechanisms for following name reactions mediated preparation of coumarin motifs (**Figure 2**).

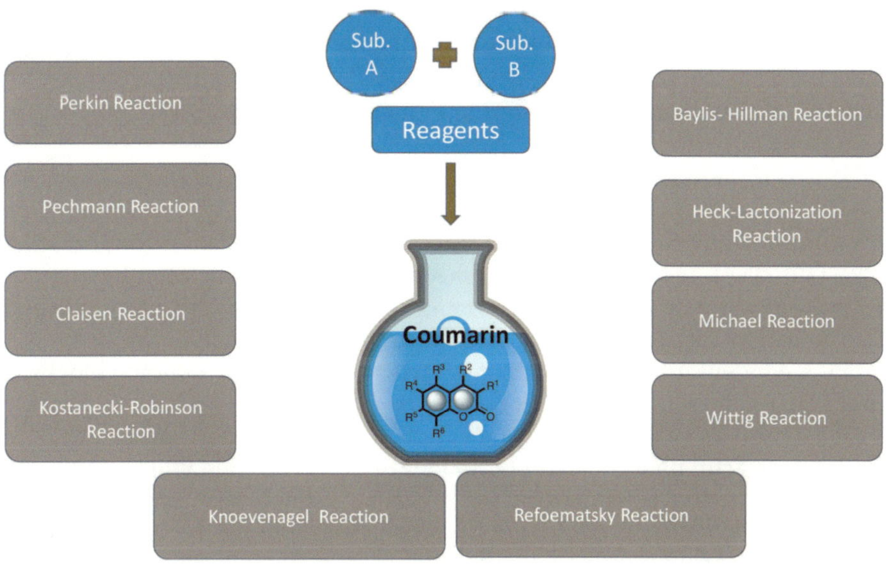

Figure 2.
Name reactions mediated synthesis of coumarin.

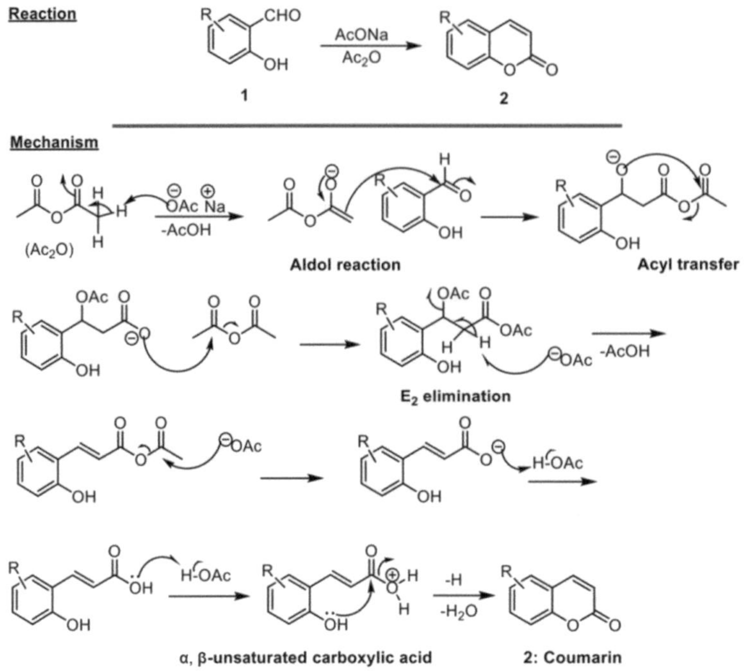

Figure 3.
Synthesis of coumarin via Perkin reaction.

2.1 Perkin reaction

In 1968, the first time Pekin demonstrated the method for the construction of coumarin by the condensation reaction of simple salicylaldehyde in the presence of acetic anhydride [37].

The Perkin reaction of salicylaldehyde **1** and an acetic anhydride are mixed together in the basic reaction condition, a chemical process that furnished, α, β-unsaturated aromatic acid in the presence of sodium acetate followed by intramolecular cyclization produced the expected substituted coumarin **2**. The proposed mechanism of this reaction is described in **Figure 3** [2, 37, 38].

2.2 Pechmann reaction

The Pechmann achieved the initial discovery of the Pechmann condensation in 1883 [39]. Typically, carbolic acid **3** reacts with a carboxylic ester having α-carbonyl group **4** in an acidic environment to produce the desire coumarins **5** [40]. The most widely reported process for producing coumarins through Pechmann condensation, which scheme starts with two basic building blocks, phenol and β-ketoester, and produces good coumarin yields most of the cases. Several catalysts were tested for this reaction, including sulfuric acid, trifluoroacetic acid, phosphorous pentoxide, $ZrCl_4$, $TiCl_4$, and ionic liquids, etc. enabled by Pechmann condensation [41]. Various groups have reported for the preparation of coumarin scaffolds *via* Pechmann methods. In the mechanism of the reaction involved following paths initial step is esterification followed by the attack of activated carbonyl group, allows to forms the ring system. The last step of the reaction involves dehydration. The proposed plausible mechanism of this

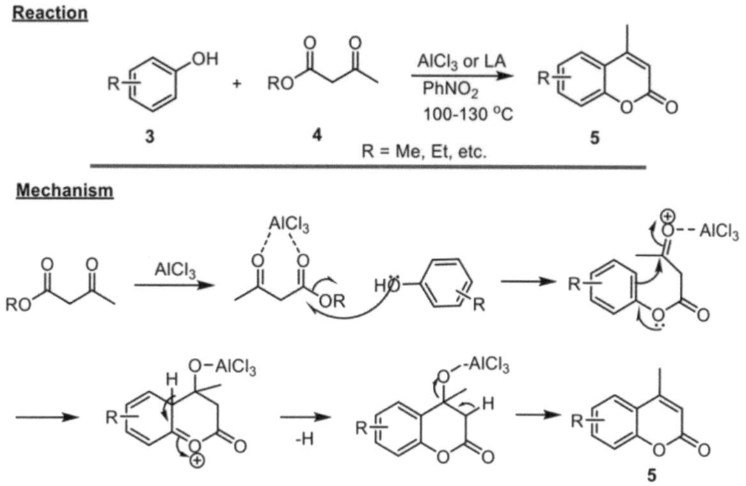

Figure 4.
Coumarin synthesis via Pechmann reaction.

reaction shown in **Figure 4**. By reacting substituted phenols and ethyl acetoacetate in the presence of a zinc-iodine mixture in refluxing toluene, a number of substituted coumarins have been produced in yields ranging from 25 to 77%. The proposed plausible mechanism of this reaction is shown in **Figure 4** [42].

2.3 Claisen rearrangement

The preparation of 3,4-substituted coumarin utilizing trifluoroacetic acid (TFA) as homogeneous promoter *via* Claisen rearrangement reaction. The reaction of phenol 3 reacts with protected allyl alcohol 6 in the presence of basic condition offers desired target compound followed by underwent 3,3 sigmatropic Claisen rearrangement in the presence TFA in moderate temperature followed by tautomerism or basic condition produced 3,4-substituted coumarin 7 in good yields. Several groups reported Claisen rearrangement mediated synthesis of coumarin analogs. The proposed plausible mechanism of this reaction is shown in **Figure 5** [43].

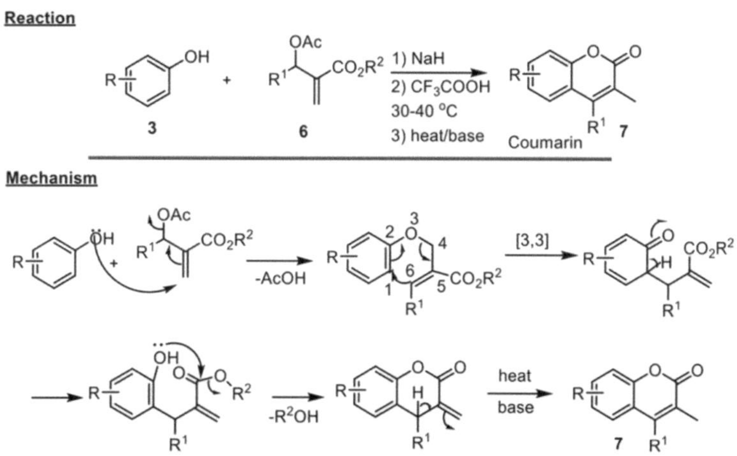

Figure 5.
Synthesis of 3,4-disubstituted coumarin via Claisen rearrangement.

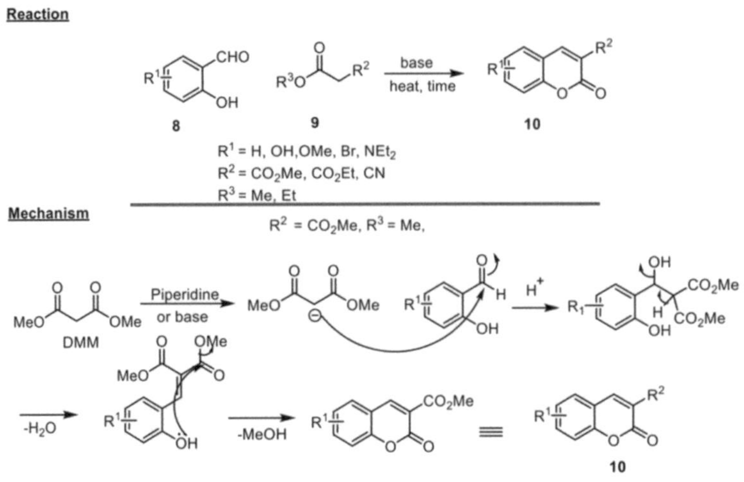

Figure 6.
Synthesis of 3-substituted coumarin via Knoevenagel reaction.

2.4 Knoevenagel reaction

Many coumarin derivatives have been derived from suitable starting materials *via* a Knoevenagel reaction. 3-substituted coumarin derivatives can be synthesized *via* base mediated process. The reaction needs to be carried out in the presence of 2-hydroxy benzaldehydes 8 and coupling partner 9 containing an active methylene group in the presence of the base under heating conditions. The yield obtained from coumarin product 10 is acceptable range [44]. The proposed plausible mechanism of this reaction is shown in **Figure 6**. There are various reports present in the literature regarding the synthesis of scaffolds of coumarin *via* Knoevenagel reaction in the presence of ultrasound solvent-free conditions [45, 46].

2.5 Kostanecki-Robinson coupling reaction

Kostanecki-Robinson coupling reaction could be utilized for the synthesis of derivatives of coumarin. The **Figure 7** shows the reaction between aliphatic anhydride **12** and aryl ketone **11** with a substitution of the hydroxyl group which gives the desired product as coumarin **13** with good to excellent yields. The proposed plausible mechanism of this reaction is described in **Figure** 7 [14, 47].

In 2004 Song et al. synthesized 4-arylcoumarins **15** from phenyl ester **14** in the presence of 4-butyl-3-methylimidazolium bromide (phase transfer catalyst), Hf $(OTf)_4$ (metallic catalyst), for 9 h at 80°C the yield of the expected product obtained was good (**Figure 8**) [48].

2.6 Reformatsky reaction

In the Reformatsky reaction of an activated acyl halide first reacts with a zinc metal to offer RZnBr followed 1,2 addition of organometallic zinc reagents to ketone **11** produced a zinc enolate after elimination of Zn [OH(Br)] to form ester. This process converts 3,4-disubstituted coumarins **13** from α, β-unsaturated ester. This synthesis protocol involving reaction steps are as shown in the **Figure 9** with the mechanism as well [49].

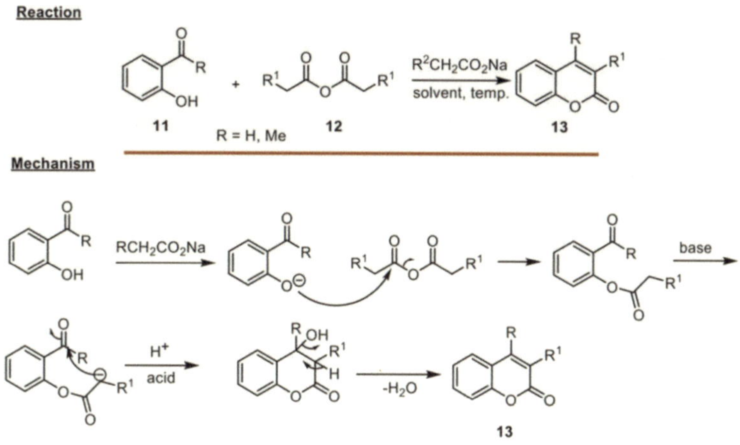

Figure 7.
Synthesis of coumarin derivatives via Kostanecki-Robinson reaction.

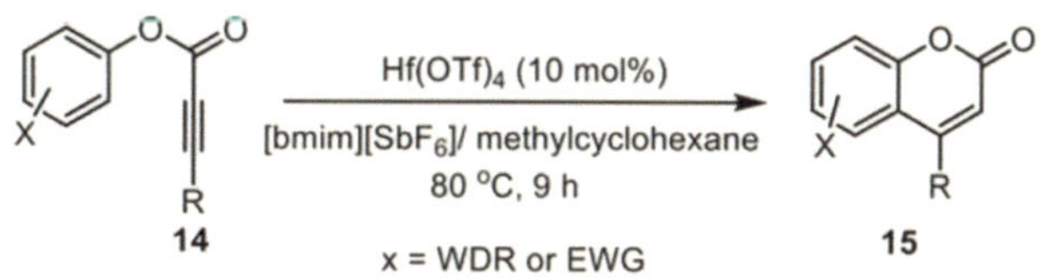

Figure 8.
Synthesis of 4-arylcoumarins.

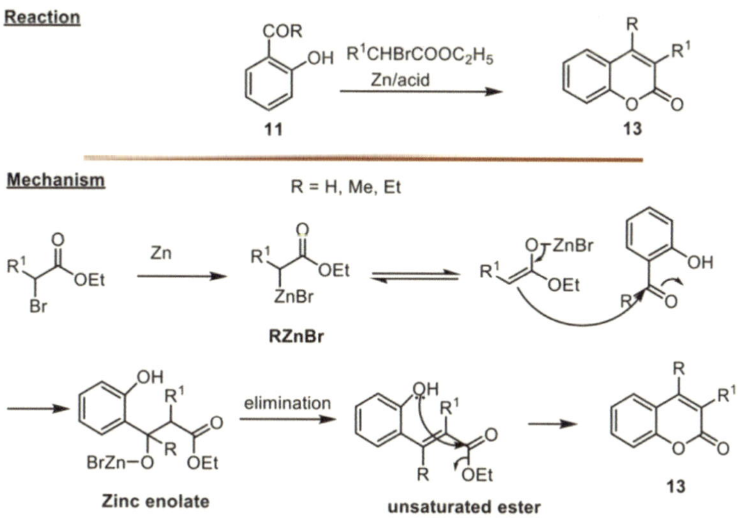

Figure 9.
Synthesis of coumarin via Reformatsky reaction.

2.7 Wittig reaction

Wittig reaction of aldehyde or a ketone **11** is mixed with a Wittig phosphine reagent (a triphenyl phosphonium ylide) to offer the expected olefin 16 in good yields along with phosphine oxide as by-product (**Figure 10**). This Wittig name reaction is

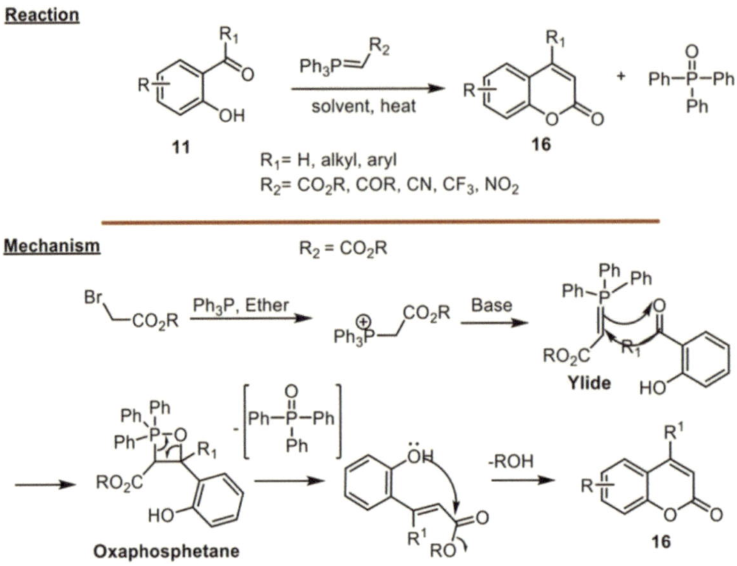

Figure 10.
Preparation of coumarin via Wittig reaction.

discovered by the German chemist Georg Wittig. He received Nobel Prize in Chemistry in 1979 his discovery of this significant olefin synthesis. This system allows the preparation of highly important natural products as well as drug molecules. The preparation of various substituted coumarin **19** derivatives obtained from corresponding phenols **17** having ortho-carbonyl group and triphenyl(α-carboxymethylene) phosphorane imidazole ylide **18** has been carried out by Kumar et al. by following applying the route of Wittig reaction, the yield of olefin reported was good. The mechanism of the reaction suggests that the reaction proceeds through the phosphorane intermediates as shown in **Figure 11** [50]. There are various reports present in the literature regarding the synthesis of coumarin scaffolds via Wittig reaction starting with an aldehyde or ketones with phosphonium ylide.

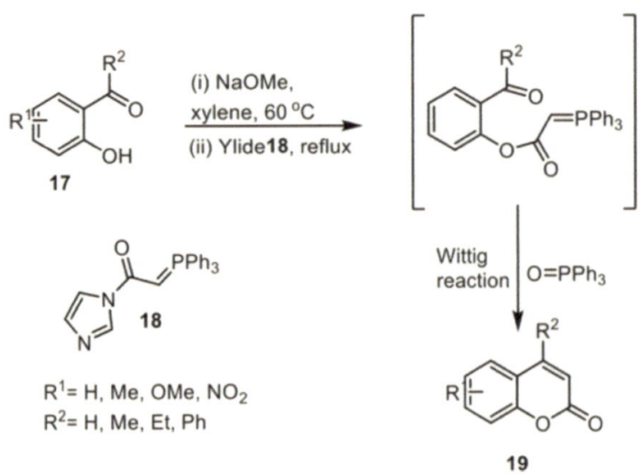

Figure 11.
Synthesis of coumarin via Wittig reaction.

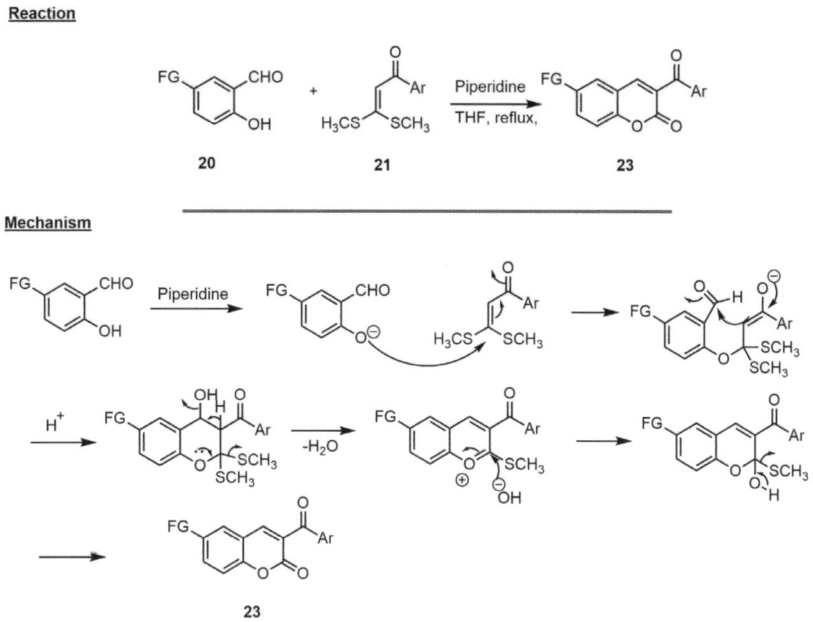

Figure 12.
Synthesis of 3-benzoyl coumarin through Michael addition approach.

2.8 Michael addition

The synthesis of 3-aroylcoumarin **23** could be carried out *via* Michael addition reaction approach from the readily present 2-hydroxybenzaldehyde **20** and α-aroylketene dithioacetals **21** in the presence of piperidine and refluxed condition in THF as a solvent. The mechanism of the reaction shows that the reaction proceeds *via* 2 steps, (i) initially *via* Michael addition (ii) aldol condensation (iii) elimination of water and -SCH_3 as shown in **Figure 12** [51].

2.9 Heck-lactonization reaction

The Heck-Lactonization reaction can be carried out for the synthesis of coumarin analogues **26** *via* Pd catalysis and were examined different reaction conditions tested using aqueous water and organic solvent conditions. Even with 10 mol% of the $PdCl_2$ or Pd $(OAc)_2$-catalysts showed, enoate **25** interacted with iodo compound **24** to produce coumarin **26** in good to moderate yields under conditions A and B with water. The organic solvent condition C was indicating low chemical yield and the proposed catalytic cycle as shown in **Figure 13** [52].

2.10 Baylis-Hillman reaction

As seen in **Figure 14**, the Baylis-Hillman approach was used to create substituted coumarins. In the presence of DABCO, 2-hydroxybenzaldehydes **8** reacts with methyl acrylate **27** to produce a combination of chromene **28** and coumarin **29**. Nevertheless, related interactions between 2-hydroxybenzaldehydes and *tert*-butyl acrylate in the presence of conventional or microwave heating led to respective

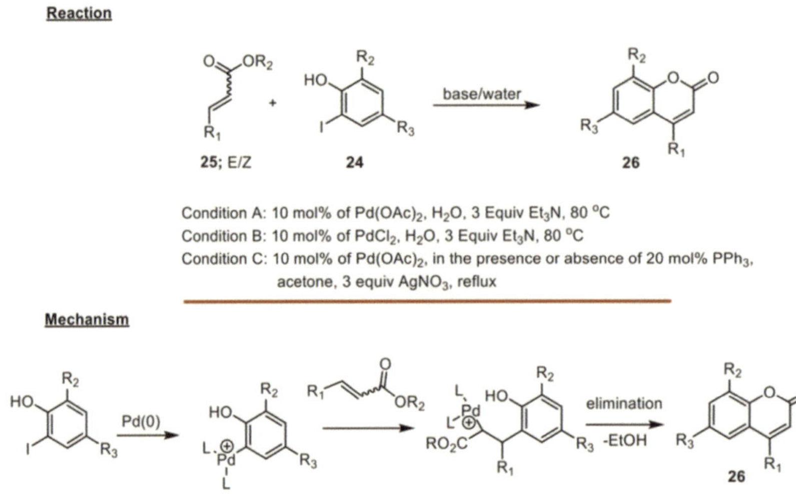

Figure 13.
Synthesis of coumarin via Heck-lactonization reaction.

Baylis-Hillman adducts, that further undergoes cyclization *via* reflux in AcOH to produce a combination of both 3-substituted chromene and coumarin. Decent quantities of 3-(chloromethyl) coumarins **33** are obtained by treating the Baylis-Hillman adducts **30** and strong acid HCl when refluxed in AcOH. Additionally, 3-methyl coumarins **31** are produced by the reaction of compound **30** with HI under reflux in a solution of Ac_2O and AcOH, and these 3-formyl coumarins **32** are produced by a subsequent reaction with SeO_2. The plausible mechanism of the reaction has been shown in the **Figure 14** [53].

3. New approaches for the synthesis of coumarins derivatives

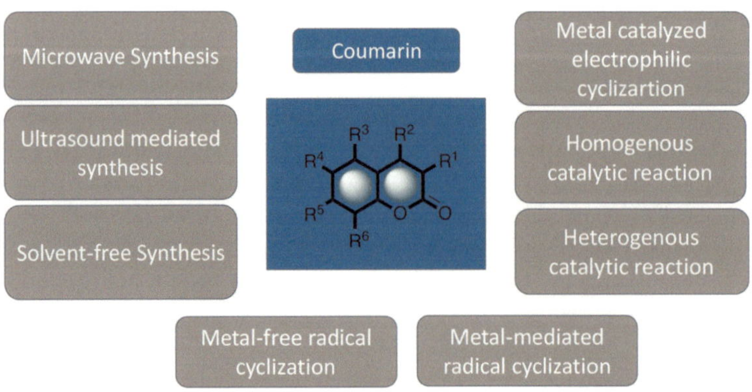

3.1 Microwave mediated innovative synthesis

Recently, microwave-mediated organic synthesis has replaced conventional heating methods. In recent years, the synthesis of organic molecules has increasingly

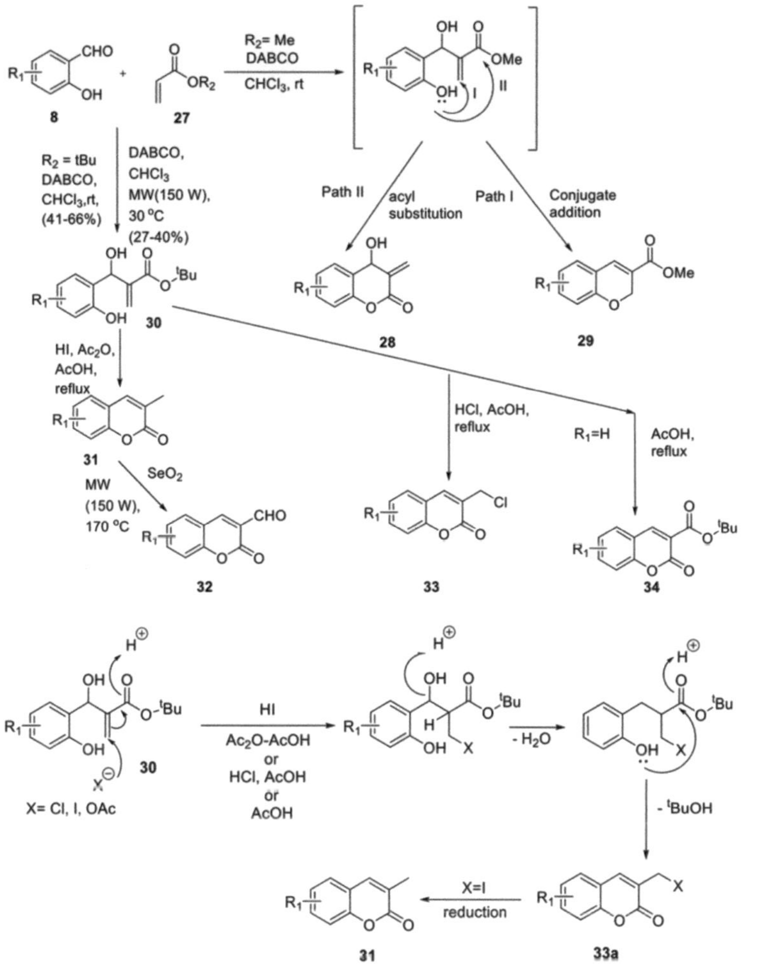

Figure 14.
Baylis Hillman reaction for the synthesis of substituted coumarin.

relied on the use of microwave energy to heat chemical reactions. In contrast to dramatically accelerating chemical reactions, direct microwave heating is known to reduce the formation of side products, increases yield, and improves the reproducibility. Various academic research institutions have already embraced microwave irradiation as a method for fast reaction in order to efficiently synthesize and discover new chemical substances [54].

In the year 2017, Brahmbatt and co-workers demonstrated the microwave- assisted preparation of 3-aryl-furo[3,2-c] coumarins **37**. The time required for the synthesis was just 2–4 minutes and the yield was also good (72–82%) as shown **Figure 15** [55].

In the same year, Desai et al. reported the preparation of 4-(substitutedphenyl)-2-(furan-2-ylmethyleneamino)-6-(2-oxo-2*H*-chromen-3-yl)nicotinonitrile derivatives **42**. The reaction was carried out in a solvent-free condition, the reaction was performed with 2-amino-4-(substitutedphenyl)-6-(2-oxo-2*H*-chromen-3-yl) nicotinenitriles **41** and 2-furfuealdehyde and the microwaves (300 W) were irradiated for 8–10 min. in the presence of acetic acid and a catalytic amount of $ZnCl_2$ (**Figure 16**) [56].

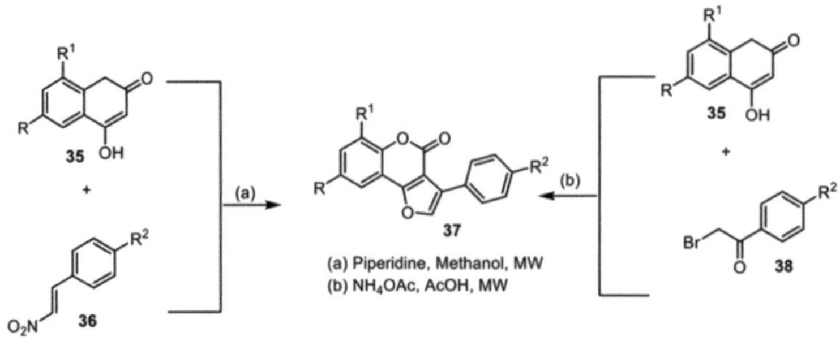

Figure 15.
Microwave assisted synthesis of 3-aryl-furo[3,2-c] coumarins.

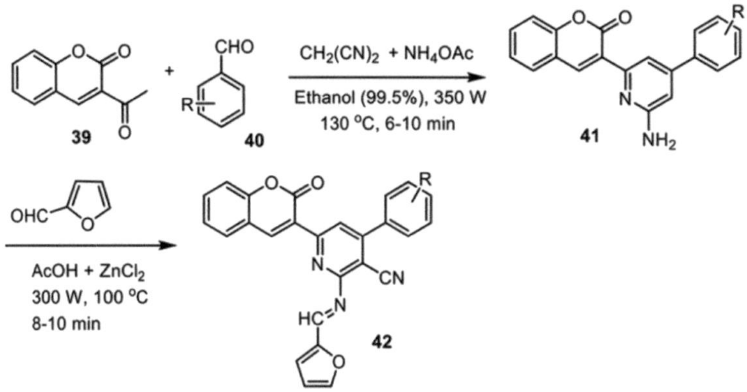

Figure 16.
Synthesis of coumarin-based nicotinonitrile.

[3,4-d]triazole-fused coumarin derivatives were synthesized by Schwendt and co-workers. The yield obtained was best (63–94%) in the presence of DMF (solvent) and at 160°C for 1 min [57].

A variety of coumarin-carbonodithioate and coumarin-maltol derivatives were synthesized, showing antibacterial activity and antitumor in a relatively short period in the presence of microwave radiations. It was stated that this approach was 24 times faster than the traditional technology [58, 59].

Pyrido[3,2-c]coumarins were synthesized in the presence of ammonium acetate, the reaction was carried out with suitable starting materials. The yield obtained was good and the time required for the reaction was about 3–4 mins [60]. Synthesis of coumarin-thiazolidine-2,4-dione was carried out recently by Mangasuli and co-workers. The reaction was performed in the presence of K_2CO_3 (catalyst), the starting material utilized was coumarin and thiazolidinedione.

3.2 Ultrasound helped synthesis

Comparing ultrasonic irradiation to traditional energy sources, there are various benefits like heat, light, and electric potential) [61, 62]. The primary cause of the chemical reactions caused by ultrasound is acoustic cavitation, which is the formation,

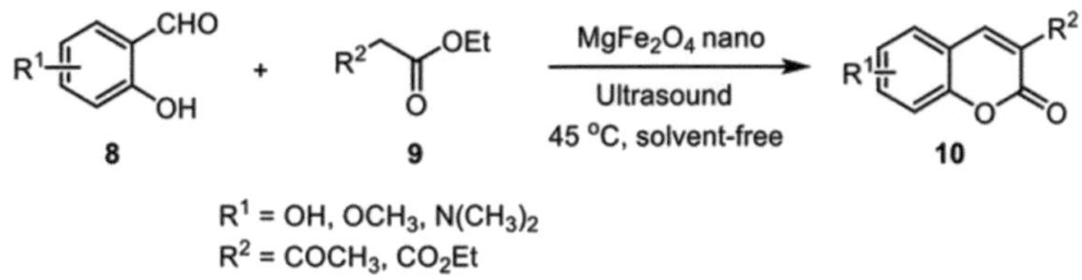

Figure 17.
Ultrasound assisted synthesis of 3-substituted coumarins.

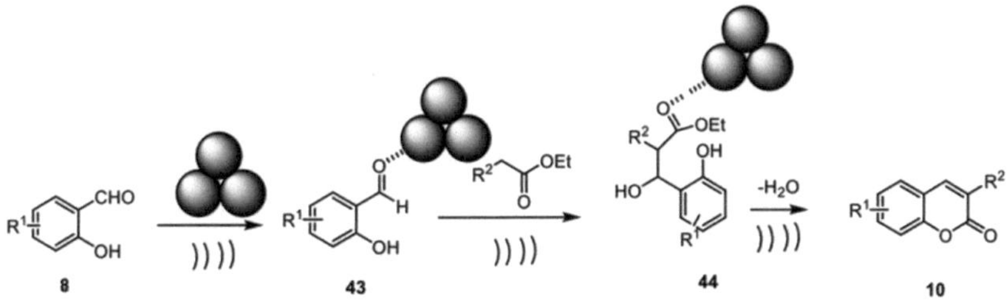

Figure 18.
Mechanism of the synthesis of 3-substituted coumarins.

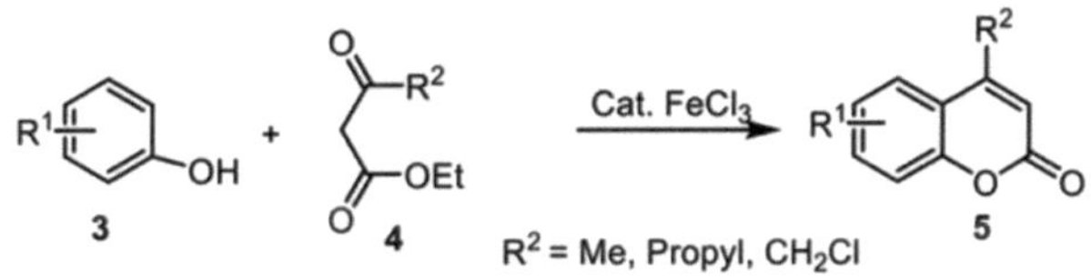

Figure 19.
Synthesis of coumarin via ultrasonic/microwave radiation.

development, and implosive bursting of bubbles. In many fields of chemistry research, including organic synthesis and solid-state materials, ultrasonic-assisted synthesis techniques have gained a lot of interest as shown **Figure 17** [63].

The 3-substituted coumarins **10** can be synthesized in the presence of $MgFe_2O_4$ nanoparticle, and the reaction can be carried out between salicylaldehyde **8** and 1,3-dicarbonyl compound **9** in the existence of the Ultrasound radiation. The mechanism for the syntheis of 3-substituted coumarin **10** was reported by Ghomi and co workers in 2018 and has been shown in the **Figure 18** [64].

The Pechmann condensation reaction for the synthesis of 3- substituted coumarin **5** in the presence of ultrasound radiation as well as in microwave radiation was carried out in the presence of catalyst $FeCl_3$, it was found that the yield reported was better with the methodology used than the conventional method as mentioned in **Figure 19** [65].

Bis-coumarin derivatives have been synthesized in the presence of ultrasound radiation, the reaction was carried out between various aromatic aldehydes and 4-hydroxycoumarin [66].

3.3 Solvent-free synthesis

Large volumes of hazardous and volatile organic solvents are used in numerous conventional chemical reactions. Green chemistry's major objective is to replace such toxic reaction solvents. The design of solvent-free green processes has attracted

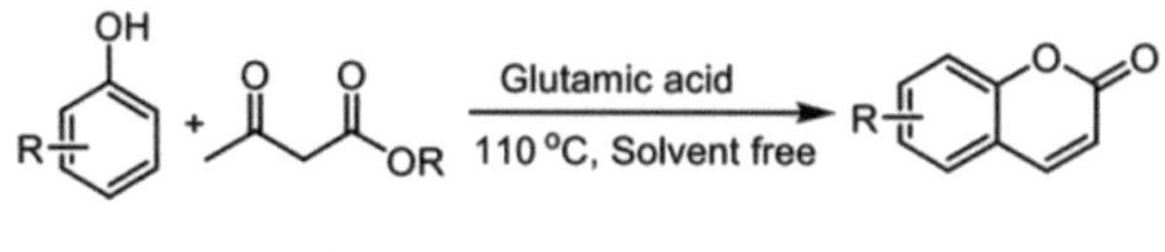

Figure 20.
Solvent-free synthesis of coumarin.

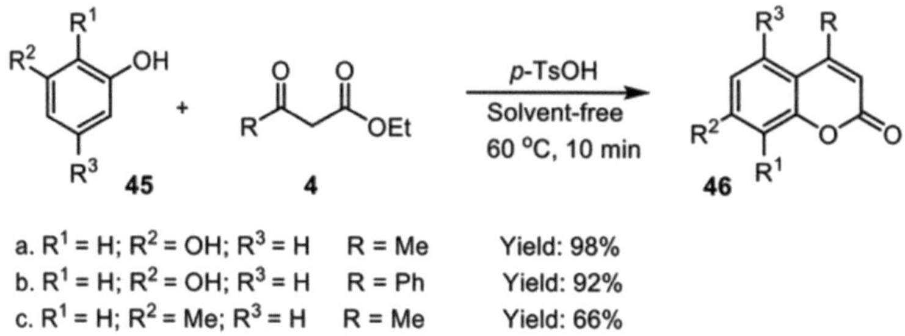

Figure 21.
Synthesis of substituted coumarin in solvent-free condition.

noticeably more interest from researchers as environmental consciousness on a worldwide scale rises. Many researchers have reported the synthesis of coumarin in a solvent free condition. In the year 2014, Sabetpoor et al. reported the synthesis of simple coumarin analogs **5** in the solvent-free conditions in the presence of glutamic acid as a catalyst. The reaction was carried out between the reactant phenol **3** and keto-ester **4** (**Figure 20**). The yield obtained of the expected product was excellent, in ranging from 83 to 93% [67].

The solvent-free Knoevenagel and Pechmann preparation of coumarin **46** has been carried out by Sugino and Tanaka. Pechmann reaction was carried out between reactant substituted phenol **45** and keto-ester **4** utilizing *p*-toluenesulfonic acid (PTSA catalyst) at 60°C. The yield obtained was good ranging 66–98% with different substituents at Resorcinol **45** and ethyl acetoacetate **4** in **Figure 21**. After heating the mixture for 10 min, it was kept for cooling and then crystalline product was obtained after adding water to the reaction mixture, followed by recrystallization from EtOH [68]. In 2011, a similar type of method has been reported by Makrandi et al. [69].

The same authors have also demonstrated the Knovenagel reaction of salicylaldehyde **47** and β-keto ester **4**, in the presence of piperidine at room temperature for 5 min. The neutralization of the reaction mixture has been carried out using aq. HCl, followed by filtration and recrystallization from EtOH. The 3-ethoxycarbonylcoumarin **48** was obtained with a great yield up to 95%. As well as the other substituted coumarin was also obtained with a high yield [68] as given **Figure 22**.

3.4 Light induced metal-free radical cyclization

In 1912, at the start of the 19th century, Ciamician created a unique technique that used light as a natural source in a chemical reaction. Moreover, utilizing the irradiating method, several organic photochemical processes based on Ultra Violet were developed [70]. In this respect, MacMillan initially investigated in 2008 how a

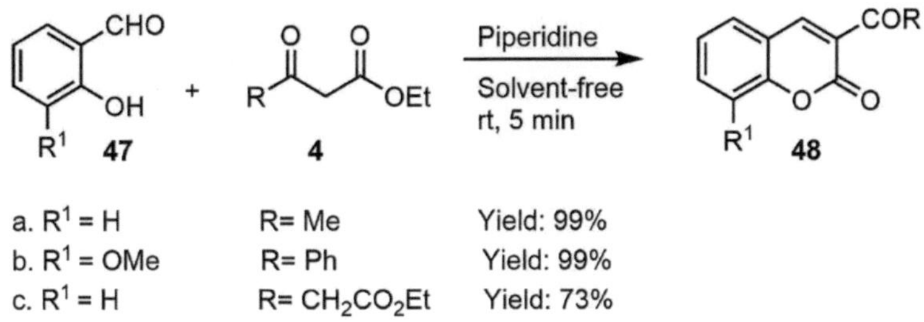

Figure 22.
Synthesis of coumarin via Knovenagel reaction in solvent-free condition.

combination of an organocatalyst and a photosensitive catalyst could enhance the asymmetrical alkylation of aldehydes. Because of its unique single electron transfer (SET) path in very mild reaction circumstances, as well as the fact that it is a secure, economical, and sustainable energy source, visible light-irradiated photoredox catalysis has recently gained a lot of interest. By pairing visible light with metals such as Ruthenium, Iridium, Copper, Nickel, and others, many C-C and carbon-heteroatom bonds can be created. Several studies about how to produce coumarins through metal-free/transition-metal catalyzed inter and intramolecular, radical and electrophilic cyclizations have indeed been reported in the literature, but their practical implementation is constrained by the requirement of a toxic metal, substance, or reagent. In order to synthesize coumarins and other compounds, it is crucial to develop simple, practical, and environmentally friendly techniques [71].

A novel photocatalytic technique to produce (3-acyl 4-arylcoumarin) **51** from aldehyde and aryl alkyne ester was reported in 2018 by Itoh et al. Using visible light, the reaction of phenyl 3-phenyl-2-propynate **49** with *p*-tolualdehyde **50** was conducted in the presence of a AQN catalyst (10 mol%), an oxidant like Bz_2O_2, and K_2CO_3. The necessary 1,2-ester compound was generated in good yields using *p*-substituted phenoxy rings carrying either electron-donating substitution (CH_3, OMe) or electron-poor groups (I, RCOOR', CH_3CO, and OAc). It's noteworthy that the team performed a few straightforward control trials to demonstrate the molecular pathway (**Figure 23**). The method's appealing qualities also include mild reaction conditions, affordable catalysts, and readily available substrates [72].

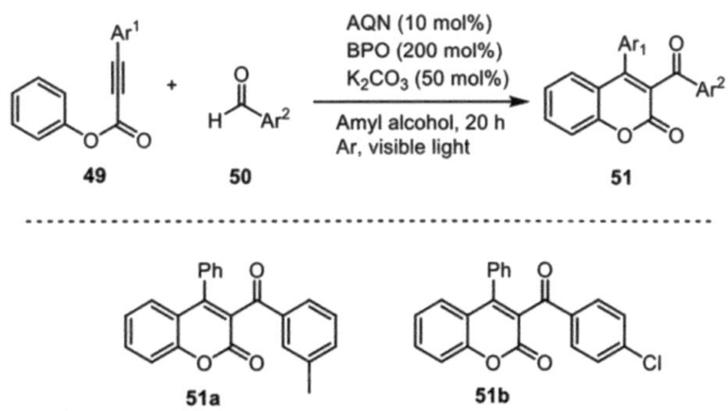

Figure 23.
Synthesis of 3-acyl 4-arylcoumarin.

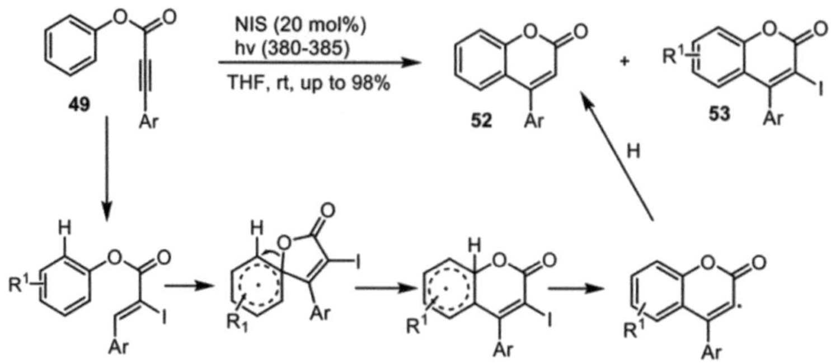

Figure 24.
Synthesis of iodo coumarins via light-assisted metal-free radical cyclization.

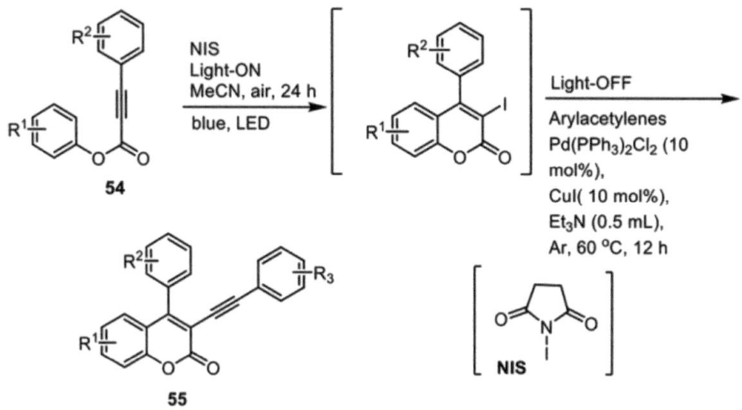

Figure 25.
Synthesis of 3-arylacetylene coumarins.

Li and colleagues described the straightforward photocatalyzed cyclization of alkynoates **49** to iodo coumarins in a THF solvent at 25°C and metal-free circumstances in 2019 (**Figure 24**). After examining the effects of various light sources on the procedure, it was discovered that visible light with a wavelength of 380–385 nm is the most effective. Iodo coumarins **53** were synthesized in high yields using substrates that had substitutions at the *p*-position of the benzene ring [71].

Wu and co-workers very recently reported, in 2020, a simple and practical one-pot reaction to synthesize 3-arylacetylene coumarins **55** from the precursor **54** using thermo-catalysis and photosensitizer-free photocatalytic activity. The basic and effective photocatalytic reactions of *p*-tolyl 3-phenylpropiolate and *N*-iodosuccinimide (NIS) have been conducted in acetonitrile to carry out the one-pot process as given **Figure 25** [73].

3.5 Metal-mediated radical cyclization

Metal-catalyzed reactions have established themselves as one of the crucial steps in the synthesis of organic compounds. There is various methodology reported by chemists for the construction of coumarin derivatives via metal-assisted Radical cyclization reaction. Sulfonyl coumarins **58** can be synthesized from phenyl 3-

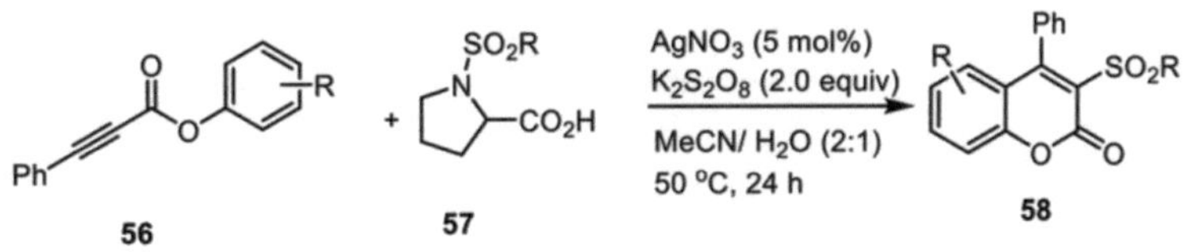

Figure 26.
Synthesis of sulfonyl coumarins.

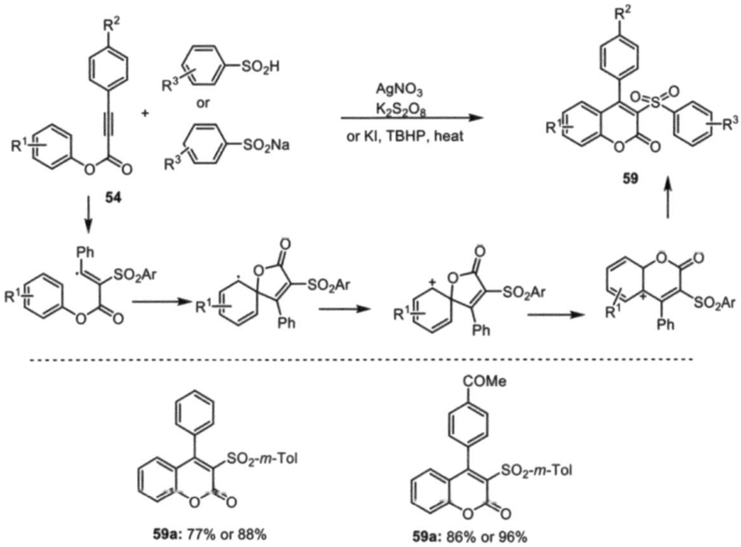

Figure 27.
Synthesis of 4-phenyl 3-sulfonylcoumarins.

phenylpropiolates **56** and tosyl prolines **57** as starting material in the presence of $AgNO_3$ (catalyst), and an oxidant like potassium persulphate in solvent $MeCN/H_2O$ (2:1) at a temperature of 50°C as showed in **Figure 26**. It is to be noticed that the yield is not good with alanine, phenylalanine, and glycine however, it is good with valine or 2-methylalanine sulfomide [74].

In the year 2018, Zhang et al. reported the synthesis of 3-phenyl sulfonyl-coumarins. The reaction is carried out with starting material **54** and sodium sulfinate/ sulfinic acids, the reagents required for the reaction are silver nitrate, TBHP, and KI, in solvent (mixture of acetonitrile and water) at 80°C, in the presence of nitrogen atmosphere (**Figure 27**). Various substituted starting materials **54** can be used for synthesizing different derivatives of coumarin **59**. The yield of the product is better with sulfinates than sulfinic acid [75].

Recently in 2019, 3-monofluoromethylated coumarins **60a-c** have been synthesized by Fu and coworkers *via* monofluoromethylation in the presence of a silver catalyst. When used as a CH_2F radical source, phenyl alkynoate **54** and monofluoromethyl benzo[*d*]thiazol-2-yl sulfone were combined with $AgNO_3$(10 mol %) and potassium persulfate (4.0 equiv.) in DMSO (3.0 mL) and heated to 60°C in an environment of nitrogen for 24 h (**Figure 28**). The reaction is compatible with both electron-rich and deficient substituents at the para position of phenyl ring [76].

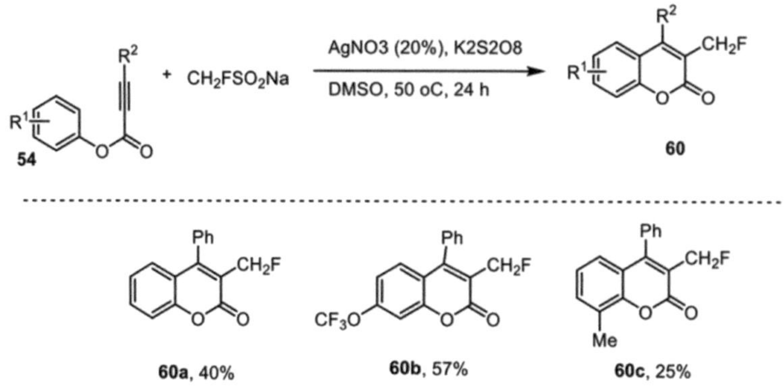

Figure 28.
Synthesis of 3-monofluoromethylated coumarins.

3.6 Metal catalyzed electrophilic cyclization

It has been a challenge for chemists and is crucial in the disciplines of agrochemicals, medicines, and healthcare to activate the C-H bond by a metal-catalyst that results in the novel and advantageous chemically synthesized reaction that creates the C-C bond. A beneficial use relates to organic compounds such as annulated arenes, carbocycles, and heterocycles. Intramolecular hydroarylation is the methodical insertion of arene C-H bonds over numerous bonds in an intermolecular approach. In 2014, it was demonstrated that Au(III) catalyzed electrophilic hydroarylation of aryl alkynoate **61** yielded its respective coumarin analogues **62** through ortho cyclization route and de-aromative ipso-cyclized product in good yield by a modest adjustment in the reaction process. The specific production of coumarin derivatives **62** results from the utilization of an Au-catalyst and AgOTf as additives in an anhydrous DCE solvent, and the presence of a little amount of water results in spirocycles **63** as illustrated in **Figure 29** [77].

Anderson and colleagues demonstrated the electrophilic cyclization-catalyzed formation of 3-organoselenyl-2*H*-coumarins **64a-e** and 3-organoselenylquinolinones from related aryl alkynoates and arylpropiolamides **54**, respectively, in the ideal

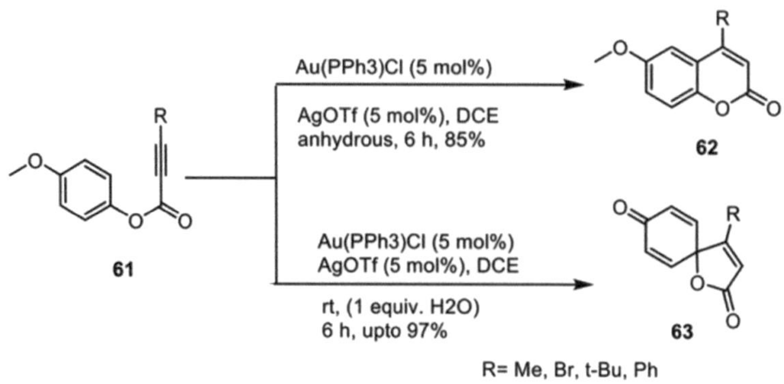

Figure 29.
Au-catalyzed coumarin synthesis.

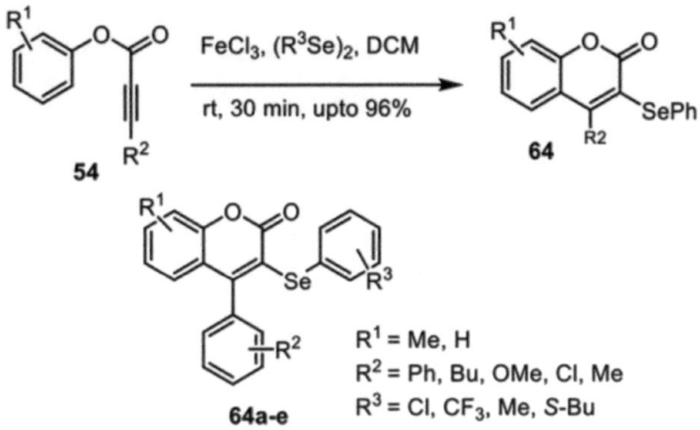

Figure 30.
Fe-catalyzed coumarin synthesis.

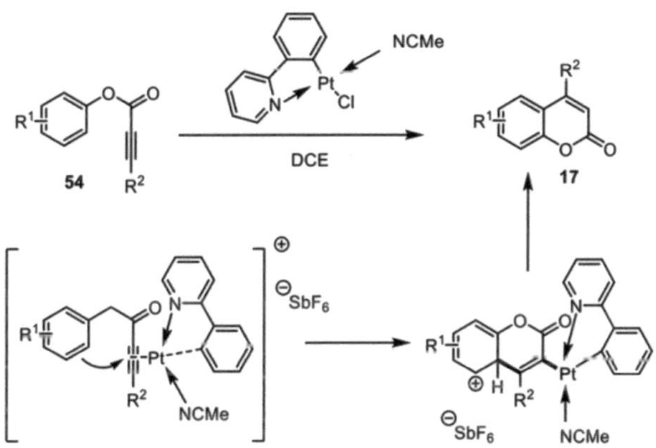

Figure 31.
Pt-catalyzed coumarin synthesis.

reaction conditions attained in a similar manner with organoselenium reagents in dichloromethane solvent as given in **Figure 30** [78].

In 2020, Zaitceva and colleagues showed that cyclometalated (ppy)Pt(II) compounds can catalyze the intramolecular cyclization of phenyl propynoates **54** to produce coumarins **17** and benzocoumarins. A considerable number of substituted coumarins and benzocoumarins were synthesized employing this catalytic approach, and a wide variety of variants were identified to be compatible with the cyclization mechanism (**Figure 31**) [79].

3.7 Homogeneous catalytic reaction

In 2014 Chang et al. proposed a gentle and metal-free approach to synthesize 3-sulfenylated coumarins **66a-c** through cyclization of aryl alkynoates **54** and corresponding *N*-sulfanylsuccinimides **65** promoted by $BF_3{\cdot}Et_2O$ as a Lewis acid (**Figure 32**) [80].

Wu et al. published a practical and flexible approach for functionalizing 3-sulfenylcoumarin **67**a-c and derivatives of 3-sulfenylquinolinone *via* iodine-catalyzed electrophilic cyclization in 2017 (**Figure 33**). The reaction was carried out with phenyl

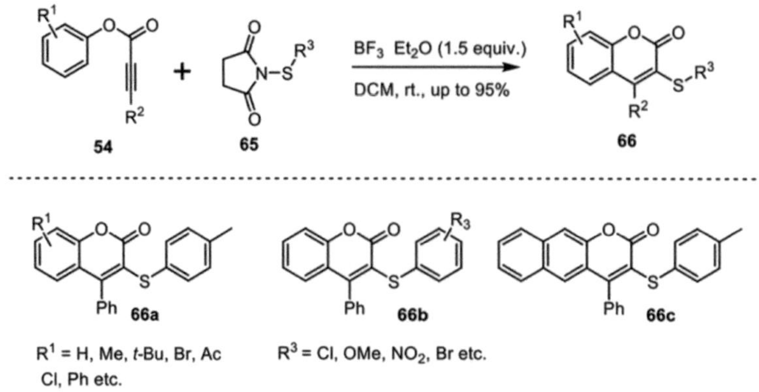

Figure 32.
Preparation of 3-sulfenylated coumarin.

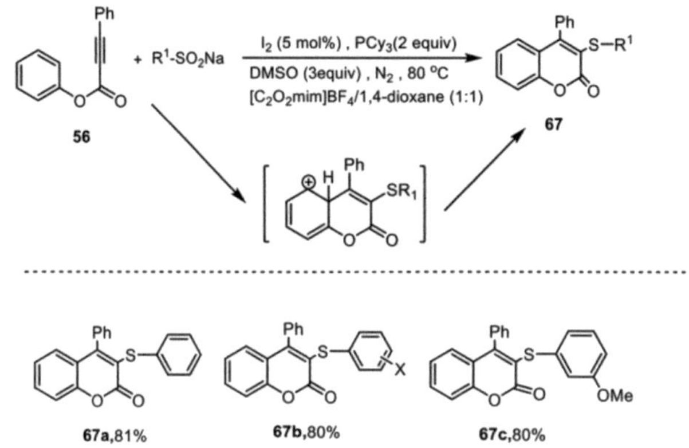

Figure 33.
Iodine-mediated synthesis of 3-sulfenylcoumarin.

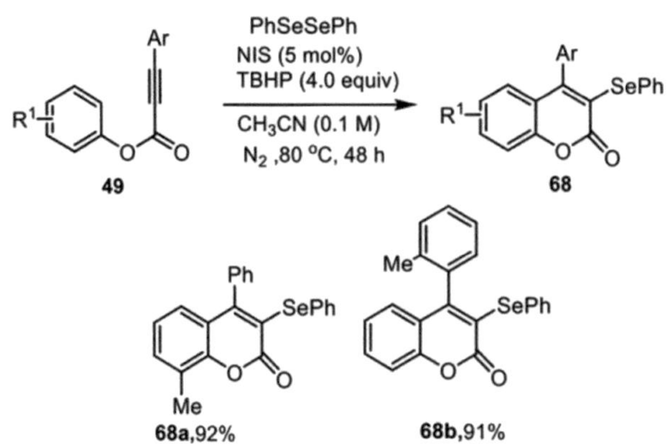

Figure 34.
Formation of 3-organoselenyl-2H-coumarins by propargylic aryl ether.

3-phenylpropiolate **56** and sodium benzenesulfinate, iodine and in DMSO and solvent mixture dioxane and [C_2O_2mim]BF_4 were used as co-solvents [81].

Later in 2019, Fang and coworkers reported a methodology for the preparation of 3-organoselenyl-2*H*-coumarins **68** via oxidative radical cyclization of propargylic aryl ethers **49** and diaryl diselenides. The reaction of (3-phenoxyprop-1-yn-1-yl)benzene and diphenyl diselenide was carried in CH_3CN with *tert*-butyl hydroperoxide(4.0 equiv.) at 80°C for 48 h yielded 85% of the expected product **68a-b** as shown in **Figure 34** [82].

3.8 Heterogeneous catalytic reaction

One of the foundational elements of the chemical and energy industries is heterogeneous catalysis will play a key role in facilitating the shift to these sectors' eventual transformation to carbon-neutral operations [83]. Nowadays, heterogenous catalysis is playing an important role in the organic synthesis, and still it is used for converting petroleum as well as natural gas into the cleaner, capable fuels [84, 85].

Researchers have long struggled with the activation of the C-H bond by metal catalysts, which results in the novel and advantageous synthetic organic reaction that creates the C—C bond and is crucial in the agrochemical, pharmaceutical, and medical fields. Intramolecular hydroarylation, which is the systematic introduction of arene C—H bonds over multiple bonds in the intermolecular path way, gives a useful organic products like annulated arene carbocycles as well as heterocycles. A novel method has been demonstrated by Yuzo Fujiwara et al. in the year 2000, for the preparation of coumarins and quinolinones *via* intermolecular hydroarylation substrate. Utilizing the procedure, different aryl alkynoates **54** and alkynanilides can be quickly converted to the required coumarin derivatives **69a-b** at a temperature between 25 and 27°C and with a catalytic quantity of Pd(OAc)2 in a mixture of solvents TFA and DCM as shown **Figure 35** [86].

Dalibor and coworkers in 2004 have also demonstrated $PtCl_4$ catalyzed intramolecular electrophilic hydroarylation of various substrates having different structures which includes alkynoate esters **70**, propargylamines, and propargyl ethers, that results into a great yield of 6-endo cyclized compounds of coumarin derivatives **71** and **72** in DCM as well as in solvent like dioxane at 25°C. This reaction was studied with a variety of transition metals and their salts, and it was determined that $PtCl_2$ and $PtCl_4$ were both successful, however $PtCl_2$ produced lower yields than $PtCl_4$ when used with appropriate organic solvents such as DCM or dioxane (**Figure 36**) [87].

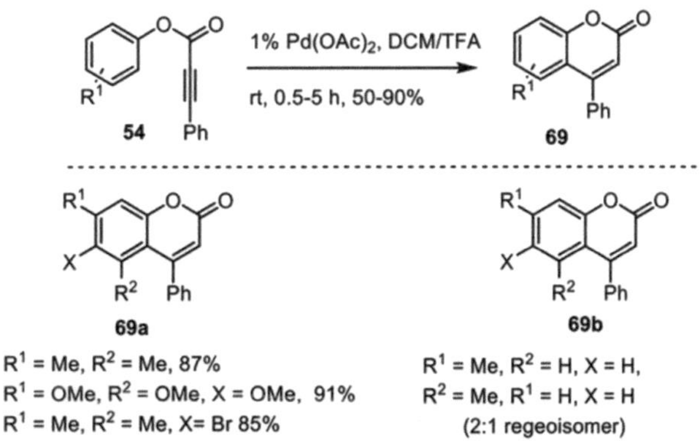

Figure 35.
Synthesis of coumarins by pd-catalyzed intramolecular hydroarylation.

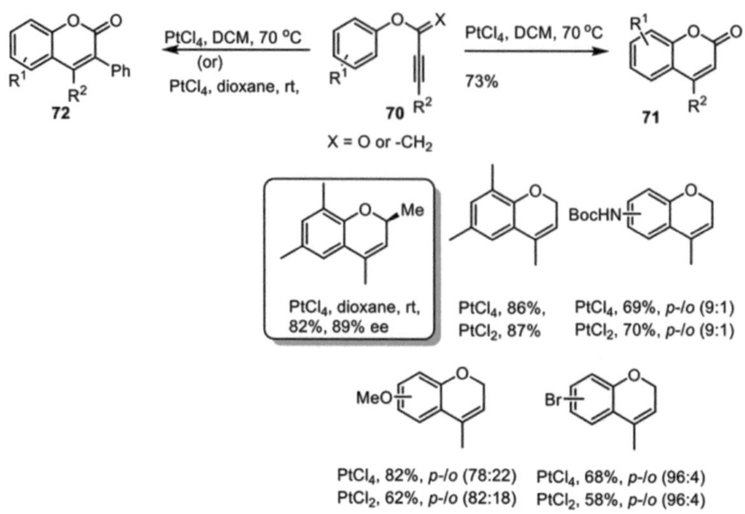

Figure 36.
Preparation of coumarins catalyzed by $PtCl_4$.

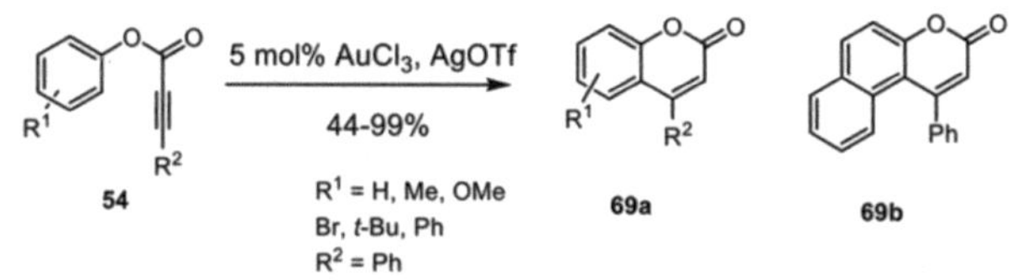

Figure 37.
Method for preparation of coumarins catalyzed by $AuCl_3$.

Later in 2004 Chuan and coworkers have described gold(III)-mediated intermolecular hydroarylation reaction of aryl alkynoates **54** under neat conditions to give coumarin derivatives **69** in high yields (**Figure 37**). In this reaction higher temperature (70°C) was needed in some cases and the process of the reaction is distinct from the same reactions which were catalyzed by Pd(II), Pt(II), Pt(IV), and Ru(II) has been described [88].

4. Recent applications of coumarin based derivatives

Both natural and synthesized coumarins have received a lot of interest recently due to their various biological and pharmacological characteristics, which expand their industrial potential (**Table 1**). The usage of coumarins in the creation of biochemical probes has also significantly grown as a result of their photosensitizing and fluorescent capabilities. Umbelliferone, esculetin, and quercetin are coumarins that have anti-oxidant effects and defend against oxidative damage to cellular DNA [89, 90]. Esculetin has been reported best for protecting cells from oxidative damages [91]. By preventing vitamin K from working, the dicumarol exhibits anti-coagulant qualities, whereas angelmarin has been utilized for pancreatic cancer. Sakaguchi and co-workers reported numerous derivatives of coumarin, including 7,7-dihydroxy-coumarins and others, which are known to have anti-oxidant effects. Free radical production results in oxidative damage to DNA, which activates p53. Recently, in 2017 and 2018, Sahu [92] and co-workers and Witaicenis [93] and coworkers respectively, reported the antioxidant property of 4-methylesculetin.

S. No.	Activity	Derivative of coumarin	Reference
1	Antioxidant [97–99]	Chrysin Scopoletin Umbelliferone	[97–99]
2	Antiinflammatory [100–102]	Esculetin 4-methyl esculetin Umbelliferone-6-carboxylic acid Scopoletin	[100–102]
3	Anticancer [103–107]	Esculetin Umbelliferone Scopoletin Cloricromene	[103–106]
4	Anticoagulant [22, 108, 109]	Warfarin Acenocoumarol	[22, 108, 109]
5	Photosensitivity [110, 111]	Furanocoumarins Psoralens	[110, 111]
6	Enzyme inhibitors [105, 112]	Daphnetin Vipirinin	[105, 112]

Table 1.
Coumarin derivatives and their activities.

Coumarin structures like Cloricromene, Warfarin shows good anti-cancer activity and there are being used in the treatment of cancer. Vipirinin, a derivative of coumarin has is used to suppress the Vpr-depended gene (viral) and thus acts as an antagonist [94]. Osthol, a coumarin is well known for its antioxidant property which prevents acetaminophen activation as well as functions in the regulation of the concentration of free calcium that exists in the intracellular space [95]. Coumarins' photosensitivity is beneficial for

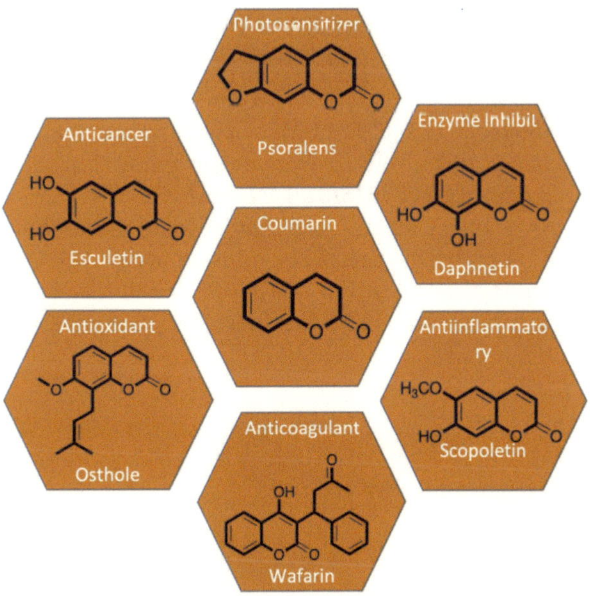

Figure 38.
Coumarin derivatives and their applications.

detecting the activation of particular enzymes as well as bio-molecules including DNA, protein, as well as lipids [96]. Also, they have been utilized in the area of pharmacology to analyze the potency of various drug molecules. Furanocoumarins acts as potent photo-sensitizing agents by killing bacteria and inactivating virus in a conjugated reaction with Ultraviolet light (**Figure 38**) [13].

5. Conclusions

Coumarin derivatives are highly valuable lactone-based organic molecules for academics and bio-pharma industries due to the unique nature of chemical properties, and flexible design system to access a broad range of functionalized coumarin scaffolds via simple synthetic chemical transformation as a result, a huge number of coumarin derivatives have been successfully developed and produced in single steps. This review permits many protocols of one-pot construction of coumarins and their derivatives, together with well-studied name reactions and other types of coupling reactions as well using different metal-based catalysis, and visible light-mediated photocatalysis. While these aforementioned traditional name reactions have been employed for several decades, new methods developments have been widely achieved in the last few years such as microwave-, ultrasound-helped, light-induced metal-free radical cyclization, metal-mediated radical cyclization, metal-catalyzed electrophilic cyclization, miscellaneous, homogeneous catalytic reaction and heterogeneous catalytic reaction and solvent-free conditions, This new system were applied in a challenge to maximize the single-pot reaction yield, diminish the reaction time, and reduce the unwanted side reactions and as well as making these reactions more bioactive candidates. This review concluded that this catalysis-based new developments can play an increasing role in the preparation of highly substituted coumarin derivatives by modifying the unpleasant typical reaction conditions associated with the standard synthetic routes.

Acknowledgements

The authors acknowledge ICT-IOC, Bhubaneswar for providing necessary support. Rambabu Dandela thanks DST-SERB for Ramanujan fellowship (SB/S2/RJN-075/2016), Core research grant (CRG/2018/000782) and ICT-IOC start-up grant, Mahender Khatravath thanks Central University of South Bihar, for providing the facilities. Vasudevan Dhayalan expresses his gratitude to the DST-SERB for the Ramanujan Fellowship (Grant No. RJF/2020/000038) and the National Institute of Technology Puducherry, Karaikal, India for their assistance with research.

Conflict of interest

The authors declare no conflict of interest.

Notes/thanks/other declarations

VD and CM Thanks to SERB for the Ramanujan Fellowship and NIT-Puducherry.

Author details

Deepika Sharma[1], Vasudevan Dhayalan[2*], Chitrarasu Manikandan[2]
and Rambabu Dandela[1]

1 Department of Industrial and Engineering Chemistry, Institute of Chemical Technology, Bhubaneswar, India

2 Department of Chemistry, National Institute of Technology Puducherry, Karaikal, India

*Address all correspondence to: dhaya.chem@nitpy.ac.in

References

[1] Matos MJ, Santana L, Uriarte E, Abreu OA, Molina E, Yordi EG. Coumarins — An Important Class of Phytochemicals. In: Rao AV, Rao LG, editors. Phytochemicals - Isolation, Characterisation and Role in Human Health. IntechOpen. 2015:113-140. DOI: 10.5772/59982

[2] Perkin WH. On the hydride of aceto-salicyl. Journal of the Chemical Society. 1868;**21**:181-186. DOI: 10.1039/JS8682100181

[3] Jain PK, Joshi H. Coumarin: Chemical and pharmacological profile. Journal of Applied Pharmaceutical Science. 2012; **2**(6):236-240. DOI: 10.7324/JAPS.2012.2643

[4] Santana L, Uriarte E, Roleira F, Milhazes N, Borges F. Furocoumarins in medicinal chemistry. Synthesis, natural occurrence and biological activity. Current Medicinal Chemistry. 2012; **11**(24):3239-3261. DOI: 10.2174/0929867043363721

[5] Signore G, Nifosì R, Albertazzi L, Storti B, Bizzarri R. Polarity-sensitive coumarins tailored to live cell imaging. Journal American Chemical Society. 2010;**132**(4):1276-1288. DOI: 10.1021/ja9050444

[6] Balewski Ł, Szulta S, Jalińska A, Kornicka A. A mini-review: Recent advances in coumarin-metal complexes with biological properties. Frontiers in Chemistry. 2021;**9**:1-18. DOI: 10.3389/fchem.2021.781779

[7] Nasab NH, Azimian F, Kruger HG, Kim SJ. Coumarin-chalcones generated from 3-acetylcoumarin as a promising agent: Synthesis and pharmacological properties. Chemistry Select. 2022;**7**(11):1-13. DOI: 10.1002/slct.202200238

[8] Nakamura Y, Sakata Y, Hosoya T, Yoshida S. Synthesis of functionalized benzopyran/coumarin-derived aryne precursors and their applications. Organic Letter. 2020;**22**(21):8505-8510. DOI: 10.1021/acs.orglett.0c03106

[9] Zhao YR, Zheng Q, Dakin K, Xu K, Martinez ML, Li WH. New caged coumarin fluorophores with extraordinary uncaging cross sections suitable for biological imaging applications. Journal of American Chemistry Society. 2004;**126**(14): 4653-4663. DOI: 10.1021/ja036958m

[10] Kulkarni M, Kulkarni G, Lin C-H, Sun C-M. Recent advances in coumarins and 1-azacoumarins as versatile biodynamic agents. Current Medicinal Chemistry. 2006;**13**(23):2795-2818. DOI: 10.2174/092986706778521968

[11] Prahadeesh N, Sithambaresan M, Mathiventhan U. A study on hydrogen peroxide scavenging activity and ferric reducing ability of simple coumarins. Emerging Science Journal. 2018;**2**(6):417-427. DOI: 10.28991/esj-2018-01161

[12] Bhatnagar A, Sharma PK, Kumar N, Dhudhe RA. Review on recent advances in coumarin derivatives with their multidisciplinary actions. Der Pharmacia Lettre. 2010;**2**(4):297-306

[13] Sandhu S, Bansal Y, Silakari O, Bansal G. Coumarin hybrids as novel therapeutic agents. Bioorganic Medicinal Chemistry. 2014;**22**(15):3806-3814. DOI: 10.1016/j.bmc.2014.05.032

[14] Calcio Gaudino E, Tagliapietra S, Martina K, Palmisano G, Cravotto G. Recent advances and perspectives in the synthesis of bioactive coumarins. Royal Society of Chemistry Advance. 2016;

6(52):46394-46405. DOI: 10.1039/c6ra07071j

[15] Abdallah M, Hijazi A, Dumur F, Lalevée J. Coumarins as powerful photosensitizers for the cationic polymerization of epoxy-silicones under near-UV and visible light and applications for 3D printing technology. Molecules. 2020;**25**(9):1-12. DOI: 10.3390/molecules25092063

[16] Zhang G, Zheng H, Guo M, Du L, Liu G, Wang P. Synthesis of polymeric fluorescent brightener based on coumarin and its performances on paper as light stabilizer, fluorescent brightener and surface sizing agent. Applied Surface Science. 2016;**367**:167-173. DOI: 10.1016/j.apsusc.2016.01.110

[17] Bakhtiari G, Moradi S, Soltanali S. A novel method for the synthesis of coumarin laser dyes derived from 3-(1H-Benzoimidazol-2-Yl) coumarin-2-one under microwave irradiation. Arabian. Journal of Chemistry. 2014;**7**(6):972-975. DOI: 10.1016/j.arabjc.2010.12.012

[18] Liu X, Cole JM, Waddell PG, Lin TC, Radia J, Zeidler A. Molecular origins of optoelectronic properties in coumarin dyes: Toward designer solar cell and laser applications. Journal of Physical Chemistry. 2012;**116**(1):727-737. DOI: 10.1021/jp209925y

[19] Chen T, Ma L, Tang Z, Yu LX. Identification of coumarin-based food additives using terahertz spectroscopy combined with manifold learning and improved support vector machine. Journal of Science. 2022;**87**(3):1108-1118. DOI: 10.1111/1750-3841.16064

[20] Wong TC, Sultana CM, Vosburg DA. A green, enantioselective synthesis of warfarin for the undergraduate organic laboratory. Journal of Chemical Education. 2010;**87**(2):194-195. DOI: 10.1021/ed800040m

[21] Bipat R. From rat poison to medicine: Medical applications of coumarin derivatives. Phytochemicals in Human Health. 2020:1-14. DOI: 10.5772/intechopen.89765

[22] Verhoef TI, Redekop WK, Daly AK, Van Schie RMF, De Boer A, Maitland-Van Der Zee AH. Pharmacogenetic-guided dosing of coumarin anticoagulants: Algorithms for Warfarin, acenocoumarol and phenprocoumon. British Journal of Clinical Pharmacology. 2014;**77**(4):626-641. DOI: 10.1111/bcp.12220

[23] Rastij V, Vrandečić K, Ćosić J, Šarić GK, Majić I, Agić D, et al. Effects of coumarinyl schiff bases against phytopathogenic fungi, the soil-beneficial bacteria and entomopathogenic nematodes: Deeper insight into the mechanism of action. Molecules. 2022;**27**(7):1-17. DOI: 10.3390/molecules27072196

[24] Choi TJ, Song J, Park HJ, Kang SS, Lee SK. Anti-Inflammatory activity of glabralactone, a coumarin compound from angelica sinensis, via suppression of TRIF-Dependent IRF-3 signaling and NF- κ B pathways. Mediators of Inflammation. 2022;**2022**:1-11. DOI: 10.1155/2022/5985255

[25] Rawat A, Vijaya Bhaskar Reddy A. Recent advances on anticancer activity of coumarin derivatives. European Journal Medicinal Chemistry Reports. 2022;**5**:100038. DOI: 10.1016/j.ejmcr.2022.100038

[26] Metwally NH, Elgemeie GH, Fahmy FG. Green synthesis: Antimicrobial activity of novel benzothiazole-bearing coumarin derivatives and their fluorescence

properties. Egyptian Journal of Chemistry. 2022;**65**(2):679-686. DOI: 10.21608/EJCHEM.2021.91887.4365

[27] Dianhar H, Tristiyana QR, Purba AWA, Handayani S, Sugita P, Rahayu DUC. Antibacterial activities of 3-substituted coumarin-scaffolds synthesized under microwave irradiation. Journal Hunan University Natural Science. 2022;**49**(1):38-46. DOI: 10.55463/issn.1674-2974.49.1.6

[28] Patil SM, Martiz RM, Satish AM, Shbeer AM, Ageel M, Al-Ghorbani M, et al. Discovery of novel coumarin derivatives as potential dual inhibitors against α-glucosidase and α-amylase for the management of post-prandial hyperglycemia via molecular modelling approaches. Molecules. 2022;**27**(12):1-38. DOI: 10.3390/molecules27123888

[29] Musa MA, Cooperwood JS, Khan MO. A review of coumarin derivatives in pharmacotherapy of breast cancer. Current Medicinal Chemistry. 2008;**15**(26):2664-2679. DOI: 10.2174/092986708786242877

[30] Tejedor D, Delgado-Hernandez S, Diana-Rivero R, Diaz-Diaz A, Garcia-Tellado F. Recent advances in the synthesis of 2H-Pyrans. Molecules. 2019; **24**(16):2119-2132. DOI: 10.3390/molecules24162904

[31] Venugopala KN, Rashmi V, Odhav B. Review on natural coumarin lead compounds for their pharmacological activity. Biomed Research International. 2013:1-14. DOI: 10.1155/2013/963248

[32] Sharma D, Dhayalan V, Chatterjee R, Khatravath M, Dandela R. Recent advances in the synthesis of coumarin and its derivatives by using aryl propiolates. Chemistry Select. 2022;**7**(4): 1-25. DOI: 10.1002/slct.202104299

[33] Cao D, Liu Z, Verwilst P, Koo S, Jangjili P, Kim JS, et al. Coumarin-based small-molecule fluorescent chemosensors. Chemical Reviews. 2019; **119**(18):10403-10519. DOI: 10.1021/acs.chemrev.9b00145

[34] Dreyer DL, Jones KC, Jurd L, Campbell BC. Feeding deterrency of some 4-hydroxycoumarins and related compounds: Relationship to host-plant resistance of alfalfa towards pea aphid (acyrthosiphon pisum). Journal of Chemical Ecology. 1987;**13**(4):925-930. DOI: 10.1007/BF01020171

[35] Ganeshapillai D, Woo LWL, Thomas MP, Purohit A, Potter BVL. C-3- and C-4-substituted bicyclic coumarin sulfamates as potent steroid sulfatase inhibitors. ACS Omega. 2018;**3**(9): 10748-10772. DOI: 10.1021/acsomega.8b01383

[36] Bozdag M, Alafeefy AM, Altamimi AM, Vullo D, Carta F, Supuran CT. Coumarins and other fused bicyclic heterocycles with selective tumor-associated carbonic anhydrase isoforms inhibitory activity. Bioorganic Medicinal Chemistry. 2017;**25**(2): 677-683. DOI: 10.1016/j.bmc.2016.11.039

[37] Rabbani GA. Concise introduction of perkin reaction. Organic Chemistry: Current Research. 2018;**07**(02):7-10. DOI: 10.4172/2161-0401.1000191

[38] Buckles RE. The use of the perkin reaction in organic laboratory classes. Journal of Chemical Education. 1950; **27**(4):210-211. DOI: 10.1021/ed027p210

[39] Lončarić M, Sokač DG, Jokić S, Molnar M. Recent advances in the synthesis of coumarin derivatives from different starting materials. Biomolecules. 2020;**10**(1):1-36. DOI: 10.3390/biom10010151

[40] Zambare AS, Kalam Khan FA, Zambare SP, Shinde SD, Sangshetti JN. Recent advances in the synthesis of coumarin derivatives via pechmann condensation. Current Organic Chemistry. 2015;**20**(7):798-828. DOI: 10.2174/1385272820666151026224227

[41] Gulati S, Singh R, Sangwan S. A review on convenient synthesis of substituted coumarins using reuseable solid acid catalysts. Royal Society of Chemistry Advances. 2021;**11**(47): 29130-29155. DOI: 10.1039/d1ra04887b

[42] Chavan SP, Shivasankar K, Sivappa R, Kale R. Zinc mediated transesterification of β-ketoesters and coumarin synthesis. Tetrahedron Letter. 2002;**43**(47):8583-8586. DOI: 10.1016/S0040-4039(02)02006-3

[43] Mustafa YF. Classical approaches and their creative advances in the synthesis of coumarins: A brief review. Journal of Medicinal Chemistry. 2021; **4**(6):612-625. DOI: 10.26655/JMCHEMSCI.2021.6.10

[44] Valizadeh H, Gholipour H. Imidazolium-based phosphinite ionic liquid (IL-OPPh2) as reusable catalyst and solvent for the knoevenagel condensation reaction. Synthetic Communication. 2010;**40**(10): 1477-1485. DOI: 10.1080/00397910903097310

[45] Sripathi SK, Logeeswari K. Synthesis of 3-aryl coumarin derivatives using ultrasound. International Journal of Organic Chemistry. 2013;**03**(01):42-47. DOI: 10.4236/ijoc.2013.31004

[46] Gholap SS, Deshmukh UP, Tambe MS. Synthesis and in-vitro antimicrobial screening of 3-cinnamoyl coumarin and 3-[3-(1H-Indol-2-Yl)-3-Aryl-Propanoyl]-2H-Chromen-2-Ones. Iranian Journal of Catalysis. 2013; **3**(3):171-176

[47] Hwang IT, Lee SA, Hwang JS, Lee KI. A facile synthesis of highly functionalized 4-arylcoumarins via kostanecki reactions mediated by DBU. Molecules. 2011;**16**(8):6313-6321. DOI: 10.3390/molecules16086313

[48] Song CE, Jung DU, Choung SY, Roh EJ, Lee SG. Dramatic enhancement of catalytic activity in an ionic liquid: Highly practical friedel-crafts alkenylation of arenes with alkynes catalyzed by metal triflates. Angewandte Chemie - Inernational Edition. 2004; **43**(45):6183-6185. DOI: 10.1002/anie.200460292

[49] Hintermann L. Comprehensive organic name reactions and reagents. By Zerong Wang. Angewandte Chemie - Inernational Edition. 2010;**49**(15): 2659-2660. DOI: 10.1002/anie.201000292

[50] Upadhyay PK, Kumar PA. Novel synthesis of coumarins employing triphenyl(α-carboxymethylene) phosphorane imidazolide as a C-2 synthon. Tetrahedron Letter. 2009; **50**(2):236-238. DOI: 10.1016/j.tetlet.2008.10.133

[51] Rao HSP, Sivakumar S. Condensation of α-aroylketene dithioacetals and 2-hydroxyarylaldehydes results in facile synthesis of a combinatorial library of 3-aroylcoumarins. Journal of Organic Chemistry. 2006;**71**(23):8715-8723. DOI: 10.1021/jo061372e

[52] Fernandes TDA, Gontijo Vaz B, Eberlin MN, Da Silva AJM, Costa PRR. Palladium-catalyzed tandem heck-lactonization from o-iodophenols and enoates: Synthesis of coumarins and the

study of the mechanism by electrospray ionization mass spectrometry. The Journal of Organic Chemistry. 2010; **75**(21):7085-7091. DOI: 10.1021/jo1010922

[53] Kaye PT, Robinson RS. Dabco-catalysed reactions of salicylaldehydes with acrylate derivatives. Synthetic Communication. 1996;**26**(11): 2085-2097. DOI: 10.1080/00397919608003567

[54] Nüchter M, Ondruschka B, Bonrath W, Gum A. Microwave assisted synthesis – A critical technology overview. Green Chemistry. 2004;**6**(3): 128-141. DOI: 10.1039/b310502d

[55] Kaneria AR, Giri RR, Bhila VG, Prajapati HJ, Brahmbhatt DI. Microwave assisted synthesis and biological activity of 3-Aryl-Furo[3,2-c]Coumarins. Arabian Journal of Chemistry. 2017;**10**: S1100-S1104. DOI: 10.1016/j.arabjc.2013.01.017

[56] Desai NC, Satodiya HM, Rajpara KM, Joshi VV, Vaghani HV. A microwave-assisted facile synthesis of novel coumarin derivatives containing cyanopyridine and furan as antimicrobial agents. Journal of Saudi Chemical Society. 2017;**21**:S153-S162. DOI: 10.1016/j.jscs.2013.12.005

[57] Schwendt G, Glasnov T. Intensified synthesis of [3,4-d]triazole-fused chromenes, coumarins, and quinolones. Monatshefte fur Chemie. 2017;**148**(1): 69-75. DOI: 10.1007/s00706-016-1885-5

[58] Mangasuli SN, Hosamani KM, Satapute P, Joshi SD. Synthesis, molecular docking studies and biological evaluation of potent coumarin–Carbonodithioate hybrids via microwave irradiation. Chemical Data Collection. 2018;**15–16**:115-125. DOI: 10.1016/j.cdc.2018.04.001

[59] Koparde S, Hosamani KM, Barretto DA, Joshi SD. Microwave synthesis of coumarin-maltol hybrids as potent antitumor and anti-microbial drugs: An approach to molecular docking and DNA cleavage studies. Chemical Data Collection. 2018;**15–16**:41-53. DOI: 10.1016/j.cdc.2018.03.004

[60] Kolita S, Bhuyan PJ. An efficient synthesis of pyrido[3, 2-c]coumarins under microwave irradiation in solvent-free conditions. Chemistry Select. 2018; **3**(5):1411-1414. DOI: 10.1002/slct.201702957

[61] Kaur N. Ultrasound-assisted green synthesis of five-membered O- and S-Heterocycles. Synthetic Communication. 2018;**48**(14):1715-1738. DOI: 10.1080/00397911.2018.1460671

[62] Banerjee B. Recent developments on ultrasound assisted catalyst-free organic synthesis. Ultrasonics Sonochemistry. 2017;**35**:1-14. DOI: 10.1016/j.ultsonch.2016.09.023

[63] Liu Y, Myers EJ, Rydahl SA, Wang X. Ultrasonic-assisted synthesis, characterization, and application of a metal-organic Framework: A green general chemistry laboratory project. Journal of Chemical Education. 2019; **96**(10):2286-2291. DOI: 10.1021/acs.jchemed.9b00267

[64] Ghomi JS, Akbarzadeh Z. Ultrasonic accelerated knoevenagel condensation by magnetically recoverable MgFe2O4 nanocatalyst: A rapid and green synthesis of coumarins under solvent-free conditions. Ultrasonics Sonochemistry. 2018;**40**:78-83. DOI: 10.1016/j.ultsonch.2017.06.022

[65] Prousis KC, Avlonitis N, Heropoulos GA, Calogeropoulou T. FeCl3-Catalysed ultrasonic-assisted, solvent-free synthesis of 4-substituted

coumarins. A useful complement to the pechmann reaction. Ultrasonics Sonochemistry. 2014;**21**(3):937-942. DOI: 10.1016/j.ultsonch.2013.10.018

[66] Al-Kadasi AMA, Nazeruddi GM. Ultrasound assisted catalyst-free one - pot synthesis of Bis-Coumarins in neat water. International Journal of Chemical Science. 2012;**10**(1):324-330

[67] Sabetpoor S, Hatamjafari F. Synthesis of coumarin derivatives using glutamic acid under solvent-free conditions. Oriental Journal of Chemistry. 2014;**30**(2):863-865. DOI: 10.13005/ojc/300265

[68] Sugino T, Tanaka K. Solvent-free coumarin synthesis. Chemistry Letters. 2001;**2**:110-111. DOI: 10.1246/cl.2001.110

[69] Sharma D, Kumar S, Makrandi JK. Modified pechmann condensation using grinding technique under solvent-free condition at roomtemperature. Green Chemistry Letters and Reviews. 2011; **4**(2):127-129. DOI: 10.1080/17518253.2010.517785

[70] Ciamician G. The Photochemistry of the future. Science. 1912;**36**(926): 385-394. DOI: 10.1126/science.36.926.385

[71] Wang Z, Li X, Wang L, Li P. Photoinduced cyclization of alkynoates to coumarins with n-iodosuccinimide as a free-radical initiator under ambient and metal-free conditions. Tetrahedron. 2019;**75**(8):1044-1051. DOI: 10.1016/j.tet.2019.01.013

[72] Kawaai K, Yamaguchi T, Yamaguchi E, Endo S, Tada N, Ikari A, et al. Photoinduced generation of acyl radicals from simple aldehydes, access to 3-Acyl-4-arylcoumarin derivatives, and evaluation of their antiandrogenic activities. Journal of Organic Chemistry. 2018;**83**(4):1988-1996. DOI: 10.1021/acs.joc.7b02933

[73] Wu X, Jia M, Huang M, Kim JK, Zhao Z, Liu J, et al. A visible-light-induced "on-off" one-pot synthesis of 3-arylacetylene coumarins with AIE Properties. Organic and Biomolecular Chemistry. 2020;**18**(17):3346-3353. DOI: 10.1039/d0ob00479k

[74] Kanyiva KS, Hamada D, Makino S, Takano H, Shibata T. α-Amino acid sulfonamides as versatile sulfonylation reagents: Silver-catalyzed synthesis of coumarins and oxindoles by radical cyclization. European Journal of Organic Chemistry. 2018;**2018**(43):5905-5909. DOI: 10.1002/ejoc.201800901

[75] Ren H, Mu Y, Zhang M, Zhang AQ. Synthesis of 3-phenylsulfonylcoumarins by cyclisation of phenyl propiolates with sulfinic acids or sodium sulfinates. Journal of Chemical Research. 2018; **42**(10):515-520. DOI: 10.3184/174751918X15385227785338

[76] Fu W, Sun Y, Li X. Silver-catalyzed monofluoromethylation of alkynoates with sodium monofluoroalkanesulfinate (CH2FSO2Na) to Access 3-monofluoromethylated coumarins. Synthetic. Communication. 2020;**50**(3): 388-398. DOI: 10.1080/00397911.2019.1697452

[77] Aparece MD, Vadola PA. Gold-catalyzed dearomative spirocyclization of aryl alkynoate esters. Organic Letter. 2014;**16**(22):6008-6011. DOI: 10.1021/ol503022h

[78] Mantovani AC, Goulart TAC, Back DF, Menezes PH, Zeni G. Iron(III) chloride and diorganyl diselenides-mediated 6- endo-dig cyclization of Arylpropiolates and Arylpropiolamides leading to 3-organoselenyl-2 h -

Coumarins and 3-Organoselenyl-Quinolinones. Journal of Organic Chemistry. 2014;**79**(21):10526-10536. DOI: 10.1021/jo502199q

[79] Zaitceva O, Bénéteau V, Ryabukhin DS, Eliseev II, Kinzhalov MA, Louis B, et al. Cyclization of aryl 3-aryl propynoates into 4-arylcoumarins catalyzed by cyclometalated platinum(II) complexes. Tetrahedron. 2020;**76**(14):1-9. DOI: 10.1016/j.tet.2020.131029

[80] Gao WC, Liu T, Zhang B, Li X, Wei WL, Liu Q, et al. Synthesis of 3-sulfenylated coumarins: BF3·Et2O-mediated electrophilic cyclization of aryl alkynoates using N-sulfanylsuccinimides. Journal of Organic Chemistry. 2016;**81**(22):11297-11304. DOI: 10.1021/acs.joc.6b02271

[81] Wu W, An Y, Li J, Yang S, Zhu Z, Jiang H. Iodine-catalyzed cascade annulation of alkynes with sodium arylsulfinates: Assembly of 3-sulfenylcoumarin and 3-sulfenylquinolinone derivatives. Organic Chemistry Frontiers. 2017; **4**(9):1751-1756. DOI: 10.1039/c7qo00326a

[82] Fang JD, Yan XB, Zhou L, Wang YZ, Liu XY. Synthesis of 3-organoselenyl-2H-coumarins from propargylic aryl ethers via oxidative radical cyclization. Advanced Synthesis and Catalysis. 2019; **361**(9):1985-1990. DOI: 10.1002/adsc.201801565

[83] Friend CM, Xu B. Heterogeneous catalysis: A central science for a sustainable future. Accounts of Chemical Research. 2017;**50**(3):517-521. DOI: 10.1021/acs.accounts.6b00510

[84] Descorme C, Gallezot P, Geantet C, George C. Heterogeneous catalysis: A key tool toward sustainability. ChemCatChem. 2012;**4**(12):1897-1906. DOI: 10.1002/cctc.201200483

[85] Maheswara M, Siddaiah V, Damu GLV, Rao YK, Rao CV. A Solvent-free synthesis of coumarins via pechmann condensation using heterogeneous catalyst. Journal of Molecular Catalysis A: Chemical. 2006; **255**:49-52. DOI: 10.1016/j.molcata.2006.03.051

[86] Jia C, Piao D, Kitamura T, Fujiwara Y. New method for preparation of coumarins and quinolinones via Pd-catalyzed intramolecular hydroarylation of C-C triple bonds. Journal of Organic Chemistry. 2000;**65**(22):7516-7522. DOI: 10.1021/jo000861q

[87] Pastine SJ, Youn SW, Sames D. PtIV-catalyzed cyclization of arene-alkyne substrates via intramolecular electrophilic hydroarylation. Organic Letter. 2003;**5**(7):1055-1058. DOI: 10.1021/ol034177k

[88] Shi Z, He C. Efficient functionalization of aromatic C-H bonds catalyzed by Gold(III) under mild and solvent-free conditions. Journal of Organic Chemistry. 2004; **69**(11):3669-3671. DOI: 10.1021/jo0497353

[89] Garg SS, Gupta J, Sharma S, Sahu D. An insight into the therapeutic applications of coumarin compounds and their mechanisms of action. European Journal of Pharmaceutical Sciences. 2020;**152**:105424. DOI: 10.1016/j.ejps.2020.105424

[90] Ren QC, Gao C, Xu Z, Feng LS, Liu ML, Wu X, et al. Bis-coumarin derivatives and their biological activities. Current Topics in Medicinal Chemistry. 2018;**18**(2):101-113. DOI: 10.2174/1568026618666180221114515

[91] Liang C, Ju W, Pei S, Tang Y, Xiao Y. Pharmacological activities and synthesis of esculetin and its derivatives: A mini-review. Journal of Lipid Research. 2017; **58**(3):519-528. DOI: 10.3390/molecules22030387

[92] Sahu D, Sharma S, Singla RK, Panda AK. Antioxidant activity and protective effect of suramin against oxidative stress in collagen induced arthritis. European Journal of Pharmaceutical Sciences. 2017;**101**: 125-139. DOI: 10.1016/j.ejps.2017.02.013

[93] Witaicenis A, de Oliveira ECS, Tanimoto A, Zorzella-Pezavento SFG, de Oliveira SL, Sartori A, et al. 4-Methylesculetin, a coumarin derivative, ameliorates dextran sulfate sodium-induced intestinal inflammation. Chemical Biological Interactions. 2018; **280**:59-63. DOI: 10.1016/j.cbi.2017.12.006

[94] Ong EBB, Watanabe N, Saito A, Futamura Y, Abd El Galil KH, Koito A, et al. Vipirinin, a coumarin-based HIV-1 Vpr inhibitor, interacts with a hydrophobic region of Vpr. Journal of Biological Chemistry. 2011;**286**(16): 14049-14056. DOI: 10.1074/jbc.M110.185397

[95] Cai Y, Sun W, Zhang XX, Lin YD, Chen H, Li H. Osthole prevents acetaminophen-induced liver injury in mice. Acta Pharmaceutica Sinica B. 2018; **39**(1):74-84. DOI: 10.1038/aps.2017.129

[96] Deasy B, Bogan DP, Smyth MR, O'Kennedy R, Fuhr U. Study of coumarin metabolism by human liver microsomes using capillary electrophoresis. Analytica Chimica Acta. 1995;**310**(1):101-107. DOI: 10.1016/0003-2670(95)00136-N

[97] Pushpavalli G, Kalaiarasi P, Veeramani C, Pugalendi KV. Effect of chrysin on hepatoprotective and antioxidant status in D-galactosamine-induced hepatitis in rats. European. Journal of Pharmacology. 2010;**631**: 36-41. DOI: 10.1016/j.ejphar.2009.12.031

[98] Shaw CY, Chen CH, Hsu CC, Chen CC, Tsai YC. Antioxidant properties of scopoletin isolated from sinomonium acutum. Phythotherapy Research. 2003;**17**(7):823-825. DOI: 10.1002/ptr.1170

[99] Luzi F, Puglia D, Dominici F, Fortunati E, Giovanale G, Balestra GM, et al. Effect of gallic acid and umbelliferone on thermal, mechanical, antioxidant and antimicrobial properties of poly (vinyl alcohol-co-ethylene) films. Polymer Degradation and Stability. 2018;**152**:162-176. DOI: 10.1016/j.polymdegradstab.2018.04.015

[100] Witaicenis A, Seito LN, Di Stasi LC. Intestinal anti-inflammatory activity of esculetin and 4-methylesculetin in the trinitrobenzenesulphonic acid model of rat colitis. Chemical Biological Interaction. 2010;**186**(2):211-218. DOI: 10.1016/j.cbi.2010.03.045

[101] Nurul Islam M, Joo Choi R, Eun Jin S, Shik Kim Y, Ra Ahn B, Zhao D, et al. Mechanism of anti-inflammatory activity of umbelliferone 6-carboxylic acid isolated from angelica decursiva. Journal of Ethnopharmacology. 2012; **144**(1):175-181. DOI: 10.1016/j.jep.2012.08.048

[102] Ding Z, Dai Y, Hao H, Pan R, Yao X, Wang Z. Anti-inflammatory effects of scopoletin and underlying mechanisms. Pharmaceutical Biology. 2008;**46**(12):854-860. DOI: 10.1080/13880200802367155

[103] Turkekul K, Colpan RD, Baykul T, Ozdemir MD, Erdogan S. Esculetin